Systolic Signal Processing Systems

ELECTRICAL ENGINEERING AND ELECTRONICS

A Series of Reference Books and Textbooks

Editors

Marlin O. Thurston
Department of Electrical
Engineering
The Ohio State University
Columbus, Ohio

William Middendorf
Department of Electrical
and Computer Engineering
University of Cincinnati
Cincinnati, Ohio

1. Rational Fault Analysis, *edited by Richard Saeks and S. R. Liberty*

2. Nonparametric Methods in Communications, *edited by P. Papantoni-Kazakos and Dimitri Kazakos*

3. Interactive Pattern Recognition, *Yi-tzuu Chien*

4. Solid-State Electronics, *Lawrence E. Murr*

5. Electronic, Magnetic, and Thermal Properties of Solid Materials, *Klaus Schröder*

6. Magnetic-Bubble Memory Technology, *Hsu Chang*

7. Transformer and Inductor Design Handbook, *Colonel Wm. T. McLyman*

8. Electromagnetics: Classical and Modern Theory and Applications, *Samuel Seely and Alexander D. Poularikas*

9. One-Dimensional Digital Signal Processing, *Chi-Tsong Chen*

10. Interconnected Dynamical Systems, *Raymond A. DeCarlo and Richard Saeks*

11. Modern Digital Control Systems, *Raymond G. Jacquot*

12. Hybrid Circuit Design and Manufacture, *Roydn D. Jones*

13. Magnetic Core Selection for Transformers and Inductors: A User's Guide to Practice and Specification, *Colonel Wm. T. McLyman*

14. Static and Rotating Electromagnetic Devices, *Richard H. Engelmann*

15. Energy-Efficient Electric Motors: Selection and Application, *John C. Andreas*

16. Electromagnetic Compossibility, *Heinz M. Schlicke*

17. Electronics: Models, Analysis, and Systems, *James G. Gottling*

Electrical Engineering-Electronics Software

Systolic Signal Processing Systems

edited by

Earl E. Swartzlander, Jr.

TRW Inc.
Redondo Beach, California

Marcel Dekker, Inc.　　　　　　New York • Basel

Library of Congress Cataloging-in-Publication Data

Systolic signal processing systems.

    Includes index.
    1. Signal processing--Digital techniques.
I. Swartzlander, Earl E.
TK5102.5.S663   1987   621.38'043   87-3467
ISBN 0-8247-7717-4

MARCEL DEKKER, INC.
270 Madison Avenue, New York, New York 10016

Current printing (last digit):
10 9 8 7 6 5 4 3 2 1

PRINTED IN THE UNITED STATES OF AMERICA

Preface

The field of special purpose digital processor development is evolving rapidly. This results in part from advances in VLSI circuit design and fabrication technology. Specifically, through computer aided design (CAD), silicon compilation, and VLSI fabrication advances it has become practical to create custom digital processors for a wide range of military, industrial, and commercial applications. The processor architecture must respect the constraints imposed by the design and fabrication technologies. The repetitive use of common structures (i.e., modules) and the minimization of "random" interconnections are crucial to reducing the design effort and to creating producable integrated circuits.

This book is about systolic signal processing systems: networks of signal processors with extremely efficient data flow between the processors. The terminology is new: systolic systems were first proposed about a decade ago by Professor H. T. Kung and his students, including Charles Leiserson. This concept represents a significant evolution from earlier pipelined digital signal processors.

Pipeline architectures have been employed for several decades to implement analog filters and high speed radar signal processors. The basic concept is that a cascade chain of processors can be provided with storage between the processors so that multiple data enter "into the pipeline" while previous data are being processed. The resulting data flow is similar to a firefighters' bucket brigade; in both cases the rate of flow is determined by the speed of the individual processors and not by the number of stages in the pipeline.

Systolic systems are so named because data flows through them in a rhythmic fashion similar to the flow of blood through a body. They feature data connections only between neighboring processors, use

multiple copies of a few types of processors, and avoid the use of
global memory. These attributes are well matched to the constraints
of VLSI technology. Localized interconnection (i.e., data flow only
between neighboring processors and no global memory) reduces the
need for exotic high-pin-count packaging, and allows increasing
numbers of processors to be placed on a chip (or wafer) as technology
evolves. Use of a few types of processing elements greatly simplifies
the VLSI design challenge.

Although much attention has been paid to the architecture of
general-purpose digital computers, relatively little has been devoted
to special purpose signal processing systems where the benefit from
the use of systolic architectures is greatest. This book represents
a joint effort by about a dozen internationally recognized experts to
improve the understanding of such systems. Chapters are included
describing the basic concepts, current systems, new approaches, and
extensions to the future. Our ultimate purpose is to foster a greater
awareness of the techniques, technologies, and benefits of systolic
processor architectures for signal processing applications.

This book is written for advanced students, engineers, and
managers who wish a concise introduction to the key concepts and
future directions of systolic processor architectures. Working engineers
and managers may find it useful to attend workshops and symposia,
such as those sponsored by the IEEE, to extend their understanding.

The first chapter provides an introduction to the use of mathematical modeling for the design of pipelined signal processors. Dr.
Danny Cohen shows the relationship between the functions performed
by circuit elements (and their interconnection) and the resulting
processor's operation, allowing the designer to better understand
what a given processor will do prior to its construction.

In chapter 2, Professor Per E. Danielsson describes systolic
convolvers. Since the convolution operation is the cornerstone of
modern signal processing, the concepts are of great importance. A
wide variety of serial/parallel multiplier cells (which are cascaded to
perform convolution) are described. This chapter introduces many of
the concepts of circuit level pipelining and systolic processing.

Professor H. T. Kung is credited with developing the concept
of systolic processing. In chapter 3, he describes the CMU Warp
processor (the term Warp denoting ultra high speed is derived from
the "Star Trek" television series). Warp is a 32-bit floating point
systolic array capable of computing at rates of 10 million floating point
operations per cell. This chapter shows how various algorithms are
implemented on it.

In chapter 4, Professor S. Y. Kung describes the wavefront
array concept. This represents an extension to systolic architecture
in that data move in self-timed "waves" across the array. This chapter

develops the concept and shows how algorithms are implemented on arrays using data flow graphs.

In the next chapter, Professor Peter Cappello provides a methodology for making trades between space and time in processor arrays. In general, adding more processors (i.e., increasing space) does not produce a proportionate reduction in processing time. This chapter shows ways to design arrays that achieve more nearly optimum performance.

The development of a methodology to optimize the array implementation for a given algorithm is extended in chapter 6 by Dr. Sailesh Rao and Professor Thomas Kailath. Their approach is based on expressing the problem as a Regular Iterative Algorithm (RIA) and developing procedures to implement the RIA efficiently. Their work provides a mathematical framework for architecture development and comparison.

In chapter 7, Professor Tom Leighton and Charles E. Leiserson show how systolic algorithms apply to the development of defect tolerant architectures for Wafer Scale Integration (WSI). In WSI, circuits are implemented on integrated circuit wafers (which may contain many defective circuit elements). As shown in this chapter, systolic architectures facilitate reconfiguration to bypass defective circuits.

Finally, in chapter 8, Professors Renato Stefanelli and Mariagiovanna Sami examine fault tolerance in the context of a two-dimensional array. This work leads to a protocol for dynamic fault circumvention. Although developed for a two-dimensional array, it should also be applicable to other regular structures.

Some users may be troubled by variations in the style of presentation; this is one of the disadvantages of a multiauthor book. Exposure to the pioneers who are developing this field should help to compensate for the inconsistencies.

This book was developed with significant support from the staff at Marcel Dekker, Inc., and my secretary, Lauren Hall. Thanks are due to all of them for their help. On behalf of the technical community, it is a pleasure to extend special thanks to the authors who made time in their already busy schedules to document their work, thereby sharing it with all of us.

EARL E. SWARTZLANDER, JR.

Contributors

PETER R. CAPPELLO *Computer Science Department, University of California, Santa Barbara, California*

DANNY COHEN *University of Southern California Information Sciences Institute, Marina del Rey, California*

PER E. DANIELSSON *Department of Electrical Engineering, Linköping University, Linköping, Sweden*

THOMAS KAILATH *Information Systems Laboratory, Stanford University, Stanford, California*

H. T. KUNG *Department of Computer Science, Carnegie-Mellon University, Pittsburgh, Pennsylvania*

SUN-YUAN KUNG *Signal & Image Processing Institute, Department of Electrical Engineering, University of Southern California, Los Angeles, California*

TOM LEIGHTON *Mathematics Department and Laboratory for Computer Science, Massachusetts Institute of Technology, Cambridge, Massachusetts*

CHARLES E. LEISERSON *Laboratory for Computer Science, Massachusetts Institute of Technology, Cambridge, Massachusetts*

SAILESH K. RAO *AT&T Bell Laboratories, Holmdel, New Jersey*

MARIAGIOVANNA SAMI *Department of Electronics, Politecnico di Milano, Milano, Italy*

RENATO STEFANELLI *Department of Electronics, Politecnico di Milano, Milano Italy*

Contents

1

A Mathematical Approach to Computational Network Design

DANNY COHEN *University of Southern California Information Sciences Institute, Marina del Rey, California*

1 INTRODUCTION

In this chapter we demonstrate the application of a precise mathematical notation to the design and manipulation of computational networks. This notation captures the concepts of arithmetic operations (such as addition and multiplication) and of timing (e.g., delay). Once a design is expressed by means of such a mathematical notation, it can be evaluated objectively against a predefined set of design objectives, such as performance and cost. Throughout this chapter the term "design" means the structure/architecture of the computational network. This term is the hardware equivalent of the software term "algorithm."

In Section 2 we define the design objectives that guide the examples in this chapter. Obviously, other sets of design objectives may be used without deviating from the spirit of the chapter.

In Section 3 we deal with the implementation of a finite impulse response (FIR) filter, a typical signal processing problem. In that section, several designs are suggested and evaluated objectively, and the mathematical notation to express them is developed, in parallel. Here we consider first a design that follows closely the mathematical definition of the FIR filter. Later this design is transformed several times in order to improve it with respect to the predefined design objectives. Here the graphic representations of the designs are the source of the intuition for the synthesis, and their mathematical representations are mainly a means for the analysis, such as verifying the correctness of the various transformations of the design.

In Section 4 the same technique is applied to synthetic aperture radar (SAR) processing. We consider and evaluate several designs that result directly from the mathematical definition and the notation.

1

In Section 5 the same techniques and notation are applied to polynomial multiplication, division, and simultaneous multiplication and division. In that section the mathematical representation is the guiding force for the synthesis, and the graphic representations are used only for demonstration.

In Section 6 we discuss a special type of filter, the spectral filter. Operational calculus is used on the mathematical notation for deriving the most efficient implementation network for that filter. This mathematical notation is a very powerful tool, complementing the intuition based on conventional graphic representation. It could be used for both the synthesis and analysis of computational networks.

2 THE DESIGN GOALS

To achieve an optimal design, it is necessary to define the design objectives. The following objectives are typically considered to be important:

1. Correctness and accuracy
2. High computation rate
3. Low delay
4. Low parts count
5. Modularity, simplicity, etc.
6. Low power
7. Small size
8. Low cost

Obviously, this is only a partial list. For different applications the relative weights of these objectives may vary. It is generally accepted that (1) is the most important, even though there is evidence that this is not always the case. In some cases (8) is the dominant factor; in others, it is (6) and (7). Here (1) through (5) are considered to be the most important, in that order.

3 FIR-FILTER EXAMPLE

Consider the finite impulse response (FIR) filter, defined by

$$y_n = \sum_{i=1}^{N} a_i x_{n-i} \tag{1.1}$$

This is a nonrecursive filter of the Nth order. Each output (Y) is a weighted average of the previous N inputs (X). Typically, the X sequence is a time series, and the x's are available sequentially, with x_i preceding x_{i+1} by one "cycle."

3.1 Z Operator

Let Z be the "delay-by-one-cycle" operator such that $Zx_i = x_{i-1}$. In systems controlled by a central master clock, this Z operator may be implemented by a simple register. Z^0 is defined to be the unit operator and Z^n is defined by $Z^n = ZZ^{n-1}$. It may be implemented by an n-stage first-in, first-out (FIFO) shift register.

We use the following properties of the Z operator:

1. $Z^n x_i = x_{i-n}$.
2. $ZF(x) = F(Zx)$.
3. If C is time invariant, $ZC = C$.

Z^n with $n < 0$ is a prediction by $|n|$ steps into the future. Since prediction of external input is not easy to implement with conventional logic, it is advisable to use only $n \geqslant 0$ when applying the Z^n operator to external input.

3.2 FIR-Filter Implementation

The expression

$$y_n = \sum_{i=1}^{N} a_i x_{n-i} \qquad (1.1)$$

may also be written as

$$y_n = \sum_{i=1}^{N} a_i Z^i x_n$$

Using operator-calculus notation, this may be written as

$$Y = \left(\sum_{i=1}^{N} a_i Z^i \right) X$$

For N = 4 this yields

$$Y = (a_1 Z + a_2 Z^2 + a_3 Z^3 + a_4 Z^4)X \qquad\qquad (1.2)$$

Equation (1.2) can be implemented by the network shown in Fig. 1.1. The circles in the figure with the a_i's represent the multiplications by the constants written inside them. Checking this network against the design objectives reveals that:

1. *Correctness*: The correct expression is indeed computed, since the values at P_4, P_7, P_9, and P_{10} are Zx_n, $Z^2 x_n$, $Z^3 x_n$, and $Z^4 x_n$, respectively.
2. *Computation rate*: The computation rate is the reciprocal of the computation period, the time needed for one multiplication and for adding N numbers.
3. *Delay*: The delay is the time from giving x_n to the network until y_n is computed. The delay is one cycle.

It is not simple to quantify the parts count (4) and the modularity objective (5). However, the parts count (4) can be improved. The values at P_1, P_2, P_3, and P_4 are all equal to Zx. Therefore, these points could be unified. Similarly, P_5, P_6, and P_7 could be unified, and so can P_8 and P_9. This does not change (1), (2), and (3), but it does improve (4). The new network is shown in Fig. 1.2. Hence the parts count objective (4) is improved by eliminating six delay operators, or $N(N-1)/2$ in the general case. The modularity objective (5) is also improved, as seen from the repeated modules, marked by dashed lines in Fig. 1.2.

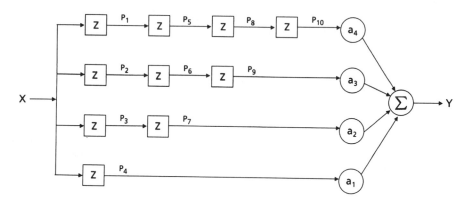

FIGURE 1.1 The implementation of Eq. (1.2).

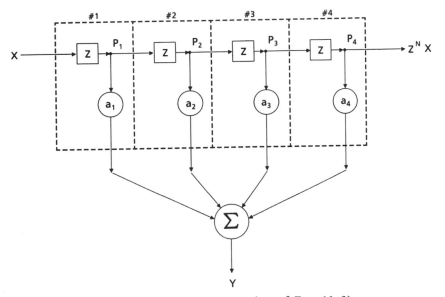

FIGURE 1.2 The improved implementation of Eq. (1.2).

3.3 Improving This Design

The N-term summation (represented by the sigma) is the Achilles heel of this design, because it does not comply with the modularity requirement. In addition, the direction of the information flow from the repeated modules into the summation is perpendicular to the direction in which these modules are arranged. This may cause problems with the geometry of the wiring, both in very large scale integration (VLSI) and discrete implementations, and also on and between printed circuit boards.

In addition, the required number of output lines from any set of modules is proportional to the number of modules, and this may pose severe problems for implementation at any scale. To alleviate this problem, the N-term summation is implemented by $(N-1)$ two-input additions, divided between the modules, as shown in Fig. 1.3. The network shown in the figure is composed of N identical modules. This is a great improvement for design objective (5), modularity. The leftmost adder, the one in the first module (with a_1), a does not perform any real addition operation, because one of its inputs is always zero. Its purpose is to improve the modularity. Obviously, in discrete implementations, there is no need to include it. Eliminating it improves the parts count slightly. On the other hand, keeping it is a small price for reducing the number of module types.

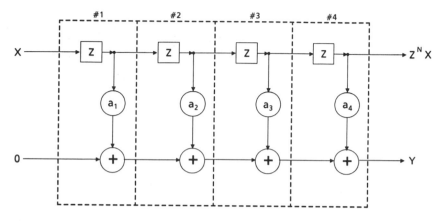

FIGURE 1.3 Using sequential addition.

This implementation is represented by

$$Y = \left(\sum_{i=1}^{N} a_i Z^i \right) X$$

To improve the delay involved in this computation, notice that

$$Z^{-1}Y = \left(\sum_{i=1}^{N} a_i Z^{i-1} \right) X = \left(\sum_{i=0}^{N-1} a_{i+1} Z^i \right) X \tag{1.3}$$

Only nonnegative powers of Z are used for the input values (X), on the right-hand side of Eq. (1.3). On the left-hand side there is a Z^{-1}, which is a prediction, applied to the output (Y). This indicates that at the nth cycle, when x_n is given, y_{n+1} is available. Equation (1.3) yields the implementation shown in Fig. 1.4. In the figure the leftmost adder (in module 1) is redundant. So is the rightmost delay (in the last module). They do not tax the performance, but they improve the modularity.

3.4 Improving the Operation Rate

The major deficiency of all the networks considered so far is their operation rate, objective (2). The operation period ("cycle") cannot be shorter than the time required for multiplication and addition of N quantities. Even when the multipliers are arranged such that the multiplication time overlaps the addition time, the addition must still propagate through N (or N − 1) stages. Since N may be very large,

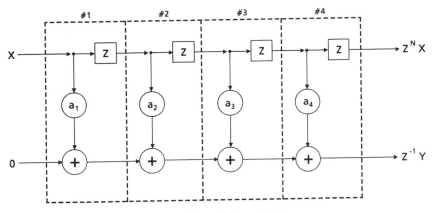

FIGURE 1.4 The implementation of Eq. (1.3).

it is desirable to eliminate the need for this long propagation. This
may be achieved by the pipeline idea: using storage units (hence
causing some delay) to improve the computation rate, by allowing
various steps of the computations to overlap, hence to be executed
simultaneously. This is done here by using Z units between the
modules, which delay the output but improve the computation rate.

The resulting network is shown in Fig. 1.5. The proof that
the output is correct is its computation. Let S_j be the output of such
a network such as that shown in Fig. 1.5, but with j modules. We
will prove that in general the output of an N-module network, S_N,
is a delayed Y as defined in Eq. (1.1).

From the structure of the network and the modules, as shown in
Fig. 1.5, we get the following relation:

$$S_j = (a_j Z^{2j-2})X + ZS_{j-1} \quad \text{for } j > 0 \text{ and } S_0 = 0 \tag{1.4}$$

First we will show that

$$S_N = \left[\sum_{i=1}^{N} Z^{N-i}(a_i Z^{2i-2}) \right] X \tag{1.5}$$

and then we will show that $S_N = Z^{N-2}Y$.

Equation (1.5) is proven by induction. Assume that it holds
for S_{N-1} and use Eq. (1.4) to evaluate S_N:

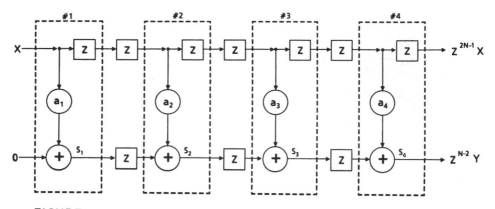

FIGURE 1.5 Implementing a pipelined adder.

$$S_N = (a_N z^{2N-2})X + ZS_{N-1}$$

$$= (a_N z^{2N-2})X + z\left[\sum_{i=1}^{N-1} z^{N-1-i}(a_i z^{2i-2})\right]X$$

$$= \left[(a_N z^{2N-2}) + \sum_{i=1}^{N-1} z^{N-i}(a_i z^{2i-2})\right]X$$

$$= \left[\sum_{i=1}^{N} z^{N-i}(a_i z^{2i-2})\right]X \quad \text{Q.E.D.} \tag{1.6}$$

This proves Eq. (1.5). Expand Eq. (1.6) and get

$$S_N = \left[\sum_{i=1}^{N} z^{N-i}(a_i z^{2i-2})\right]X = \sum_{i=1}^{N} (a_i z^{N+i-2})X$$

$$= z^{N-2}\left[\sum_{i=1}^{N} (a_i z^i)\right]X = z^{N-2}Y \quad \text{Q.E.D.} \tag{1.7}$$

This proves that the network shown in Fig. 1.5 computes Y delayed by $N - 2$ cycles. Check this network against the design objectives:

1. *Correctness*: The correct expression is computed, as proven by Eq. (1.7).
2. *Computation rate*: The computation period is now the time required

for a single multiplication followed by a single addition, independent of the value of N. Since it is easy to overlap the execution of the multiplication and the addition, we do not attempt to separate them, even though this may slightly improve the computation rate.

3. *Delay*: The computation delay is equal to (N − 2) cycles.
4. *Parts count*: The number of adders and multipliers needed is as before. Howver, three delays are needed in each module. Hence the total parts count is higher (i.e., worse) than before.
5. *Modularity*: The modularity is not as "nice" as in the network shown in Fig. 1.4, which includes only components included in the repeated modules.

    To improve the modularity, we merge the new delays into the old modules. In order not to introduce additional delays, we include in each module the delay from its upper right and the delay from its lower left. Hence the network implementation is now composed of N modules, each as shown in Fig. 1.6, without the need for any additional components. By using a network consisting of N modules as shown here, the rate is the best achievable (without separating the multiplication from the addition).

3.5 Another Look

We would like to point out that a very important design decision was made before without any justification or even discussion. This design decision is the direction of the sequentialization of the summation operator. We introduced it as a left-to-right sequence of adders without considering other possibilities. We can use a tree structure with $\log_2 N$ depth. Its carry chain is only $\log_2 N$ long. This is better than N, but still might be too long. The same pipeline approach may be used again, by using the delay operation, Z, between every pair of successive adders.

    How does this design check against the objectives?

1. *Correctness*: The correct expression is indeed computed, as shown before.
2. *Computation rate*: The rate is optimal. As before, we wish not to split the addition from the multiplications.
3. *Delay*: The delay is only $\log_2 N$.
4. *Parts count*: The total number of adders required for adding N numbers is N − 1, whether they are arranged in linear order or in a tree structure. Hence no change in the number of adders is needed. The multipliers count is also the same as before.
5. *Modularity*: The binary tree of adders is again perpendicular to the data flow and may impose geometrical problems.

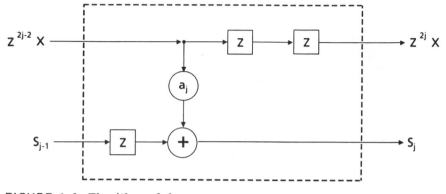

FIGURE 1.6 The jth module.

3.6 And Another Look

We have considered the left-to-right and binary tree arrangements.
Next we consider the right-to-left option. At first, it does not appear
to be different from the left-to-right option, but this point is worth
verifying. Let us look at the network shown in Fig. 1.4, with the
direction of the addition reversed. The resulting network is shown in
Fig. 1.7. The networks shown in Fig. 1.4 and 1.7 are equivalent.
Therefore, the latter suffers from the same problem that the former
does—the need for simultaneous addition of N numbers.

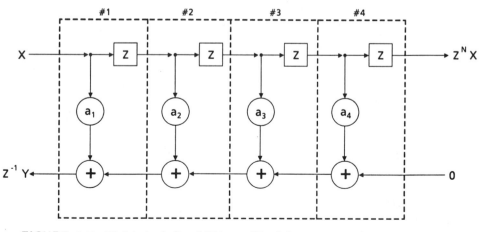

FIGURE 1.7 Right-to-left addition with delays.

The same pipeline idea can be used again, by adding delays. This results in the network shown in Fig. 1.8, similar to Fig. 1.5. This new network also has to be checked against the design objectives. Starting from (1), the correctness, we compute the value of the output S (which is actually S_n). Now we get

$$S = \left[\sum_{i=1}^{N} z^{i-1}(a_i z^{2i-2}) \right] X \tag{1.8}$$

This is very similar to Eq. (1.5), except that the output of the ith module is delayed now by z^{i-1} instead of by z^{N-i} as before, when the addition was left-to-right. The simplification of Eq. (1.8) yields

$$S = \left[\sum_{i=1}^{N} z^{i-1}(a_i z^{2i-2}) \right] X = z^{-3} \left[\sum_{i=1}^{N} (a_i z^{3i}) \right] X$$

This is obviously not the desired Y. Therefore, the network shown in Fig. 1.8 does not perform the correct computation.

Why does the very same approach that worked so well in the network shown in Fig. 1.5 fail now? The reason is that in both cases the delays between the adders on the "lower" line are needed in order to make the computation period independent of N. The purpose of the delays on the "upper" line is to compensate for the delays on the lower line such that the addition is performed coherently. Since in the left-to-right network in Fig. 1.5, data flow on both lines (the lower and the upper) in the same direction, the same delays have to be introduced in both, to keep the data "in step." However, in the right-to-left network of Fig. 1.8, data flow on these lines in opposite directions. Hence to compensate for a delay on the lower line, data should be accelerated on the upper line. Since Z is used on the lower, Z^{-1} should be used on the upper.

It is unfortunate that the Z^{-1} operation is a prediction that cannot be implemented in the general case. However, in this case each Z^{-1} happens to follow a Z, such that each cancels the effect of the other and no delays are needed on this line. Figure 1.9 shows the modified network. A careful examination shows that this network computes

$$S = \sum_{i=1}^{N} z^{i-1}(a_i X) \tag{1.9}$$

Again, the new design has to be checked against all the design objectives. Starting with (1), the correctness, we get

Danny Cohen

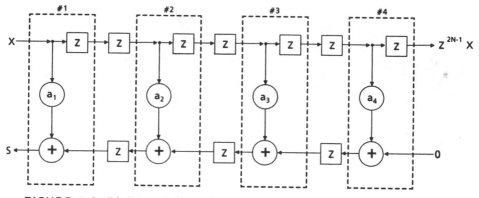

FIGURE 1.8 Right-to-left addition with delays.

$$S = \sum_{i=1}^{N} Z^{i-1}(a_i X) = Z^{-1}\left[\sum_{i=1}^{N} Z^i(a_i X)\right]$$

$$= Z^{-1}\left[\sum_{i=1}^{N} (a_i Z^i)\right]X = Z^{-1}Y \quad \text{Q.E.D.}$$

This proves the correctness and also shows that the delay is as before. We also know that the computation period is minimal, since it is equal to the longest "atomic" operation. The parts count is lower than in any other design, and the network is modular.

Based on the above, this design is optimal with respect to correctness (1), computation rate (2), and delay (3), and it also scores highly in the parts count (4) and the modularity (5) categories. An alternative way to draw this network is shown in Fig. 1.10. The addition is performed, again, in the left-to-right direction, because the order of the a_i's is reversed.

3.7 Design Evaluation

The Z notation can be used for the evaluation of all the networks shown before, from Fig. 1.1 to Fig. 1.10. The transformation between the various alternatives can (and should) be performed without the aid of figures and intuition.

Let us review the systems discussed so far. System (A) is the one resulting directly from the definition, and is shown in Fig. 1.1 through 1.3. It computes

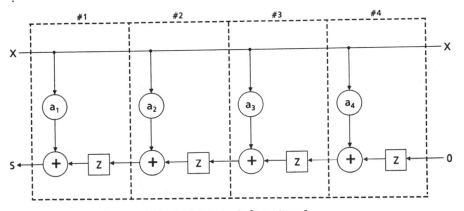

FIGURE 1.9 The modified right-to-left network.

System (A): $Y = \displaystyle\sum_{i=1}^{N} (a_i z^i)\, X$

Since one delay could be saved this network was transformed into system (B), the one shown in Fig. 1.4. It computes

System (B): $z^{-1}Y = \left[\displaystyle\sum_{i=0}^{N-1} (a_{i+1} z^i) \right] X$

Then, to improve the rate, (B) was transformed into the pipelined system (C), as shown in Fig. 1.5. It computes

System (C): $z^{N-2}Y = \left[\displaystyle\sum_{i=1}^{N} z^{N-i}(a_i z^{2i-2}) \right] X$

Then the right-to-left addition was introduced, and (C) was transformed into system (D), the one shown in Fig. 1.9 and Fig. 1.10. This system computes

System (D): $z^{-1}Y = \displaystyle\sum_{i=0}^{N-1} z^i(a_{i+1} X)$

Compare and evaluate these systems, by using their representations, without referring to the figures.

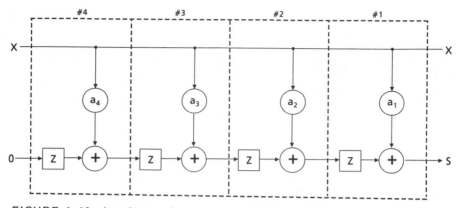

FIGURE 1.10 An alternative drawing of Figure 1.9.

1. *Correctness*: From the representation above it is evident that all
    of these systems perform the correct computation because all of
    them were shown to compute $Z^k Y$, for some k.
2. *Computation rate*: Both (A) and (B) require adding N quantities
    at once. Therefore, their computation period is equal to the time
    required for a multiplication followed by the addition of N numbers,
    where (C) and (D) require only the time needed for a multiplication
    and a single addition.
3. *Delay*: In system (A) y_n is available in the same cycle as x_n. We
    use this for delay reference and denote it as zero. System (B)
    computes $Z^{-1}Y$. Hence, the Y is one cycle earlier than in system
    (A), and its delay is -1. System (C) computes $Z^{N-2}Y$. Hence the
    Y is N $-$ 2 cyles later than in system (A), and its delay is N $-$ 2.
    System (D) has the same delay of -1 as system (B). In summary,
    in the general case, the delays are:

System implementation:	A	B	C	D
Delay (cycles):	0	-1	N $-$ 2	-1

However, even though both systems (B) and (D) have the same delay
in cycles, (D) requires a shorter time from the availability of the
input x_n until the output y_{n+1} is computed, because (D) uses a
shorter cycle than (B).

4. *Parts count*: All four implementaions require N multipliers and N
    (or N $-$ 1) adders. They differ only in the requirements for delay
    units. Both systems (A) and (B) require N delays for X. System
    (C) requires 2N delays for X and N delays for the partial sums of
    the products. System (D) requires N delays for the partial sums
    of the products. If fixed-point arithmetic is used, these delays
    units may require more capability (bits) than for delaying X.

5. *Modularity*: All four implementations are equally modular, with
    the same level of complexity.

    The rating of these system is summarized in the following table.
S > T means that S is better than T.

1. Correctness: $(A) = (B) = (C) = (D)$
2. Computation rate: $(C) = (D) > (A) = (B)$
3. Delay: $(D) > (B) > (A) = (C)$
4. Parts count: $(A) = (B) > (D) > (C)$
5. Modularity: $(A) = (B) = (C) = (D)$

This shows that system (D) is the best design if performance (rate
and delay) is the major objective. In some cases, such as when fixed-
point arithmetic is used, system (B) may be the best design if
reducing the parts count is the major design objective.

4 SYNTHETIC APERTURE RADAR

In this section we discuss an example taken from synthetic aperture
radar (SAR) data processing. This SAR application is introduced
first; then a design for its implementation is discussed.

4.1 SAR Problem

Consider a moving platform, such as an aircraft or a spacecraft,
traveling along a straight line. Every period of (NT) time it transmits a radar burst, whose echo is recorded N times, T periods apart.
Typical numbers are N = 1000 and T = 100 nsec, corresponding to
F = 10 MHz.

Let i be the serial number of a given burst and let j be the serial
number of a given echo return inside it. The value of j varies between
0 and N − 1. The value of i increases as long as the platform is in
motion. The data D(i,j) is recorded at the time $t = (Ni + j)T$ in a one-
dimensional serial sequence. We omit the T from the notation. Similarly,
the Z operator is a delay by this period.

The input D(k) precedes the input D(k + 1) by one cycle. Hence
the Z operator is a delay by one cycle, and Z^{-1} is a prediction by one
cycle which cannot be applied to external input data. We refer to
(i,*) as columns and to (*,j) as rows. Hence there are N rows,
parallel to the platform trajectory, and the columns which correspond
to the radar bursts are perpendicular to the trajectory.

The purpose of collecting the data set {D(i,j)} is to use it for the
computation of the "surface function" F(i,j), defined by

$$F(i,j) = \sum_{k=-m}^{m} a_k D(i - k, j) \qquad (1.10)$$

for the fixed set of coefficients $\{a_k | -m \leqslant k \leqslant +m\}$. This is a weighted average of $D(i,j)$ with its neighbors, of the same jth row, up to m columns on each side.

The definition (1.10) is an extreme simplification of the actual SAR problem. For simplicity many crucial details are omitted. Among these complex details are the dependence of the a's on the position inside the burst (the j value), the surface elevation, and the effects of the angle between the trajectory of the platform and the motion of the planet. These and other details are very important for the actual SAR process, but are not necessary for understanding the concept.

4.2 Design of the Network

When applying the Z operator to the data we get

$$ZD(i,j) = D(i, j - 1) \quad \text{for } j > 0 \quad \text{and} \quad ZD(i,0) = D(i - 1, N - 1)$$

and

$$Z^N D(i,j) = D(i - 1, j) \qquad (1.11)$$

Substituting Eq. (1.11) into Eq. (1.10) gives

$$F(i,j) = \left[\sum_{k=-m}^{m} (a_k Z^{kN}) \, D \right] = \left[Z^{-mN} \sum_{k=0}^{2m} (a_{k-m} Z^{kN}) \right] D \qquad (1.12)$$

Since we cannot implement the Z^{-mN} operation, we must compute $Z^{mN}F$, which is F lagging mN cycles (m columns) behind the input data sequence D. This is expected, since the definition of $F(i,j)$, Eq. (1.10), requires data m bursts on each side, past and future.

Since Eq. (1.12) is very similar to Eq. (1.1), we already know how to compute it. Equation (1.12) may be written as

$$Z^{mN}F = \left[\sum_{k=0}^{2m} (a_{k-m} Z^{kN}) \right] D \qquad (1.13)$$

Figure 1.11 shows the implementations, for m = 2, of Eq. (1.13). Equation (1.13) is basically like the FIR example, say Eq. (1.3), except that Z^N is used here instead of the Z there. Hence, following

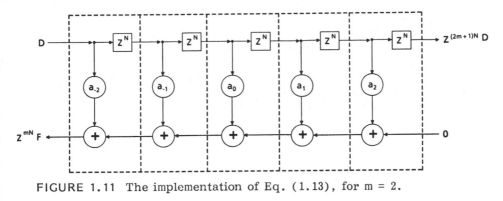

FIGURE 1.11 The implementation of Eq. (1.13), for m = 2.

what we did for the FIR filter example, we get the fastest implementation represented by a pipeline implementation (1.14), similar to
Eq. (1.9):

$$Z^{mN}F = \left[\sum_{k=0}^{2m} Z^{kN}(a_{k-m}D) \right] \qquad (1.14)$$

It is left for the reader to check this design against the design objectives, (1) through (5). Figure 1.12 shows the implementation,
for m = 2, of Eq. (1.14). As mentioned earlier , when fixed-point
arithmetic is used, the Z^N operators in Eq. (1.14) may be more expensive than those of Eq. (1.13), because they store products having
more bits than the raw data signal, D.

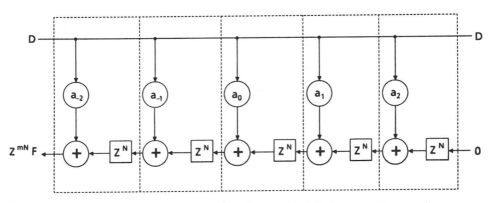

FIGURE 1.12 The implementation of Eq. (1.14) for m = 2.

The reason for moving the Z operators from the raw data to the partial sum of the products, where it is more expensive, is to pipe-line the addition (by avoiding the long carry-chain propagation and overlapping the addition) in order to improve the computation rate. However, this separation can be achieved by a single Z, for any value of N. Therefore, to achieve the improved computation rate, without "overpaying" in parts, the following implementation can be used:

$$Z^{mN}F = \left[\sum_{k=0}^{2m} Z^k (a_{k-m} Z^{k(N-1)}) \right] D \qquad (1.15)$$

The three occurrences of Z in Eq. (1.15) correspond to three different meanings. The first, on the left-hand side, represents the delay in the computation of F (relative to D) and does not represent any device. The second Z, in Z^k, represents the registers used for holding partial sums of products, and the third, in $Z^{K(N-1)}$, represents the FIFO used for delaying the input signal, D. This is an optimal spread of the delays; the bulk of the delay is implemented in the cheaper way, on the raw data, whereas the more expensive delay (for the sum of products) needed for the pipeline is implemented in the minimal quantity, only one delay per unit. Figure 1.13 shows the implemen-tation, for m = 2, of Eq. (1.15).

5 MULTIPLICATION AND DIVISION OF POLYNOMIALS

The FIR filter and the SAR examples were synthesized by using intuition on computational networks represented by drawings. The Z notation was used primarily for analysis (e.g., correctness verifi-cation and performance evaluation) and not for synthesis. In this section we compute multiplication and division of polynomials, and design computational networks to implement these operations. How-ever, here the Z notation is also used for the synthesis of the net-works, and the diagrams are used only to demonstrate the design.

5.1 Polynomial Multiplication

Let A(t) and X(t) be polynomials in t, of degrees c and m, respectively:

$$A(t) = \sum_{i=0}^{c} a_i t^i$$

$$X(t) = \sum_{i=0}^{m} x_i t^i$$

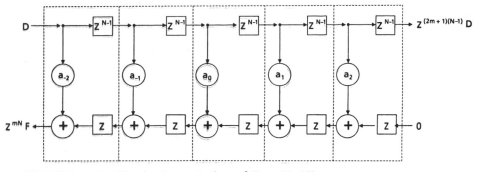

FIGURE 1.13 The implementation of Eq. (1.15).

Let $Y(t)$ be the product polynomial of $A(t)$ and $X(t)$:

$$Y(t) = \sum_{n=0}^{m+c} y_n t^n = \left(\sum_{i=0}^{c} a_i t^i\right)\left(\sum_{j=0}^{m} x_j t^j\right) = \sum_{i=0}^{c} \sum_{j=0}^{m} a_i x_j t^{i+j}$$

By equating the coefficients of t, we get $j = n - i$ and

$$y_n = \sum_{i=0}^{c} a_i x_{n-i} \qquad \text{(where } x_i = 0 \text{ for } i < 0 \text{ and for } i > m) \qquad (1.16)$$

We are interested in finding the coefficient set of the polynomial $Y(t)$ from the given coefficient sets of $A(t)$ and $X(t)$ rather than evaluating any of these polynomials for particular values of t. In many applications $A(t)$ is a fixed polynomial and $X(t)$ is variable. The computation problem is to compute the $m + c + 1$ coefficients of $Y(t)$ from the fixed $c + 1$ coefficients of $A(t)$ and the given $m + 1$ variable coefficients of $X(t)$.

   Since Eq. (1.16) is similar to Eq. (1.1) except for the boundary conditions, the same networks that compute the FIR filter can also perform this polynomial multiplication. Since Eq. (1.16) contains a_0, one more stage is needed, and the computation is performed such that y_n is available in the cycle when x_n is given. Hence the delay now is 0 instead of the -1 cycle as we had before.

   Figure 1.14 shows the network for this computation. It starts with a_0 (compared with a_1 in the previous network) and its output is Y (compared with $Z^{-1}Y$ before). Because of the boundary conditions it is necessary to clear all the delay units before starting the operation and to provide $x_i = 0$ for $i = m + 1, m + 2, \ldots, m + c$. When these values are given, the last c coefficients of Y are obtained. Since there are $m + c + 1$ coefficients of Y and only $m + 1$ coefficients of X, this "runout" operation is indeed expected.

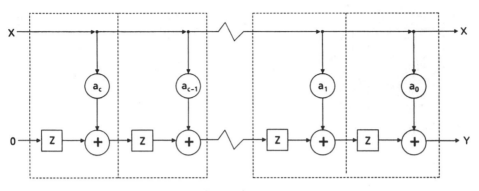

FIGURE 1.14 Polynomial multiplication.

The initial clearing can be performed, just like the runout operation, by proving the network with c zero values for X. During this period the obtained Y values are invalid. Following the FIR example, we use the network shown in Fig. 1.14 for polynomial multiplication. It is represented by

$$Y = \sum_{i=0}^{c} Z^i(a_i X) \tag{1.17}$$

5.2 Reversing the Order of X

In several applications it is preferred that x_n is available before x_{n-1}. In these cases x_m is leading and x_0 trailing. If this order is used the operator Z has a predicting role and Z^{-1} is a delay. Since Eq. (1.17) is implemented with positive powers of Z, we need another implementation with only negative powers of Z. Multiply Eq. (1.17) by Z^{-c} and get

$$Z^{-c}Y = \sum_{i=0}^{c} Z^{i-c}(a_i X) = \sum_{i=0}^{c} Z^{-(c-i)}(a_i X) = \sum_{i=0}^{c} Z^{-i}(a_{c-i} X) \tag{1.18}$$

Since this has the same structure as Eq. (1.17), the same network (Fig. 1.14) can be used to perform this operation, except for the following:

1. Z^{-1} is used instead of Z. However, since Z^{-1} here is a delay just as Z is there, this is not a change of function, only of notation.

2. The order of the a_i's is reversed, because we now have a_{c-i} where we had a_i before.
3. The output is $Z^{-c}Y$ instead of Y.

Therefore, when x_n is given to the network, y_{n+c} is computed. When the leading coefficient of X, x_m, is given to the network, the leading coefficient Y, y_{m+c}, is computed. The resulting network is shown in Fig. 1.15.

5.3 Summing Polynomial Products

Consider the problem of computing W(t), defined by

$$W(t) = A(t) * X(t) + B(t) * Y(t)$$

where A(t) and B(t) are of degree c and X(t) and Y(t) are of degree m. Obviously, W(t) is of degree m + c. By using Eq. (1.18), we get

$$Z^{-c}W = \sum_{j=0}^{c} Z^{-j}(a_{c-j}X) + \sum_{j=0}^{c} Z^{-j}(b_{c-j}Y) \qquad (1.19)$$

For c = 3 this is implemented by the network shown in Fig. 1.16. However Eq. (1.19) may be better written as

$$Z^{-c}W = \sum_{j=0}^{c} Z^{-j}[(a_{c-j}X) + (b_{c-j}Y)]$$

FIGURE 1.15 Polynomial multiplication (most significant term leading).

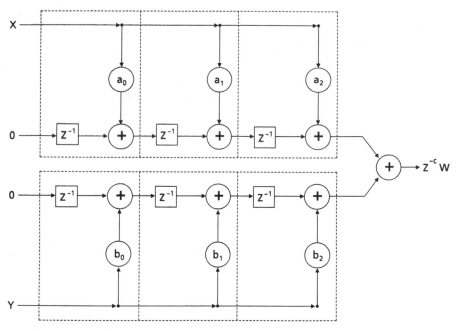

FIGURE 1.16 Sum of polynomial products.

This suggests the network shown in Fig. 1.7. Since a three-input adder is twice as complex as a two-input adder, this does not reduce the number of adders, but it reduces the number of required delay units.

5.4 Polynomial Division

Polynomial division is the inverse of polynomial multiplication. The division is defined in the usual way, finding $X(t)$ from the relation

$$Y(t) = A(t) * X(t) \qquad \text{where } a_c \neq 0$$

$A(t)$ and $Y(t)$ are given polynomials of degree c and $m + c$, respectively. $X(t)$, the polynomial to be determined, is of degree m.

Division, unlike multiplication, can be performed only by starting with the most significant (highest power) of Y. This nonsymmetry is due to requiring that the leading coefficient of $A(t)$ must not be zero. Therefore, we use Eq. (1.18) and not Eq. (1.17) in order to invert the multiplication. Equation (1.18) stated that

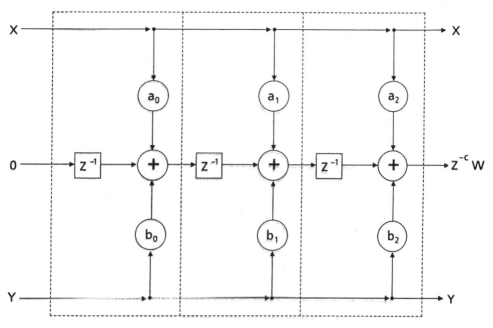

FIGURE 1.17 Sum of polynomial products, combined.

$$Z^{-c}Y = \sum_{i=0}^{c} Z^{-i}(a_{c-i}X)$$

Since the operation has to be performed from the most significant to the least significant term, at any stage in the computation of $X(t)$, the higher-order terms of $X(t)$ may already be known. Therefore, we seek to express X by using A, Y, and $Z^{-i}X$ for $i > 0$.

Extract the term with $i = 0$ from Eq. (1.18) and get

$$Z^{-c}Y = a_c X + \sum_{i=1}^{c} Z^{-i}(a_{c-i}X)$$

Isolate this term and get

$$a_c X = Z^{-c}Y - \sum_{i=1}^{c} Z^{-i}(a_{c-i}X) \tag{1.20}$$

To share the Z^{-c} operation, Eq. (1.20) is transformed into

$$a_c X = \sum_{i=1}^{c} Z^{-i}[E(i,c)Y - (a_{c-i}X)]$$

where $E(c,c) = 1$ and $E(i,c) = 0$ if $i \neq c$. Since $a_c \neq 0$, X can be expressed explicitly by

$$X = a_c^{-1} \sum_{i=1}^{c} Z^{-i}[E(i,c)Y - (a_{c-i}X)] \tag{1.21}$$

The network for implementing Eq. (1.21) is shown in Fig. 1.18.

Before starting this operation, all the Z units are cleared. Since X is synchronized with Y, x_i is computed at the same cycle when y_i is given. Since the first coefficient of Y is y_{m+c}, during the first c cycles the output is equal to zero, corresponding to x_i for $m < i \leqslant m + c$. During the next $m + 1$ cycles the coefficients of X are available, with x_m leading and x_0 trailing. After all, the $m + c + 1$ cycles the Z units include the same data that were present in the Z units of the network shown in Fig. 1.15, just before the multiplication process started.

Since all the Z units in this network were cleared before the multiplication, all the Z units should contain zeros after the division. If they contain any nonzero value, Y(t) was not an exact product of A(t). In fact, the values in the c delay units are the coefficients of the remainder polynomial, R(t), whose degree is less then m. This polynomial is defined by

$$R(t) = Y(t) - A(t) * X(t)$$

From observing Fig. 1.18 one may notice that y_0 does not have the opportunity to affect any value of x_i since it is given to the network at the same time as x_0 is computed without having the chance to propagate through the c delay units. This may be counterintuitive; however, it is true. $y_{c-1}, y_{c-2}, \ldots, y_0$ affect only R(t), not X(t).

5.5 Checking the Operations

To check (somewhat weaker than "verify") these operations, we prove that if we use these networks first to perfrom the multiplication of any arbitrary polynomial X(t) by the given polynomial A(t), by following Eq. (1.18), and then to perform the division of this product by the same given polynomial, A(t), following Eq. (1.21), the same arbitrary polynomial X(t) results.

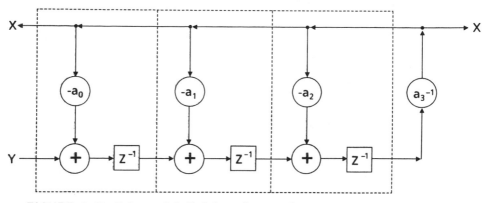

FIGURE 1.18 Polynomial division, for c = 3.

Let $Y(t)$ be the result of the multiplication of $X(t)$ by $A(t)$, and let $S(t)$ be the result of the division of $Y(t)$ by $A(t)$. We will prove that $S(t) = X(t)$. From Eq. (1.20) we get

$$S = a_c^{-1}\left[z^{-c}Y - \sum_{i=1}^{c} z^{-i})a_{c-i}S)\right]$$

Use Eq. (1.18) to substitute for $Z^{-c}Y$ and get

$$S = a_c^{-1}\left[\sum_{i=0}^{c} Z^{-i}(a_{c-i}X) - \sum_{i=1}^{c} Z^{-i}(a_{c-i}S)\right]$$

$$= a_c^{-1}\left[a_cX + \sum_{i=1}^{c} Z^{-i}(a_{c-i}X) - \sum_{i=1}^{c} Z^{-i}(a_{c-i}S)\right]$$

$$= X + a_c^{-1}\sum_{i=1}^{c} Z^{-i}[a_{c-i}(X - S)]$$

Substitute:

$$X - S = a_c^{-1}Z^0 a_c(X - S)$$

And get

$$\sum_{i=0}^{c} z^{-i}[a_{c-i}(X - S)] = 0 \text{ and } A(t) * [X(t) - S(t)] = 0$$

Since $A(t)$ is not zero, $S(t)$ must be equal to $X(t)$.   Q.E.D.

5.6 Simultaneous Multiplication and Division of Polynomials

Define $S(t)$ to be the polynomial obtained by multiplying the arbitrary polynomial $X(t)$ by the given polynomial $A(t)$, and then by dividing this product, $Y(t)$, by another given polynomial, $B(t)$, also of degree c, such that $b_c \neq 0$.

By following Eq. (1.18) and Eq. (1.20), we get

$$S = b_c^{-1}\left[\sum_{i=0}^{c} z^{-i}(a_{c-i}X) - \sum_{i=1}^{c} z^{-i}(b_{c-i}S)\right]$$

$$= b_c^{-1}\left[a_c X + \sum_{i=1}^{c} z^{-i}(a_{c-i}X - b_{c-i}S)\right]$$

The network to perform this computation (for $c = 3$) is shown in Fig. 1.19. This network was synthesized here by using formal manipulation of the mathematical expressions. It is not clear that conventional intuition and graphical diagrams lead to it, unlike the FIR examples, which could be derived as easily without the mathematical expressions.

6 SPECTRAL FILTERS

The mth element of the discrete Fourier transform (DFT) is computed (at the time n) by evaluating the following spectral filter:

$$y_n = \sum_{j=0}^{N-1} w^{jm} x_{n-j} \quad \text{where } w = e^{-2\pi i/N} \text{ and } i^2 = -1$$

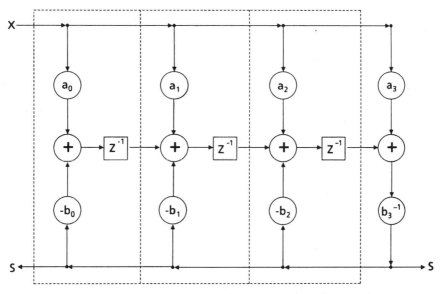

FIGURE 1.19 The implementation of $S = (AX)/B$, for $c = 3$.

Apply the Z operator and get

$$y_n = \sum_{j=0}^{N-1} w^{jm} x_{n-j} = \sum_{j=0}^{N-1} w^{jm} z^j x_n$$

This may be written as

$$Y = \left[\sum_{j=0}^{N-1} (w^m z)^j \right] X$$

This is a geometric series; its sum is

$$Y = \frac{1 - (w^m z)^N}{1 - w^m z} X = \frac{1 - w^{mN} z^N}{1 - w^m z} X$$

This is the product of the operators $(1 - w^{mN} z^N)$ and $(1 - w^m z)^{-1}$ in either order. The implementation of the latter operator may not look intuitive to the uninitiated.

Let $U = (1 - w^m z)^{-1} X$; hence $X = (1 - w^m z) U = U - w^m z U$. Therefore, $U = X + w^m z U$, which is easy to implement. This operator has an infinite response, whereas $(1 - w^{mN}Z^N)$ has a finite response. Therefore, it is better to choose the following order:

$$Y = (1 - w^{mN}Z^N)(1 - w^m Z)^{-1}X$$

Even though $w^{mN} = 1$ (from the definition of w), one may choose not to eliminate the multiplication by it but to implement it with the value of $(w^m)^N$ as computed in the precision used by that particular system, rather than the theoretical value of 1. This alleviates the round-off problems caused by the finite precision of the system.

We also have a choice between the orders wZ and Zw in both operators. In the former operator the ZwZ and wZw orders can be used, too. Since it is easy to perform multiply-add, we use the wZ order and understand that this may be better implemented otherwise under certain circumstances. The corresponding network is therefore represented by

$$Y = (1 - w^{mN}Z^N)U \qquad \text{where } U = X + w^m ZU$$

This requires that two successive additions are computed in each cycle, one for the evaluation of U and one for Y. However, these two additions can overlap (pipeline) by delaying Y for one cycle:

$$ZY = (1 - w^{mN}Z^N)ZU \qquad \text{where } U = X + w^m ZU$$

Since the term ZU appears twice, we combine them and save a Z unit:

$$ZY = (1 - w^{mN}Z^N)T \qquad \text{where } T = ZU \text{ and } U = X + w^m T \qquad (1.22)$$

Figure 1.20 shows a network implementing Eq. (1.22)

7 SUMMARY AND CONCLUSIONS

Computational networks can (and should) be represented by precise mathematical expressions modeling both operations and storage (hence, delay). Formal manipulation of these expressions may be used both for the synthesis and the analysis of various networks, with different properties. In particular, it may be used for exploring the variety of networks for implementing a given function.

Furthermore, these representations may be transformed symbolically (as opposed to graphically) in order to generate alternative networks, and to evaluate them according to the design objectives.

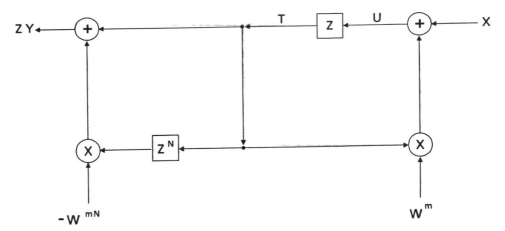

FIGURE 1.20 The implementation of Eq. (1.22), a spectral filter.

These transformations may continue until no further improvement is achieved. This mathematical tool can be used for proving the correctness and the properties of networks, including networks designed by other means.

In particular, networks as these shown in Fig. 1.19 and Fig. 1.20 are beyond the intuitive graphic presentations, but are well within the capabilities of the mathematical tool presented in this chapter.

2

Serial/Parallel Convolvers

PER E. DANIELSSON *Department of Electrical Engineering, Linköping University, Linköping, Sweden*

1 INTRODUCTION

Convolution is probably the most common form of signal processing performed, both by nature itself and by man-made systems. In this chapter we deal with convolution of time-discrete and digital signals. Furthermore, we restrict the operation to be linear, very often also nonrecursive. Thus the discrete computation we consider is defined by

$$Y^{(i)} = \sum_{l=1}^{L} A_l X_l^{(i)} \tag{2.1}$$

where $X_l^{(i)}$ is one of L input samples, so that

$$X_{l+1}^{(i)} = X_l^{(i-1)}$$

$Y^{(i)}$ is the output at point (i), and A_l is one of L coefficients, the weights of an L-point convolution kernel.

Both X_l and A_l are binary numbers which we assume to be fractions. In 2's-complement representation we may then write

$$X_l = -x_{l0}2^0 + \sum_{n=1}^{N-1} x_{ln}2^{-n} = -x_{l0}2^0 + \ldots x_{ln}2^{-n} + \ldots + x_{l,N-1}2^{-N+1}$$

$$A_l = - a_{l0} 2^0 \sum_{k=1}^{K-1} a_{lk} 2^{-k} = - a_{l0} 2^0 + \ldots a_{lk} 2^{-k} + \ldots + a_{l,K-1} 2^{-K+1}$$

A surprisingly large number of algorithms and methods can be used for the computation of Eq. (2.1). The basic form of this expression indicates that one should do L multiplications of a K-bit number with an N-bit number. However, many variations are possible, for the following two reasons.

1. The total sum consists of bit contributions along three "index axes," l,k,n. Different computation schemes arise simply by doing the accumulation in an order prescribed by different index permutations.
2. Since the operands A_l are constants, their values are a kind of a priori knowledge that can be used to bypass certain computation steps.

A good example of exploitation of (2) is shown in [1], where the constants A_l are compiled into the microprogram for an ALU. A zero bit a_{lk} of a constant A_l means that the partial sum $X_l 2^{-k}$ should not be accumulated in a certain cycle. Thus it is possible to create a microprogram that is depleted of such cycles and has an execution time proportional to the number of nonzero bits in the constants.

The so-called distributed arithmetic method [2,3] draws from both (1) and (2). In this case all the possible subtotals of the constants A_l are precalculated and stored in a table. At execution time, the X_l's are fetched by bit column, least significant bit first. The bit columns are used to index the table and bring out a ready-made subtotal. Thus the total accumulation takes N cycles and does not involve traditional multiplication at all.

A rather different approach is to use pipelining and exploit the fact that the production of output results employs overlapping windows of the input data. This involves the superscript index (i) in Eq. (2.1), which is understood to be incremented in the outermost computation loop in the two methods just described.

If we allow several results to be under computation at the same time, feedback of results to the input becomes more difficult. This is why we temporarily restrict the discussion to nonrecursive operations only given by Eq. (2.1). This finite impulse response (FIR) filter is converted to an infinite impulse response (IIR) filter by extending Eq. (1.1) to

$$Y^{(i)} = \sum_{l=1}^{L} A_l X_l^{(i)} + \sum_{r=1}^{R} B_r Y_r^{(i)} \qquad (2.2)$$

where $Y_r^{(i)}$ is one of R output values so that

$$Y_1^{(i)} = Y^{(i-1)} \qquad Y_r^{(i)} = Y^{(i-r)}$$

and B_r is one of R coefficients.

Thus the feedback part of Eq. (2.2) has the same basic structure as Eq. (1.1). IIR implementations are discussed in Sec. 6.

2 SYSTOLIC CONVOLVERS

Devices for extensive pipelining have become known as systolic arrays [4]. The word "systolic" implies a heartbeat and a pulsed flow. Translated to the digital design domain, it means a clocked and pipelined system, the final output result being computed (accumulated) in a number of processor stages. In addition, the input data and control parameters are also allowed to move from input to different stages. No buses are allowed, which means that the only global signals are ground, power, and clock. In fact, in a systolic array different input and output data streams may very well flow in separate and/or opposite directions forming a linearly, orthogonally, or hexagonally connected network. The moving of 1's and 0's in different directions resembles the activity in the cardiovascular system in a living organism.

Several convolvers of the systolic type have been proposed. One type utilizes a cell, the basic function of which is multiplication [5,6]. This design is systolic on the word level; that is, the basic cycle is a word operation and the data path width is one word (input data or output result). The cell memory is used to store constants and the tag memory to evoke reset of the accumulator when appropriate.

Figure 2.1 represents a design where each cell approaches the complexity of a full-fledged processor. In the approach shown by Figure 2.2 we take the more radical step of using iterative cells at the bit level [7]. Each cell (see Fig. 2.3) performs the full-adder function, which is the basic accumulation step in the convolution operation. The number of cells in the array is $N \times L \times (K + \log L)$; in this case $L = 3, K = 3, N = 5$. The throughput rate is equal to the clock rate, which can be extremely high, since the gate depth of all cells are only two (or four, depending on actual design of the full adder).

Per E. Danielsson

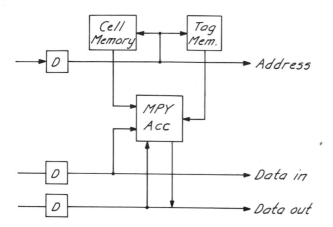

FIGURE 2.1 Programmable cell for a word-level systolic convolver.
(From Ref. 6.)

The accumulation in Fig. 2.2 does not take place in the order
prescribed by multiplication. In fact, one multiplication in Eq. (2.1),
say, $A_1 \cdot X_1$, is distributed in Fig. 2.2 and is found as the top row
of the five blocks.

The basic cell of Fig. 2.2 is described in detail by Fig. 2.3. It
is seen to have a static 1-bit storage for the coefficient bits a_{lk},
and these are distributed over the array. The leftmost data bit x_0
is the sign bit of the 2's-complement representation, which explains
why the negative counterparts of the coefficients are stored in the
three last rows of the array.

The actual loading of the coefficients can be simplified if the
memory cells for the bits a_{lk} are chained together into one long
meandering shift register. However, to improve the graphical overview, these connections have been omitted from Fig. 2.2

Now, the task of the device in Fig. 2.2 is to compute Eq. (2.1)
in a running-window fashion over the one-dimensional string of samples
$X(i)$. New X values continuously enter from the top at each clock
cycle, setting in motion a flow of Y values (Y waves) down the array.
All output values $Y(i)$ start to accumulate at the uppermost right-
hand side cell with the contribution

$$a_{12} x_{14}^{(i)}$$

In the vertical direction, this value is propagated downward so that
in the next clock cycle

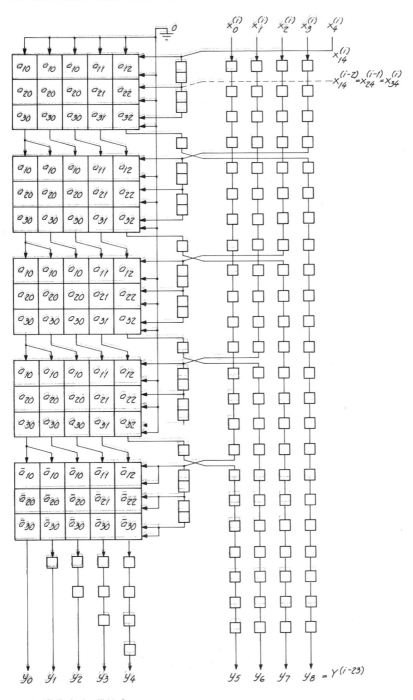

FIGURE 2.2 Bit-level systolic convolver.

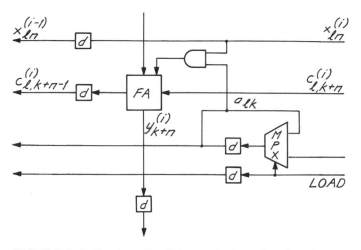

FIGURE 2.3 Basic cell of the convolver in Fig. 22.

$$a_{12}x_{14}^{(i)} + a_{22}x_{14}^{(i-1)}$$

is produced, which is equal to

$$a_{12}x_{14}^{(i)} + a_{22}x_{24}^{(i)}$$

since

$$x_{1n}^{(i)} = x_{1+1,n}^{(i+1)}$$

So, to let each correct x bit coincide with its proper wave of accumulating Y value, the x bits at the border of the array are vertically delayed two units before being propagated horizontally into the array. The vertical x pace is half the pace of the Y wave.

In the horizontal direction, the carry signals belonging to an accumulating Y value are propagated with one unit delay over the cells to produce, for example, in the first row

$$a_{12}x_{14} + 2a_{11}x_{14}$$

So, for the correct x bit to coincide with the proper Y wave, the x bits should be propagated horizontally at the same pace. The front of the Y waves will make a 45° slope; the x waves will make a slope of arctan $(1/2)$. Each x bit combines with all the a bits before its wave dies out at the lefthand side boundary of the block. Note that the vertical flow of data is shifted right one step after every Lth row.

In Fig. 2.2 we have $L = 3$. This requires $x_3^{(i)}$ to be delayed $3 + 1 = 4$ units before it gets into play since it starts its contribution three levels down and one step left. For the same reason $x_2^{(i)}$ should be delayed $6 + 2 = 8$ time units. More generally, bit $x_n^{(i)}$ has to be delayed $(N - 1 - n)(L + 1)$ time units before entering its proper block in the array. These delays make up for approximately half of the delay elements to the right of the Fig. 2.2. The rest of these elements are used for time alignment of the output bits. The last bit $y_0^{(i)}$ leaves the array 23 time units after its computation started at the upper right-hand side corner. Thanks to the shift-register system of delay elements, all the other bits in $Y^{(i)}$ leave at the same time.

A closer look at the delay conditions reveals that what is labeled $Y^{(i)}$ in Fig. 2.2 is actually composed of

$$A_1 X^{(i)} + A_2 X^{(i-1)} + A_3 X^{(i-2)}$$

where, again, (i) is the sample number of the entering data.

A similar approach is presented by Denyer and Myers [8], the difference being that their array accumulates in a different index order. Actually, an even more obvious solution would be to introduce internal pipelining in a combinatorial multiplier and cascade such units to cover the wanted convolution kernel. Another variation of this pipelined approach has been presented by Swartzlander and Gilbert [17] using the carry-save technique. However, the net result seems less regular than in the proposals mentioned previously.

All these designs, including Fig. 2.2 and [2.8], have the intrinsic drawback that IIR filters, according to Eq. (2.2), are not possible to implement. Output bits in $Y^{(i)}$ are not available until several cycles after they are needed for computation of $Y^{(i+1)}$.

The systolic designs presented so far are not very pin saving in terms of (VLSI) design. Very low pin count is only obtained by bit-serial data streams. This is the main theme of the following sections. The same approach has been used in Refs. 9 and 10, but detailed reference to this method is postponed to Sec. 7.

3 SERIAL/PARALLEL MULTIPLIERS

Serial/parallel multipliers involve three bit strings for the input variable

$$(x_0, x_1, x_2, \ldots, x_{N-1})$$

the input constant

$$(a_0, a_1, a_2, \ldots, a_{K-1})$$

and the output

$$(y_0, y_1, y_2, \ldots, y_{D-1})$$

representing three binary numbers in, say, 2's-complement form. Our a priori assumption is that the x string and the a string are serially fed into the computational unit and that the resulting y string is likewise serially shifted out. The y string is successively computed so that an original value of $0, 0, \ldots, 0$ during motion of the string is converted to the final value $y_0, y_1, \ldots, y_{D-1}$.

Thus we have three strings that move into and out from the computational unit. Since the a string consists of constants, we might assume that this string is preloaded and static. However, we will postpone such assumptions to retain generality. The computational unit is assumed to consist of one linear array of cells and can be visualized as shown by Fig. 2.4, although the three strings do not necessarily move in the same direction. Note that we do not assume any globally distributed signals besides the clock, which is not shown explicitly in the forthcoming figures but is assumed to control every d element.

The four serial/parallel multipliers of Fig. 2.5 are known from the technical literature [11–16]. For simplicity, at this point we assume that all quantities are positive numbers.

In Fig. 2.5a we see that by feeding the x string with least significant bit x_2 first, we can produce the y string serially. Figure 2.5b is a simple derivation of Fig. 2.5a by introducing extra delay elements in both the x and the y data paths. Figure 2.5c, on the other hand, is completely different since only the x string is delayed. Note that the ordering of the a bits is reversed. Without any pipelining in the y path the clock frequency will suffer from long signal propagation times in Fig. 2.5c. This defect is compensated for in Fig. 2.5d, where a delay is inserted for each cell in the data paths for both x and y.

The serial/parallel multipliers of Fig. 2.5 are only a few examples from a much larger family that will be identified as follows. For each of the three strings we define the velocities

$$v_x, v_a, v_y$$

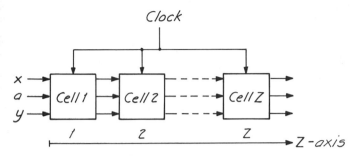

FIGURE 2.4 Bit-serial approach.

to be the number of cell distances which each string is displaced in one single time unit (clock cycle). Let us call the spatial coordinate along the linear array of cells the z axis. For each string we have the index values

$$i_x(z,t), i_a(z,t), i_y(z,t)$$

that for any given instant t equals the weight (negative power of 2 = index) of the bit at cell position z along the array. Actually, the indices i_x, i_a, and i_y are identical to the previously used indices n, k, and d, respectively.

We also define three slopes

$$w_x, w_a, w_y$$

for the static strings that equal the average increase of index per cell. The moving strings are then described by

$$i_x(z,t) = w_x(z - v_x t) \tag{2.3}$$

$$i_a(z,t) = w_a(z - v_a t) \tag{2.4}$$

$$i_y(z,t) = w_y(z - v_y t) \tag{2.5}$$

Note that for the moment we treat these new variables as if they were defined in continuous space z and continuous time t. One example of moving strings is shown in Fig. 2.6, which also introduces some notations that will be used in the following.

Now as soon as the x string overlaps the a string over a cell z, we will have a contribution to the y string. The index number of this contribution equals the sum of the a index and the x index, so that

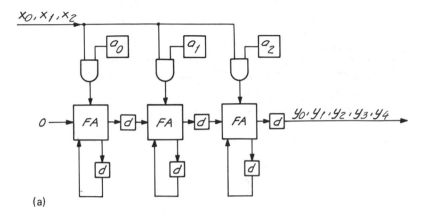

(a)

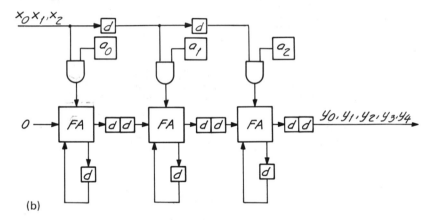

(b)

FIGURE 2.5 Four variations of serial/parallel multipliers.

$$i_y(z,t) = i_x(z,t) + i_a(z,t) \qquad (2.6)$$

Equation (2.6) must hold for all z and all t. Identification of parameters in Eq. (2.3), (2.4), (2.5), and (2.6) gives us the two basic relations between the string-defining parameters:

$$w_y = w_a + w_x > 0 \qquad (2.7)$$

$$w_y v_y = w_a v_a + w_x v_x \qquad (2.8)$$

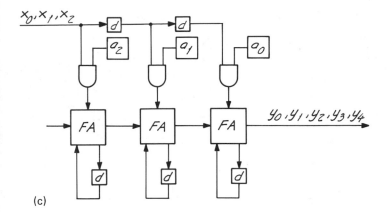

(c)

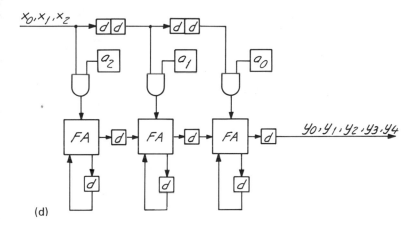

(d)

FIGURE 2.5 (Continued)

Equations (2.7) and (2.8) combine to yield

$$v_y = \frac{w_a v_a + w_x v_x}{w_a + w_x} \tag{2.9}$$

For obvious reasons, we must have $v_y > 0$; otherwise, the bits would stay forever inside the array. Also, we must have $w_y > 0$; otherwise, a signal has to propagate over a long string of cells as in Fig. 2.5c. Furthermore, v_y and w_y must have the same sign so that the y string propagates, least significant bit (LSB) first. Only then can the carries

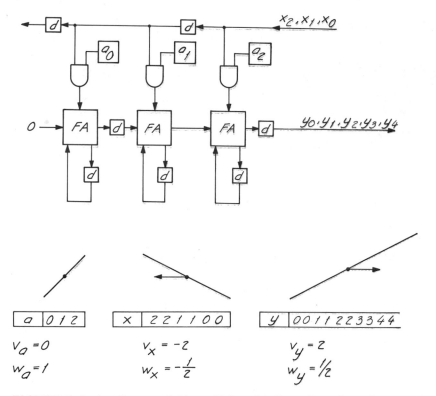

FIGURE 2.6 Another serial/parallel multiplier. Notations for the three strings.

be accumulated within the cell. Without loss of generality and avoiding mirrored equivalent solutions, we assume that the y string moves to the right.

The speeds and the slopes of the x string and the a string have to meet the special criterion

$$|v_x - v_a| < \left|\frac{1}{w_a w_x}\right| \tag{2.10}$$

since the relative speeds of the two strings must be low enough to allow all the bits of one string to combine with all the bits of the other string. In fact, if the "strictly less than" condition of Eq. (2.10) is valid, we will have a situation where several bits of one string combine with the same bit in the other string more than once. To avoid such inefficiencies we change Eq. (2.10) to the equality

$$\left| v_x - v_a \right| = \left| \frac{1}{w_a w_x} \right| \qquad (2.11)$$

There is also a lower limit for the sum of the slopes

$$\left| w_a \right| + \left| w_x \right| > 1 \qquad (2.12)$$

below which two neighboring cells may do the same computation. For example, $w_a = 1/2$, $w_x = 1/2$ and $w_a = 1/3$, $w_x = 2/3$ meet the criterion (2.12), while $w_a = 1/3$, $w_x = 1/3$ does not.

Let use now introduce the restriction that the string of constants is preloaded and static. Then $v_a = 0$ and Eq. (2.8) simplifies to

$$w_y v_y = w_x v_x > 0 \qquad (2.13)$$

while Eq. (2.11) simplifies to

$$\left| v_x \right| = \left| \frac{1}{w_a w_x} \right| \qquad (2.14)$$

Furthermore, the static string of a-bits must not coneceal any of its bits from the logic cells. Nor should it display its bits to the full adders in an irregular fashion. If one bit in the static a-vector is more exposed than another to the computational part, the design will be irregular without purpose.

The quantities $w_y v_y$ and $w_x v_x$ are equivalent to the bit rate of the output and the input string, respectively. From Eqs. (2.13) and (2.14) we get

$$\left| w_y v_y \right| = \left| w_x v_x \right| = \left| \frac{1}{w_a} \right|$$

If we want to use the simple cell of Fig. 2.3 with internal accumulation of the carry signal and keep the strict bit-serial approach, these bit rates have to equal 1. Therefore, Eq. (2.13) simplifies to

$$w_y v_y = w_x v_x = \left| \frac{1}{w_a} \right|$$

Thus the a bits are to be distributed along the array, one bit after the other in ascending or descending order. In this way, Eq. (2.12) simplifies to

$$|w_a| = 1 \quad \text{and} \quad |w_x| > 0$$

which is trivial.

The totality of equations and constraints is then reduced to

$$w_y = w_a + w_x > 0 \tag{2.7}$$

$$w_y v_y = w_x v_x = \left| \frac{1}{w_a} \right| = 1 \tag{2.15}$$

A catalog of some integer and rational number solutions to the set Eqs. (2.7) and (2.15) is given by Table 2.1. The number of solutions are, of course, unlimited and the table includes only the simplest ones. The more complex the ratios for the parameters v and w, the less regular is the distribution of the delay elements. Figure 2.7 shows implementations in a stylized form for two of the solutions in the table.

It is relatively easy to regard the various solutions simply as movements of the delay elements, at least for the simple soultions Fig. 2.6, Fig. 2.5b, and Fig. 2.5a, respectively. The value of the formalism presented here is not only that it demonstrates the complete set of possible solutions, but also that it is indispensible for finding optimal solutions for serial/parallel convolvers in next sections.

4 PROGRAMMABLE SERIAL/PARALLEL CONVOLVER

The serial/parallel multiplier is in fact a convolution of two bit strings in the sense that a specific bit of the Y string

$$y_i = \sum_{k=0}^{K-1} x_{i-k} a_k + \text{carries}$$

$$y_{i+1} = \sum_{k=0}^{K-1} x_{i+1-k} a_k + \text{carries}$$

and so on.

This indicates that the whole serial/parallel concept can be carried over from the bit level to the word level of our computational problem. Thus we should be able to compute the sums

TABLE 2.1

Solution	w_a	w_y	v_y	w_x	v_x	
A1	+1	+1/4	+4	−3/4	−4/3	
A2	+1	+1/3	+3	−2/3	−3/2	Fig. 2.7a
A3	+1	+2/5	+5/2	−3/5	−5/3	
A4	+1	+1/2	+2	−1/2	−2	Fig. 2.6, 2.15
A5	+1	+3/5	+5/3	−2/5	−5/2	
A6	+1	+2/3	+3/2	−1/3	−3	
A7	+1	+3/4	+4/3	−1/4	−4	Fig. 2.11b
B1	+1	+5/4	+4/5	+1/4	+4	
B2	+1	+4/3	+3/4	+1/3	+3	
B3	+1	+3/2	+2/3	+1/2	+2	Figs. 2.14, 2.16
B4	+1	+5/3	+3/5	+2/3	+3/2	
B5	+1	+2	+1/2	+1	+1	Figs. 2.5b, 2.10
B6	+1	+3	+1/3	+2	+1/2	Fig. 2.11a
B7	+1	+4	+1/4	+3	+1/3	
C1	−1	+1/4	+4	+5/4	+4/5	
C2	−1	+1/3	+3	+4/3	+3/4	
C3	−1	+2/5	+5/2	+7/5	+5/7	
C4	−1	+1/2	+2	+3/2	+2/3	Fig. 2.7b
C5	−1	+2/3	+3/2	+5/3	+3/5	
C6	−1	+3/4	+4/3	+7/4	+4/7	
C7	−1	+1	+1	+2	+1/2	Fig. 2.5d
C8	−1	+5/4	+4/5	+9/4	+4/9	
C9	−1	+3/2	+2/3	+5/2	+2/5	
C10	−1	+2	+1/2	+3	+1/3	Fig. 2.11c

Per E. Danielsson

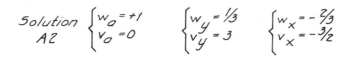

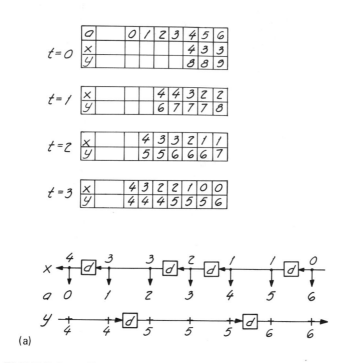

FIGURE 2.7 Two more solutions from Table 2.1.

$$Y^{(i)} = \sum_{l=1}^{L} X_l^{(i)} A_l = \sum_{l=1}^{L} X^{(i-1)} A_l$$

$$Y^{(i+1)} = \sum_{l=1}^{L} X_l^{(i+1)} A_l = \sum_{l=1}^{L} X^{(i+1-l)} A_l$$

and so on, with a structure on the word level that repeats the serial/
parallel structure of the bit level. We call such implementations
serial/parallel convolvers.

Figure 2.8 shows the same scheme as in Fig. 2.7a, the only
difference being that the bit-multiplying and gates and bit delay

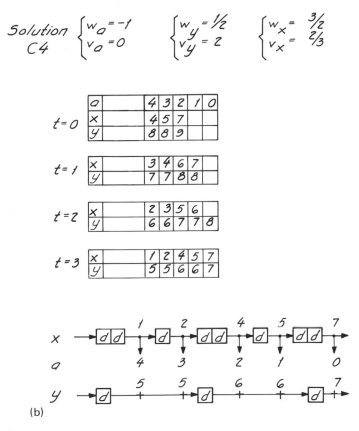

(b)

FIGURE 2.7 (Continued)

elements of Fig. 2.7a are replaced by serial/parallel multipliers (SP) and word delays (D), respectively. We also introduce the quantities $w_A, v_A, w_Y, v_Y, w_X, v_X$ for the string of words in analogy with the string of bits. It should be noted that in the Fig. 2.8 case, we are completely free to use one solution from the catalog in Table 2.1 for the serial/parallel multipliers and a completely different one for the convolver on the word level. The correctness of such designs are considered to be self-evident.

By utilizing our knowledge about the internal structure of the SP units in Fig. 2.8 we can serialize the whole convolver to the structure of Fig. 2.9. This linear array consists of identical cells and the SP multipliers are embedded in this structure at equidistant intervals. The generality of this procedure is established by the following theorem.

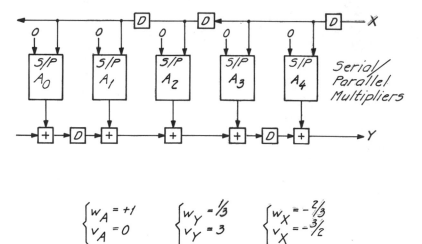

$$\begin{cases} w_A = +1 \\ v_A = 0 \end{cases} \qquad \begin{cases} w_Y = \frac{1}{3} \\ v_Y = 3 \end{cases} \qquad \begin{cases} w_X = -\frac{2}{3} \\ v_X = -\frac{3}{2} \end{cases}$$

FIGURE 2.8 Serial/parallel convolver with word paths orthogonal to bit paths.

Theorem

A (long) serial/parallel multiplier of any type having a sufficient number of cells can be used to compute a convolution sum by placing the constants $A_1, A_2, \ldots, A_L$ as bands at equidistant positions surrounded by bands of 0's. The total number of cells per pitch is D, where D is the number of bits in the output variable.

Proof: Figure 2.10 will help the reader visualize the essence of the theorem. The input variables X of this example are moving to the right twice as fast as the output variables Y. While passing A_1, Y_3 is incremented by the amout $A_1 X_2$. For correct function the distance to A_2 should be such that when X_2 enters this nonzero band at time I + t, Y_2 should then have reached the same position relative to A_2 as Y_3 and X_2 had relative to A_1 at time I. Assume that Y contains D bits. Typically, D = log L + K + N and the length

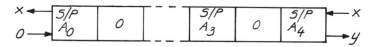

FIGURE 2.9 The programmable serialized convolver embedded in a serial/parallel multiplier structure.

of the Y string in terms of cell units is D/w_y. From Fig. 2.10 we conclude that

$$(v_x - v_y)t = \frac{D}{w_y}$$

where

$$t = \frac{D}{w_y v_x - w_y v_y} = \text{elapsed time}$$

for the X vector to move from one Y value to the next. From Eqs. (2.15) and (2.7) we get

$$v_x t = \frac{D}{w_y - \frac{w_y v_y}{v_x}} = \frac{D}{w_y - w_x} = \frac{D}{w_a} = \pm D$$

which proves the theorem since the X vector is exactly in place for the next A value if D is the number of cells per pitch.

Figure 2.11 shows several examples of events that illustrate the theorem. The important corollary is that any serial/parallel multiplier of sufficient length can be used as a full convolver by allowing for correct amount 0 space between the coefficient bands. The cellular hardware is absolutely modular and can be extended indefinitely. It is programmed for a certain convolver size L·K·N as soon as the coefficient bits are loaded into their positions.

Since the word limits for the static preloaded constants A and the moving varibles Y and X are no longer fixed, the cells of the programmable array have to be slightly more complex than the simple cell of Fig. 2.3 (see Fig. 2.12).

The y-signal is accompanied in parallel by a "clear carry" signal which, at the beginning of each Y word, resets any carry set to one by a negative Y word. The x-signal is accompanied by a word-limit signal that defines a particular bit to be the sign bit. The sign bit (if 1) converts the A constant to its 2's-complement by inverting the a bits, and at the beginning of each Y value adds +1 to the LSB position.

In Fig. 2.12, x and y are traveling in opposite directions since we are using solution A4. However, the basic cell design is the same for all serial/parallel multipliers as long as 2's-complement representation is assumed. The verification of the design through formal proofs and examples is left out for the sake of brevity. Similar cell designs are found, for example, in Refs. 15 and 16.

Per E. Danielsson

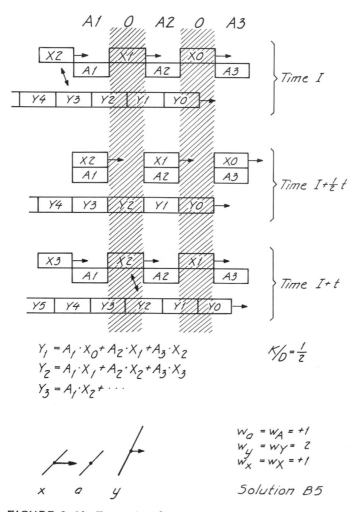

$$Y_1 = A_1 \cdot X_0 + A_2 \cdot X_1 + A_3 \cdot X_2$$
$$Y_2 = A_1 \cdot X_1 + A_2 \cdot X_2 + A_3 \cdot X_3$$
$$Y_3 = A_1 \cdot X_2 + \cdots$$

$$K/_D = \tfrac{1}{2}$$

$$w_a = w_A = +1$$
$$w_y = w_Y = 2$$
$$w_x = w_X = +1$$

Solution B5

FIGURE 2.10 Example of a serialized convolver, illustrating the theorem.

5 NONPROGRAMMABLE CONVOLVERS

A nonprogrammable convolver with serial/parallel multipliers of fixed sizes was shown in Fig. 2.8. This structure is fairly effective and should be rather easy to implement. Its regularity suffers due to its two-dimensional layout. This makes it worthwhile also to investigate the serialized model of Fig. 2.9 in this case.

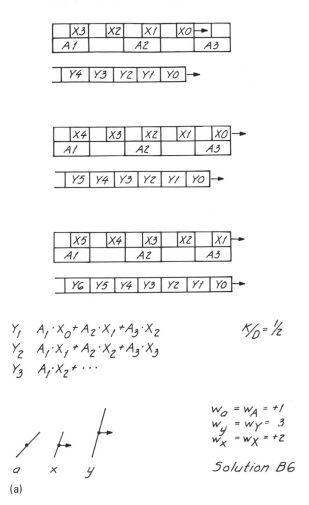

(a)

FIGURE 2.11 More examples of programmable serial/parallel convolvers.

The added complexity of the cells in Fig. 2.12 and the two extra word-limit signals are one price we pay to obtain a programmable array. Another price is the 0 bands of inactive cells. Actually, as seen from Fig. 2.11c, with K/D = 1/2, only one-fourth of the cells are really active in accumulating significant bit contributions $a_{lk}x_{ln}$. This is not unusual, however, even for nonprogrammable systolic arrays [9,10].

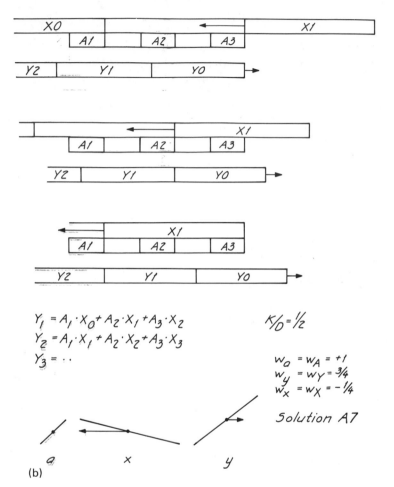

$$Y_1 = A_1 \cdot X_0 + A_2 \cdot X_1 + A_3 \cdot X_2$$
$$Y_2 = A_1 \cdot X_1 + A_2 \cdot X_2 + A_3 \cdot X_3$$
$$Y_3 = \cdots$$

$$K/_D = {}^1\!/_2$$

$$w_a = w_A = +1$$
$$w_y = w_Y = {}^3\!/_4$$
$$w_x = w_X = -{}^1\!/_4$$

Solution A7

(b)

FIGURE 2.11 (Continued)

In the present case, if we are willing to sacrifice program-mability for a convolver with a given precision and size, the whole structure can be compressed by eliminating the 0 bands. In this way, the efficiency increases with the factor D/K, which typically is close to 2. Examples of such convolvers are shown in Fig. 2.13. The data strings are moved as before over the array. However, the pace of the X words when moving from one serial/parallel multiplier to the next is accelerated in Fig. 2.13a (derived from Fig. 2.11a) by short-cuts that feed data into one multiplier from the midpoint of the previous one. In Fig. 2.13b (derived from Fig. 2.11c) the conditions

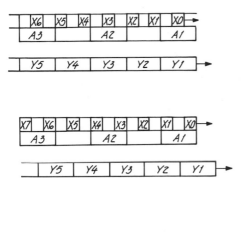

(c)

FIGURE 2.11 (Continued)

require that we instead decelerate the data with one-third of the delay of the multiplier itself. We will now explain the rationale behind these designs.

Let

$$w_A, w_X \text{ and } w_Y \text{ (with uppercase subscripts)}$$

denote the proper slopes for the word strings as before. As can be seen from Fig. 2.11a and c, these slopes are the same as the corresponding slopes of the bit strings before we cut out the 0 bands. Without the 0 bands and without accelerating or decelerating, we get the slopes

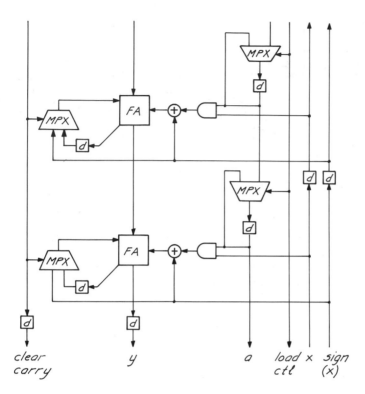

FIGURE 2.12 Cell structures for programmable serial/parallel convolvers.

$$w_A = w_a$$

$$w'_X = w_x \frac{K}{D}$$

$$w_Y = w_y \frac{K}{D}$$

since the pitch has decreased from D to K bits.
  The basic equation (2.7),

$$w_A + w'_X = w_Y$$

does not hold on the word level any more. However, the system is made correct [according to Eq. (2.7)] if we change the X slope to

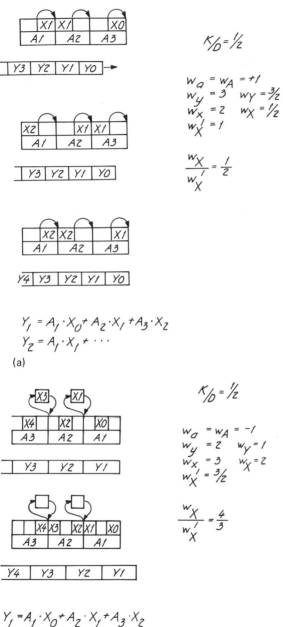

$$\frac{K}{D} = \frac{1}{2}$$

$$w_a = w_A = +1$$
$$w_y = 3 \quad w_Y = \frac{3}{2}$$
$$w_x = 2 \quad w_X = \frac{1}{2}$$
$$w_X' = 1$$

$$\frac{w_X}{w_X'} = \frac{1}{2}$$

$$Y_1 = A_1 \cdot X_0 + A_2 \cdot X_1 + A_3 \cdot X_2$$
$$Y_2 = A_1 \cdot X_1 + \cdots$$

(a)

$$\frac{K}{D} = \frac{1}{2}$$

$$w_a = w_A = -1$$
$$w_y = 2 \quad w_Y = 1$$
$$w_{x_1} = 3 \quad w_X = 2$$
$$w_X' = \frac{3}{2}$$

$$\frac{w_X}{w_X'} = \frac{4}{3}$$

$$Y_1 = A_1 \cdot X_0 + A_2 \cdot X_1 + A_3 \cdot X_2$$
$$Y_2 = A_1 \cdot X_0 + A_2 \cdot X_2 + A_3 \cdot X_3$$

(b)

FIGURE 2.13 Compressed nonprogrammable convolvers derived from Fig. 2.11(a) and (b) respectively.

$$w_X = w_Y - w_A = w_y \frac{K}{D} - w_a = w'_X - w_a \left(1 - \frac{K}{D}\right) \tag{2.16}$$

The difference between w'_X and w_X according to Eq. (2.16) constitutes the modification necessary for correctness. Normally, we may prefer acceleration instead of deceleration, the latter requiring extra delay elements.

Acceleration/deceleration conditions are not fully analyzed here for the sake of brevity. However, the net result is that solutions of type A and C in Table 2.1 all require deceleration when compressed, while solutions B5,B6, ... , require acceleration. Solutions B1 and B2 also require acceleration, although the x string on the bit level travels in the opposite direction to the X string on the word level.

$$\begin{cases} w_a = +1 \\ v_a = 0 \end{cases} \qquad \begin{cases} w_y = \frac{3}{2} \\ v_y = \frac{2}{3} \end{cases} \qquad \begin{cases} w_x = +\frac{1}{2} \\ v_x = 2 \end{cases}$$

$$\frac{K}{D} = \frac{1}{2} \qquad \begin{cases} w_Y = \frac{3}{4} \\ v_Y = \frac{4}{3} \end{cases} \qquad \begin{cases} w_X = -\frac{1}{4} \\ v_X = -4 \end{cases}$$

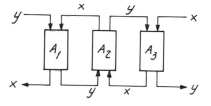

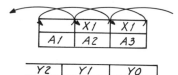

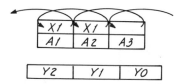

FIGURE 2.14 Unique solution B3 for compression factor $K/D = \frac{1}{2}$.

Figure 2.14 shows the solution B3 compressed with the factor $K/D = 1/2$ with folded data paths. For this case we have the unique situation that

$$w'_X = - w_X$$

This is exactly why this convolver is possible to compress without acceleration or deceleration. Also, the word strings for X and Y move in opposite directions, which makes it an excellent candidate for recursive filters, as will be shown in the next section.

This second property is shared by the solutions A1 to A7, but they all require substantial deceleration. Furthermore, the slope $w_y < 1$ for these solutions, which means that some of the cells have to propagate the y string over two cells in once clock cycle.

6 RECURSIVE FILTER DESIGN:
TWO-DIMENSIONAL CONVOLVERS

Recursive filters or IIR filters were mentioned in the introduction and the basic computation was given by Eq. (2.2). The systolic array implementation of this formula is rather self-evident using the serial/parallel convolver concept. One example, employing solution A4 in its programmable version, is shown by Fig. 2.15. Input data are fed into the middle of the array since there are four coefficients on each side. The output string is received to the right-hand side. Each output word $Y^{(i)}$ is also truncated, fed back, and used to compute the recursive part. The control signal generates the correct amount of zeros in the LS part of these data.

Since we must use each output word immediately, applied on its nearest neighbor, we cannot feed these data to the other side of the array. Instead, they have to make a quick U turn and consequently only solutions A1 to A7 in Table 2.1 are applicable.

For a compressed nonprogrammable convolver we also have an option to use the solution given by Fig. 2.14 (or the one modified to some $K/D \neq 1/2$). The recursive version is given by Fig. 2.16.

In the snapshort taken by Fig. 2.16b the MS half of $X^{(i)}$ combines with A2 accumulating to $Y^{(i+1)}$ while the LS half of $X^{(i)}$ combines with A1 accumulating to $Y^{(i)}$. The MSB of $X^{(i)}$ is, in fact, ready to jump over to the lower end of A3 and generate middle bit contributions to $Y(i+2)$. As before, half of the X data path is filled with zeros (assuming that $K/D = 1/2$). An extra delay at the feedback of the Y values is also necessary. In this case it amounts to one-third of the delay per pitch in the y data path.

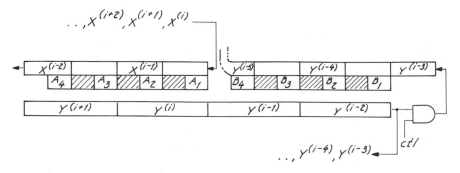

FIGURE 2.15 Recursive (IIR) filter employing solution A4.

Figure 2.17 shows parts of this convolver in detail. Note that for most of the cells we can use a design almost as simple as in Fig. 2.3. The sign bits of the constants have slightly more complex cells. The y-word limit is a control signal that travels with a fixed bit pattern 00 . . . 01 along with each y word. It clears the carries and adds a 1 in LSB if the sign bit of the constant is set.

All convolvers presented so far have been designed for one-dimensional data. Normally, a one-dimensional convolver can be extended and used also for two-dimensional data assuming that these are available in raster-scanned format, say, left-right, top-bottom.

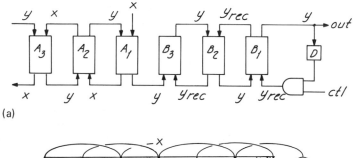

(a)

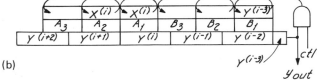

(b)

FIGURE 2.16 Recursive (IIR filter) employing the unique solution from Fig. 2.14.

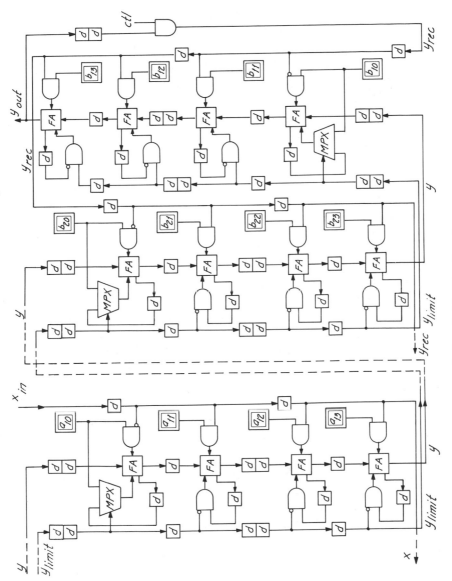

FIGURE 2.17 Detailed version of Fig. 2.16. K = R = 4.

Per E. Danielsson

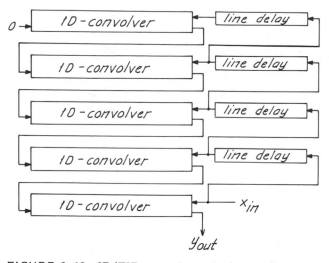

FIGURE 2.18   2D/FIR convolver designed from 1D serial/parallel
convolvers.

    Figure 2.18 shows the well-known principle for the nonrecursive
case, using a set of line delays for the input string. Hereby, we make
a two-dimensional neighborhood of data points available to the appa-
ratus that consists of chained one-dimensional convolvers. The actual
delay time should be somewhat shorter than one full line cycle (LC)
since a specific Y data at a specific time is found on one line only.
Assume that the delay of the Y string over one one-dimensional convolver is n time units. The delay per line for the X string should
then equal LC-n.
    There is only one y string that runs continuously, while there
are as many x strings as there are rows in the convolution kernel.
It would be possible to reverse the matter, that is, to use one continuous x string which delivers its contributions to different Y results.
Consequently, line delays will then be needed for the output strings
instead of the input strings as in Fig. 2.18.
    The geometry of Fig. 2.18 allows us to visualize the convolution
kernel sliding over an image of data points left-right, top-bottom.
The "oldest" x value used by the system is found at the upper left-
hand side corner.
    Any of the previously presented serial/parallel convolvers can
be used for the nonrecursive filters portrayed in Fig. 2.18. For the
recursive two-dimensional case, however, we have to make the same
U turn with the output string as we did in Fig. 2.15. Consequently,
only solutions with countermoving x and y strings can be employed.
Again, let us visualize the convolution kernel as a template moving
over the image of data points left-right, top-down. Very often the

kernel is symmetric in shape, meaning tht input data for the lower half of the template are taken from x values, while the input data for the upper half are taken from y values fed back from the output. These considerations lead to an overall structure shown by Fig. 2.19. The y string starts to the left of the center of the middle row, which is also the place of y output and the U turn.

7 MCWIRTHER-MCCANNY CONVOLVER

We will now take a closer look at the convolver design presented in Refs. 9 and 10. The basic principle is given in Fig. 2.20. It consists of cells which propagate carries to the left just as ordinary adders do; cells through which the X strings move to the right and the A constants to the left. The sum bits (the Y values) propagate downward to an accumulator made up of cells almost identical to those of the rest of the convolver. The constants are actually recirculated as shown in Fig. 2.20b. The different X strings are shown separately but could easily be chained together.

Like many other systolic array designs [5] the input data strings in Fig. 2.20 are interleaved with blanks (zero bits). This makes three-fourths of the cells nonactive or semiactive, which is the same efficiency that we achieved for programmable convolvers in Sec. 4. The convolver shown in Fig. 2.20 is nonprogrammable, but for this case the serial/parallel convolvers in Sec. 5 had an efficiency that was (typically) twice as high since actual cells were only needed at the nonzero bits of the constants.

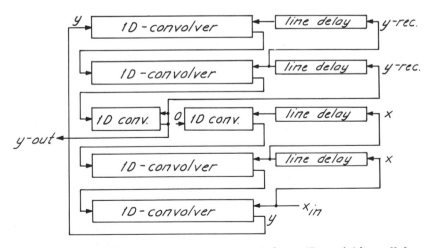

FIGURE 2.19 2D IIR convolver designed from 1D serial/parallel convolvers.

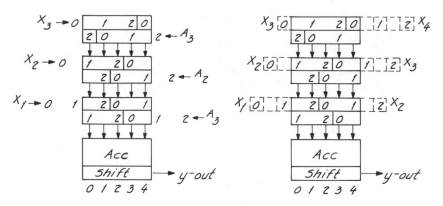

FIGURE 2.20 McWirther-McCanny convolver.

It is interesting to note that the formalism of this paper for describing moving bit strings is applicable to the McWirther-McCanny convolver. However, our condensed equation set (2.7) and (2.15), used in Sec. 4 and 5, no longer holds. Instead, we have to go back to the basic equations (2.7) and (2.8):

$$w_y = w_a + w_x \tag{2.7}$$

$$w_y v_y = w_a v_a + w_x v_x \tag{2.8}$$

In Fig. 2.20, let the horizontal axis correspond to the z axis. Then we have immediately

$$v_y = 0 \qquad w_y = + 1$$

so that

$$w_a + w_x = 1 \tag{2.17}$$

$$- w_a v_a = w_x v_x = 1 \tag{2.18}$$

$w_x v_x$ is the bit rate which, by definition, is one for bit-serial systems. Equation (2.11) is still valid:

$$|v_x - v_a| = \left| \frac{1}{w_a w_x} \right| \tag{2.11}$$

which is to say that we do not want to duplicate or miss any $a \cdot x$ bit contributions. Fig. 2.20 deliberately violates Eq. (2.11) by the inter-leaving 0's in the strings.

Finally,

$$|w_a| + |w_x| > 1 \qquad\qquad (2.12)$$

still holds.

The set of equations and inequalities (2.11), (2.12) (2.17) (2.18) has a number of solutions, some of which are given in Table 2.2. The simplest solutions D3 and E3 are shown by Fig. 2.21a and b, re-spectively. In both cases we display the behavior only. A physical implementation should fold the A paths and chain the x paths together using appropriate delays or shortcuts.

It seems that the efficiencies of these designs are the same as for the original Fig. 2.20. On average, only one-fourth of the cells are fully employed in the accumulation of new bit contributions since both the X and the A strings are followed by 0 parts of the same size as their significant bit string.

A possible way to compress the X and A strings by deleting these 0 bands is as follows. Let each X and A word be accompanied by a word-limit signal. Each cell is supplemented with an activity flip-flop the state of which is changed each time it is passed by an X or an A word limit. As shown in Fig. 2.22, an "activity window" opens and closes in each row. The net effect is that we can compress the

TABLE 2.2

	w_y	w_a	v_a	w_x	v_x
D1	+1	1/4	−4	3/4	4/3
D2	+1	1/3	−3	2/3	3/2
D3	+1	1/2	−2	1/2	2
D4	+1	2/3	−3/2	1/3	3
E1	+1	−1/3	3	4/3	3/4
E2	+1	−1/2	2	3/2	2/3
E3	+1	−1	1	2	1/2
E4	+1	−2	1/2	3	1/3

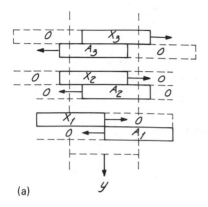

(a)

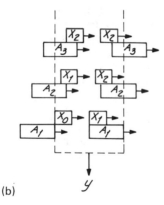

(b)

FIGURE 2.21 Two variations of the McWirther-McCanny convolver.

data streams compared to Fig. 2.21 and get rid of the 0 bands completely. On average, one-half of the cells are engaged in accumulation.

    Since we have not decreased the size of the array, it follows
that the doubled efficacy has been obtained by doubling the data
rate. Thus the accumulation of carries at the bottom of the array (not
shown in Fig. 2.20) has to be made twice as fast. Also, if the y output is wanted with doubled precision, two bits/cycle have to be output at a time.

    A final word should be said about the possibility to implement
recursion. This cannot be done in the original design of Fig. 2.20.
The last bit accumulation takes place in the middle position of the Y
word in the bottom row. But at the same time, this bit is wanted half-
way inside the very same row as the next input data.

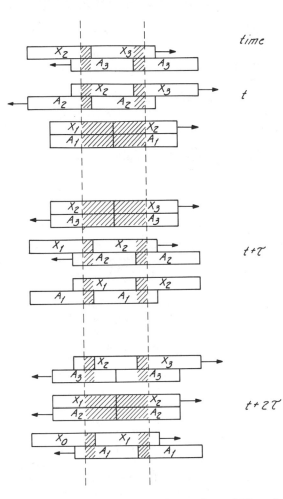

FIGURE 2.22 Compressed version of Fig. 2.21(a) with increased efficacy.

The convolver of Fig. 2.21a seems possible to use for recursion since there is enough time between production of the output bits and their possible entry in the bottom row. This is not the case in Fig. 2.21b or in the compressed version of Fig. 2.21a, Fig. 2.22. Generally, the serial/parallel convolvers of Secs. 4 and 5 with their stricter serial behavior are easier to use in recursive mode than the McWirther-McCanny convolver. This is because the output of the latter is not serial by nature but has to be serialized by an extra accumulator and shifter.

8 CONCLUSION AND DISCUSSION

A large variety of convolvers has been discussed in this chapter.
Figure 2.23 tries to summarize them in a kind of family tree with
emphasis on the systolic part.

Previously published systolic convolvers have mostly employed
word serial bit parallel techniques, very often by using rather complex cells as exemplified by Fig. 2.1. One alternative is to go down
to the bit level and use the simple full adder as the basic cell. This
will allow very high clock rates that keep 45° sloped waves of Y values
in motion. One solution (not treated here) is then obvious: Introduce
internal pipelining in combinational multipliers and cascade such
units to cover the wanted convolution kernel. A more compact solution
was shown in Fig. 2.2, where the accumulation takes place in a different order, least significant bit first.

Convolvers of the type represented by Fig. 2.2 have two major
drawbacks. They cannot be used for recursive filters, and their pin
count can become fairly high, at least compared to the strictly bit-
serial approach.

Bit-serial convolvers can utilize the serial/parallel multiplier
concept. This component is in itself a systolic design with the full
adder as its basic cell. With a certain amount of formalization it was
possible to capture the whole family of serial/parallel multipliers, and
a representative subset of solutions was collected in Table 2.1.
Thanks to the fact that convolution is on the word level a true mirror
of the multiplier behavior on the bit level, there exists a corresponding family of serial/parallel convolvers using the serial/parallel
multiplier as the basic cell instead of the full adder. This observation
lead directly to the structure pictured in Fig. 2.8.

An even more interesting possibility has also been shown in
this chapter. The one-dimensional cellular array of a (long) serial/
parallel multiplier can be used to embed a whole convolver. By making
room for bands of zeros between the cells loaded with significant A
bits, this structure also becomes programmable. One can use the
same array for many low-precision coefficients or for a few high-
precision coefficients by loading it with different coefficient bits.
The programmability requires two lines for indication of word limits
of the moving bit strings x and y. The structure can be cascaded
indefinitely.

The efficiency of the programmable serial/parallel convolver is
the same as for most systolic arrays found in the literature. However,
with few exceptions, these are not programmable but designed for a
certain problem size.

By using the formalism introduced to describe the behavior of
the serial/parallel multiplier, it was possible to transform the convolvers of Sec. 4 into compressed versions with (typically) doubled

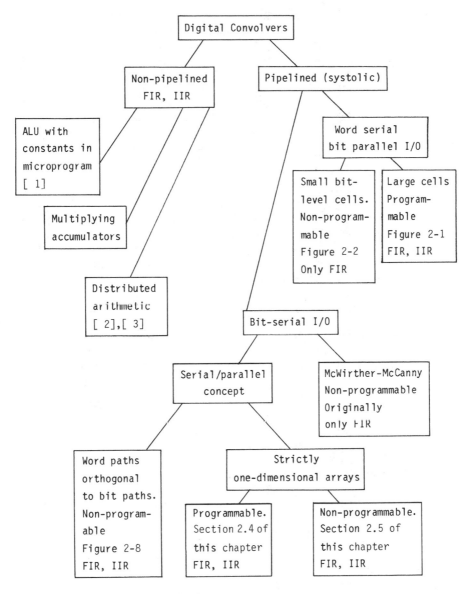

FIGURE 2.23 Convolver family tree.

efficacy. The nonprogrammability of these new convolvers is manifested by the fact that certain shortcuts or delays have to be added to the structure depending on the word lengths involved.

Recursive = IIR filters were shown to be implementable by the serial/parallel approach even if the set of possible solutions is only a subset of the ones given in Table 2.1. A necessary condition is that the input and output strings move in different directions so that the output data can make a U-turn and immediately be fed back.

Two-dimensional convolution is implemented simply by cascading a number of one-dimensional convolvers and delaying the in-data string sufficiently between each row. Recursive two-dimensional filtering was demonstrated in Fig. 2.19.

A convolver with bit-serial input/output has been described by McWirther-McCanny [9]. In this design, shown by Fig. 2.20, the A constants move in an opposite direction to the X values. They y bits also move, but orthogonally so that the whole structure becomes two-dimensional and fixed for a certain problem size. The efficiency is the same as for the programmable serial/parallel convolver. In this chapter we used the basic formalism for the serial/parallel multiplier to reveal new versions of this design. A subset of these versions, typified by Fig. 2.21a, seems to be open for some modifications that doubles its efficiency by deleting 0's which increases the bit rate with a factor of 2. A certain amount of complexity is added to the basic cell. The same set of new versions can also be used for recursive filtering. In this case, only the unmodified low-efficiency variety is applicable.

In summary, the serial/parallel convolvers of Secs. 4 to 6 seem to be advantageous in many respects:

1. They are programmable while retaining high regularity, high simplicity, and reasonable efficiency.
2. Many of them can be used for recursive filter implementation.
3. Some compressed and nonprogrammable verions, typified by Figs. 2.14, 2.16, and 2.17, have the unique properties of high efficiency, regularity, and simplicity, and can also be applicable to recursive filtering.

Beside the actual designs, the above-mentioned formalism could possibly point toward new ways to design systolic arrays in general. Many new variations are possible, which will be shown by a final example taken from [4, pp. 285—287]. The basic version presented there is shown by Fig. 2.24a.

An input data set is arriving in triangular and interleaved fashion from above. After every a-data item follows a blank (= zero). Similarly, every other y-data and x-data item is zero. The x values

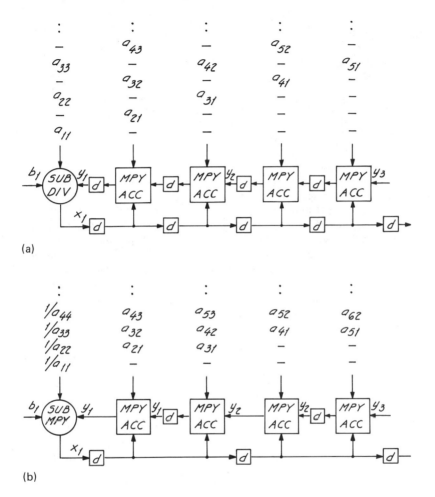

FIGURE 2.24 (a) Systolic array adopted from Ref. 4. Word-wide data paths. (b) Modification of (a) for double efficacy provided that the same clock frequency can be employed.

are produced at the left end of the array and in time sequence follow the computations

$$x_1 = \frac{b_1 - y_1}{a_{11}} \qquad t = 1$$

$$y_2 = a_{21}x_1 \qquad t = 2$$

$$x_2 = \frac{b_2 - y_2}{a_{22}} \qquad\qquad t = 3$$

$$y_3 = a_{31}x_1$$

$$y_3 = a_{31}x_1 + a_{32}x_2 \qquad\qquad t = 4$$

$$y_4 = a_{41}x_1$$

and so on.

Obviously, only half of the cells are active. Considering the time for initialization and emptying the pipeline before the next data set can be entered, the efficiency goes down to one-fourth.

A more efficient structure is shown by Fig. 2.24b. All data paths are now filled with significant data and the data rate is twice as high as that in Fig. 2.22a in terms of bit/clock cycle. The question is whether or not it is possible to keep the same cycle time in the new version. Between each clocked register set in the array, there are now two MPY/ACC. Since these processes need to be combined only in the final accumulation part, the requirement for increased cycle time will be marginal.

The singular SUB/DIV module is a greater obstacle. In Fig. 2.24a, one can allow the SUB/DIV operation to consume two clock cycles. This is not so in Fig. 2.24b. Here SUB/DIV and one MPY/ACC should be performed in one cycle time. A viable solution would then be to produce the inverted quantities $1/a_{11}, 1/a_{22}$, in an extra preceding procedure that can be pipelined over array clock cycles. The end module would then perform one SUB/MPY and one MPY/ACC operation which is comparable in complexity to two MPY/ACCs. The increase in efficiency will then (we hope) be close to a factor of 2.

ACKNOWLEDGMENTS

The main part of the material contained in this chapter was conceived and written in June 1981 while the author was a visiting scientist with IBM Research Division, San Jose. California.

REFERENCES

1.   A. Peled, On the hardware implementation of digital signal processors, *IEEE Trans. Acoust. Speech Signal Process. ASSP-24*: 76−86 (1976).

2. A Croisier, D. J. Esteban, M. E. Levilion, and V. Riso, Digital filters for PCM encoded signals, U.S. patent 3777130, (1973).

3. A Peled and B. Liu, A new hardware realization of digital filters, *IEEE Trans. Acoust. Speech Signal Process. ASSP-22*: 456–462 (1974).

4. H. T. Kung and C. E. Leiserson, in *Introduction to VLSI Systems* (C. Mead and L. Conway, eds.) Addison-Wesley, Reading, Mass., Chap. 8.3 (1980).

5. H. T. Kung and S. W. Song, A systolic 2-D convolution chip, *VLSI Document VO46* Carnegie-Mellon University, Pittsburgh, Pa. (1981).

6. J. Blackmer, G. Frank, and P. Kuekes, A 200 million operations per second (MOPS) systolic processor, *Proc. SPIE, 298*:10–18 (1981).

7. P. E. Danielsson, Iterative (systolic) arrays, *IBM Tech. Discl. Bull., 25*(8):4125–4127 (1983).

8. P. Denyer and D. Myers, Carry-save adders for VLSI signal processing, *VLSI 81* (J. P. Gray, ed.) Academic Press, New York 151–160 (1981).

9. J. G. McWirther and J. V. McCanny, A novel multibit convolver/correlator chip design based on systolic array principles, *Proc. SPIE 341*:66–74 (May 1982).

10. D. Wood, R. A. Evans, and K. W. Wood, An 8-bit serial convolver chip based on a bit level systolic array, in *Proc. Custom Integrated Circuits Conference*, Rochester, N.Y. 256–261 (May 1983).

11. L. Wanhammar, An approach to LSI implementaion of wave digital filters, dissertation, Linkoping University, Sweden (1981).

12. L. Jackson, J. Kaiser, and H. McDonald, An appraoch to the implementation of digital filters, *IEEE Trans. Aud. Electroacoust., AU-16*:413–421 (1968).

13. S. L. Freeny, Special purpose hardware for digital filtering, *Proc. IEEE, 63*:633–648.

14. D. Hampel, K. McGuire, and K. Post, CMOS/SOS serial/parallel multiplier chip, *IEEE J. Solid-State Circuits, SC-11*: 669–678 (1976).

15. R. F. Lyon, Two's complement pipeline multipliers, *IEEE Trans. Commun., COM-24*:418–425 (1976).

16. J. Kane, A low-power, bipolar, Two's complement serial pipeline multiplier chip, *IEEE J. Solid-State Circuits, SC-11* (1976).

17. E. E. Swartzlander, Jr., and B. Gilbert, Arithmetic for ultra-high-speed tomography, *IEEE Trans. Comput., C-29*:341–353 (1980).

3

Systolic Algorithms
for the CMU Warp Processor

H. T. KUNG *Department of Computer Science, Carnegie–Mellon University, Pittsburgh, Pennsylvania*

1 INTRODUCTION

CMU, with its industrial partners, has developed a 32-bit floating-point programmable systolic array for the high-speed execution of many essential computations in signal and image processing, such as the fast Fourier transform (FFT) and convolution. This is a one-dimensional systolic array that in general takes inputs from one end cell and produces outputs at the other end, with data and control all flowing in one direction. The initial version of the machine has 10 cells, each of which is capable of performing 10 million floating-point operations per second (10 MFLOPS) and is built on a single board using only off-the-shelf components. We call this machine the *Warp processor,* suggesting that it can perform various transformations at a very high spped.

We expect to have wide applications for Warp because of its high performance and high degree of programmability. A 10-cell processor, for example, can process 1024-point complex FFTs at a rate of one FFT every 600 μs. Under program control, the same processor can perform many other primitive computations in signal and image processing and in low-level vision, including two-dimensional convolution and complex matrix multiplication, at a rate of 50 to 100 MFLOPS. Together with another processor capable of performing fast divisions and square roots, the processor can also efficiently carry out a number of difficult matrix operations such as solving covariant linear systems and singular-value decomposition. In this chapter we outline the architecture of the Warp processor and describe how the signal processing tasks are implemented on the processor.

1.1 Background

Very high performance computer systems must rely heavily on parallelism, since there are severe physical and technological limits on the
ultimate speed of any single processor. Systolic array processors
offer an effective way of using a large number of processors in parallel
[1,2]. In recent years many systolic array algorithms have been
designed and several prototypes of systolic array processors have
been built [3—7]. Major efforts have now started in attempting to use
systolic array processors in large applications. Practical issues on
the implementation of systolic array processors have begun to receive
substantial attention.

   To implement systolic arrays, appropriate architectures of the
underlying processors must be developed. Architectural optimization
for the implementation of a narrow set of algorithms is usually not very
difficult, but devising an architecture that can efficiently implement
a wide class of algorithms is nontrivial. The challenge is to achieve
a balance between many conflicting goals, such as the generality of
the system versus ease of programming, flexibility versus efficiency,
and performance of the system versus design and implementation
costs. A key research issue regarding the implementation of systolic
arrays is therefore the identification of processor architectures that
optimize the trade-off between these conflicting goals for specific
applications. The CMU Programmable Systolic Chip (PSC) project [8],
for example, represents a research effort in identifying architectural
features needed in a microprocessor in order to make it an efficient
building block for the implementation of systolic arrays for a variety
of applications. The Warp processor considered in this chapter, on
the other hand, resulted from research in identifying architectures
for high-performance programmable systolic arrays for specific signal
and image processing applications.

1.2 Chapter Outline

In Sec. 2 we describe briefly the architecture of the Warp processor
and features of the CMU prototype. Main results of this chapter are
in Sec. 3 to 5, which justify the architecture of the Warp processor
by showing how it can efficiently implement convolution, interpolation,
matrix multiplication, and the FFT. Section 6 contains some concluding
remarks and a brief discussion on the use of the Warp processor in
solving systems of linear equations.

   Since January 1984 when the first version of this chapter was
written. Warp has evolved from a concept to a real machine that is
being used routinely. A postscript is included to describe Warp circa
August 1986.

2 WARP PROCESSOR ARCHITECTURE

The Warp processor has three components—the Warp processor array
or simply Warp array, the interface unit, and the host—as depicted
in Fig. 3.1. We describe the machine only briefly here; more detail
is available separately [9,10]. The Warp array performs the bulk of
the computation. The interface unit handles the input/output between
the array and the host. The host has two functions: carrying out
high-level application routines and supplying data to the Warp array.

The Warp array is a one-dimensional or linear systolic array. In
general, the array takes inputs at the leftmost cell and produces
outputs at the rightmost cell, with data and control all flowing in one
direction. There are several advantages of having this simple interconnection scheme, besides the relative ease of its design, implementation, and use. Linear arrays require the minimum-possible input/
output (I/O), in the sense that only the two end cells communicate
with the outside world. Thus an n-cell Warp processor can perform
O(n) computations for each I/O operation, even if only a constant
number of computations are performed at each cell. This property is
desirable in practice, because usually the I/O bandwidth is the major
limiting factor for achieving high performance. Linear arrays have
the additional advantage that they can always be safely synchronized
by a simple, global clock [11]. Finally, by having data and control
all flow in one direction, we can use efficient fault-tolerant techniques
to deal with faulty cells in a systolic array [12].

Each cell in the Warp array, called a Warp cell, has its own
program sequencer and has the data path shown in Fig. 3.2. The Warp
cell is currently implemented with commerically available components,
although a custom very large scale integration (VLSI) circuit implementation which will be developed in the future will be more efficient. The
following are the major features of the Warp cell data path:

Add and Mpy. These are arithmetic units for addition and multiplication operations, respectively. For the CMU prototype, the
Mpy and Add are implemented with commercial 32-bit floating-
point multiplier and adder chips, respectively [13]. To achieve
the maximum-possible throughput for the Warp cell, these chips
are used in their pipeline mode. That is, a chip starts a new
32- bit arithmetic operation every cycle (which is 200 ns for the
prototype), although the result of an operation will not emerge
from the chip's output ports until several cycles after the operation starts. The Warp array built from these pipelined arithmetic
units therfore supports pipelining at both the array and the
cell levels. These two levels of pipelining greatly enhance the
system throughput [12,14].

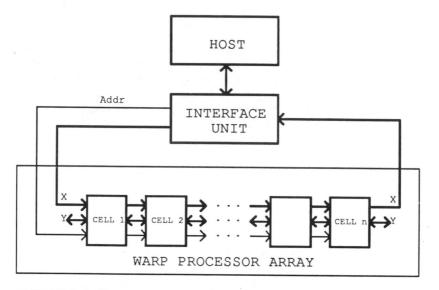

FIGURE 3.1 Warp machine overview.

A-Reg and M-Reg. These are general register files for Alu and Mpy,
respectively. Each register file contains 32 32-bit wide registers
accessible from any of six data ports. The M-Reg can also compute approximate inverse and inverse-square-root functions,
using the look-up table contained on the register file chip.
X-Queue, Y-Queue, and Addr-Queue. These queues (of 128 words
each) are provided mainly to ensure that X, Y, and Addr streams
are properly synchronized, as required by systolic algorithms.
Data random access memory (RAM). Having a memory at each cell for
buffering data, implementing look-up tables, or storing intermediate results is essential for reducing the I/O bandwidth requirement of the cells. Also by using its local memory to store
temporary data, a cell can be multiplexed to implement the
functions of multiple cells in a systolic array design. As a result,
for example, Warp can implement algorithms designed for two-
dimensional systolic arrays or one-dimensional arrays that have
more cells than the one-dimensional array of the machine. For
the CMU wire-warp prototype, the data RAM in each cell have 4K
words; for the printed circuit board implementation, the data
RAM in each cell has 32K words. The memory can perform both
a read and a write simultaneously every cycle, using addresses
selected from the Addr-Queue, crossbar, or the data RAM itself
(i.e., indirect addressing).

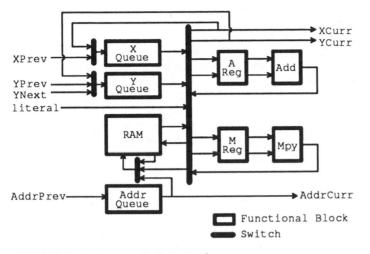

FIGURE 3.2 Warp cell data path.

Two data I/O ports (X, Y) and one address I/O port (Addr). The
 Warp cell can input as well as output two words, and a pair of
 (read/write) addresses for the data RAM, every cycle.
Crossbar. The arithmetic units, data RAM, register files, and I/O
 ports of the Warp cell are linked by a crossbar, which is shown
 as the large switch in the middle of Fig. 3.2. The crossbar has
 eight read ports, including one that accepts literals from microcode, and six write ports. The crossbar can be reconfigured
 every cycle under control of microcode to allow a read port to
 get data from any of the six write ports.
Input switches. These are used to implement computations using the
 wraparound or bidirectional data flow mode. In the wraparound
 mode the outputs of the cell is fed back to its inputs, hence
 wrapping around the cell. This mode multiplexes the use of one
 cell to implement the function of several. (The same effect can
 also be achieved through the use of other resources, such as the
 data RAM.) This increases the virtual size of the array for
 problems requiring larger array size. In the bidirectional data
 flow mode the Y input of each cell can take values from the Y
 output of the next cell, that is, the cell to the right. As to be
 discussed in Sec. 6, this feature allows the Warp array to
 implement linear systolic arrays with bidirectional data flows.

3 CONVOLUTION WITH THE WARP PROCESSOR

We first present systolic array designs for one-dimensional convolution that the Warp array can implement efficiently. Then we extend these designs to handle two-dimensional convolutions.

3.1 One-Dimensional Convolution

The one-dimensional convolution problem is defined as follows: Given a kernel as a sequence of weights $(w_1, w_2, \ldots, w_k)$ and an input sequence $(x_1, x_2, \ldots, x_n)$, compute the output sequence $(y_1, y_2, \ldots, y_{n-k+1})$, defined by

$$y_i = w_1 x_i + w_2 x_{i+1} + \cdots + w_k x_{i+k-1}$$

Depicted in Fig. 3.3 is one of the many possible systolic array designs for one-dimensional convolution, given at an abstract level [1]. Weights are preloaded into the array, one for each cell. During the computation, both inputs x_i and partial results for y_i flow from left to right, but the x_i move twice as fast as the y_i. The speed difference ensures that each y_i can meet all the k consecutive x_i that it depends on. More precisely, each y_i stays inside every cell it passes for one extra cycle; thus it takes twice as long to march through the array as does an x_i. It is an easy exercise to see that each y_i, initialized to zero before entering the leftmost cell, is indeed able to accumulate all its terms while moving to the right. For example, y_1 accumulates $w_1 x_1, w_2 x_2$, and $w_3 x_3$ in three consecutive cycles at the first, second, and third cells from the left, respectively.

Figure 3.4 depicts the internal structure of two consecutive cells, assuming that multiplication together with addition can be done in one cycle. Note that the y data stream has two latches for each cell, as

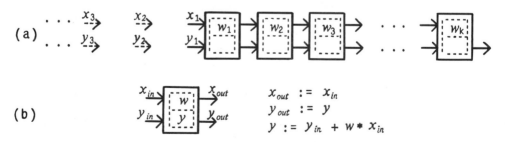

FIGURE 3.3 (a) Systolic array for one-dimensional convolution using a kernel of size k, and (b) the cell specification.

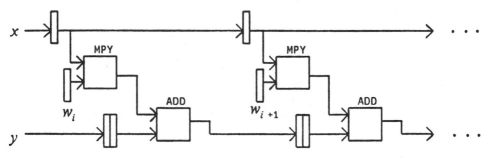

FIGURE 3.4 Cell structure for the one-dimensional convolution array of Fig. 3.3.

opposed to one latch for the case of the x data stream. Thus data on the two data streams travel at two sppeds, as required by the systolic array algorithm of Fig. 3.3.

For the Warp array, we use pipelined multiplier and adder, each having five pipeline stages. Figure 3.5 shows that it takes five instead of two steps for each y_i to pass a cell. To compensate the additional three delays on the y data stream, we use $4(= 3 + 1)$ latches on the x data stream for each cell [12].

Figure 3.6 shows another design that the Warp array can implement. In this design weight w_i used by each cell at every cycle is selected on-the-fly from the data RAM by a systolic address flowing from cell to cell on the *addr* stream. By associating each y_i with an address on the *addr* stream, this design [14] allows different sets of weights to be used to compute different y_i, as requred, for example, in interpolation and resampling of signals. Note that by going through the same delay at each cell, that is, a delay of four steps for the design of Fig. 3.6, y_i and its associated address are synchronized in the sense that they arrive at each cell at the same time.

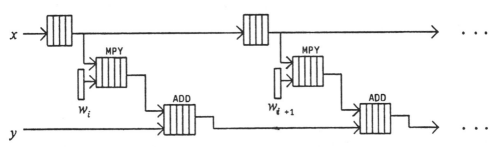

FIGURE 3.5 Two-level pipelined systolic array for one-dimensional convolution.

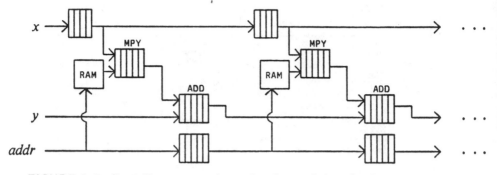

FIGURE 3.6 Systolic array using adaptive weights for interpolation and resampling.

3.2 Two-Dimensional Convolution

The two-dimensional convolution problem is defined as follows: Given the weights w_{ij} for i = 1, 2, . . . , k, and j = 1, 2, . . . ,p that form a k × p kernel, and an input image x_{ij} for i = 1, 2, . . . ,m and j = 1, 2, . . . ,n, compute the output image y_{ij} for i = 1, 2, . . .,m − k + 1 and j = 1, 2, . . . ,n − p + 1, defined by

$$y_{ij} = \sum_{h=1}^{k} \sum_{l=1}^{p} w_{hl} x_{i+h-1,j+l-1}$$

It has been shown [15] that any two-dimensional convolution problem can be converted into a one-dimensional convolution problem with the one-dimensional input sequence and kernel defined as follows. The input sequence is

$$x_1*, x_2*, \ldots , x_m*$$

where $x_i* = x_{i1}, x_{i2}, \ldots , x_{in}$. That is, the input sequence is the concatenation of the rows of the given two-dimensional image and has a total length of mn. The kernel is

$$w_1*, (n - p)!0, w_2*, \ldots , (n - p)!0, w_k*$$

where $w_i* = w_{i1}, \ldots , w_{ip}$ and (n − p)!0 stands for a string of n − p zeros. Thus the kernel is the concatenation of the rows of the given two-dimensional kernel, with a string of (n − p) zero elements inserted between each consecutive pair of rows. The length of the kernel is therefore n(k − 1) + p.

The method described here, which converts a two-dimensional prlblem into a one-dimensional one, will generate $p - 1$ invalid results for every set of n outputs [15]. Fortunately, the fraction of invalid results, which will be ignored, is very small because $p \ll n$.

If the systolic array design presented above for the one-dimensional convolution is applied directly to perform the one-dimensional convolution derived from a two-dimensional convolution, a large number of cells, that is $n(k - 1) + p$ cells would be needed in the array, and cells with zero weights would perform no useful work. From Fig. 3.5, we see that the only effects of a cell with a zero weight are to delay data on the y data stream by five cycles and those on the x data stream by four cycles. Therefore, this cell can be replaced with a cell that just introduces zero cycle delay for the x data stream and a single cycle delay for the y data stream. This degenerate cell may in turn be absorbed into the cell to the right by introducing one shift register stage (on the y data stream) to that cell. In general, if there are q non zero elements in the kernel, the systolic array needs only q cells.

Applying the foregoing cell-saving technique to one-dimensional convolution derived from two-dimensional convolution, we conclude that the convolution of an m × n image with k × p kernel can be performed on a systolic array of kp cells, where cells $ip + 1$, $i = 1$, $2, \dots, k - 1$, each have a shift register of $n - p$ stages. Figure 3.7 depicts such a two-level pipelined systolic array for two-dimensional convolution. The data RAM of each cell is used to implement the shift register on the y data stream that may be needed. The read and write addresses required at every cycle for the implementation of the shifter are supplied by the *addr* stream. It is straightforward to see that the Warp array can efficiently implement this systolic array design.

Three- or higher-dimensional convolution can also be converted into one-dimensional convolution in a similar way [15]. By using the conversion and cell-saving techniques, the architecture of the Warp array can handle convolutions of any dimensionality. But the size of the data RAM at each cell must increases as does dimensionality. For example, convolving an n × n image with a k × k kernel the data RAM must be large enough to hold $n - k$ words, as shown above, whereas convolving an n × n × n three-dimensional with a k × k × k kernel the data RAM must be large enough to hold $(n - k)(n + 1)$ words.

The design of Fig. 3.7 has a dual design for which data on the y data stream travel at a higher speed than those on the x data stream. More precisely, each cell of the dual design has six instead of four delays in the x data stream, as depicted in Fig. 3.8 [15]. The dual design has the property that the data RAM keeps data from

H. T. Kung

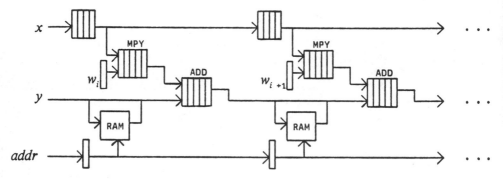

FIGURE 3.7 Two-level pipelined systolic array for two- or higher-dimensional convolution.

the x data stream instead of the y data stream. This allows a reduction of the size of the data RAM for each cell, when the word size for data on the (input) x data stream is smaller than that for data on the (output) y data stream.

For convolution with very large kernels, well-known transform methods based on the FFT should be used instead. Section 5 describes how the Warp processor can be programmed to perform the FFT.

4 MATRIX MULTIPLICATION WITH THE WARP PROCESSOR

Given $n \times n$ matrices $X = (x_{ij})$ and $W = (w_{ij})$, we want to compute their product $Y = (y_{ij})$. We present two systolic array designs for matrix multiplication that the Warp array can efficiently implement—the design of Fig. 3.11 for real matrix multiplication and the design of Fig. 3.12 for complex matrix multiplication.

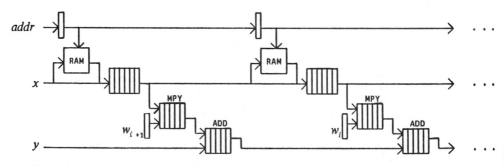

FIGURE 3.8 Another two-level pipelined systolic array for two- or higher-dimensional convolution.

Figure 3.9 depicts a simple linear systolic array at an abstract level for the matrix multiplication problem [16], with n = 8. Each cell of the array performs a multiply-accumulate operation every cycle. The jth cell from the left computes the inner product y_j of vectors $(x_{i1}, x_{i2}, \ldots, x_{in})$ and $(w_{1j}, w_{2j}, \ldots, w_{nj})$ for each i. By pumping the entries of X into the array serially in the row-major ordering and by recirculating $(w_{1j}, w_{2j}, \ldots, w_{nj})$ at cell j for each j, entries in the product matrix Y = XW will be computed and output in the row-major ordering.

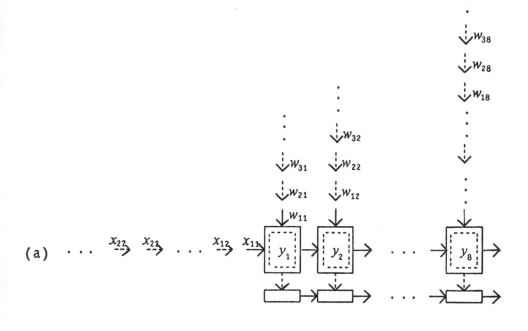

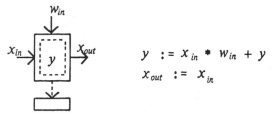

FIGURE 3.9 (a) Systolic array for matrix multiplication, and (b) cell specification.

Figure 3.10 shows the internal structure of two consecutive cells for computing y_i and y_{i+1}, assuming that multiplication together with addition can be done in one cycle. Each cell performs a multiply-accumulate operation every cycle and the result is kept in the y register. After n multiply-accumulate operations, the contents of the y register is transferred to the corresponding latch on the y data stream. These computed results on the y data stream shift to the right systolically at the rate of one cell every cycle. When they reach the right end cell of the sytolic array, they are output. Note that on the x data stream two rather than one latch is provided for each cell. This ensures that computation for y_i be computed two cycles earlier than that for y_{i+1}, and therefore the computed y_i and y_{i+1} will not collide on the y data stream.

To facilitate the recirculation of $(w_{1j}, w_{2j}, \ldots, w_{nj})$, we store at cell j these values in a shifter implemented by the data RAM. The particular w to be used in any given cycle is selected by an address (*addr*), which is input to the cell every cycle. Since the address patterns for all the cells are the same, they are passed systolically from cell to cell. To synchronize the *addr* stream with the x data stream, the same number of latches (two latches) are provided for both streams at each cell.

4.1 Interleaving Multiple Matrix Multiplications

We now consider the case that the multiplier and adder each have five pipeline stages. Since the adder has five stages, accumulations can take place only once every five cycles. To make full use of the adder and multiplier, we interleave computations for five independent matrix multiplications on the systolic array. (These matrix multiplications may actually be subtasks of a single, large matrix multiplications.) This implies that a new task can enter the adder every cycle, and that five independent accumulations can be updated simultaneously

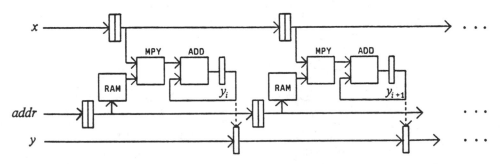

FIGURE 3.10 Cell structure for the systolic matrix multiplication array of Fig. 3.9.

at various stages of the adder at any given cycle. However, with this interleaving scheme, cells will output results in bursts. That is, at every nth cycle a cell will start outputing results for five consecutive cycles. To avoid collisions on the y data stream, among outputs from different cells, we use $6(= 5 + 1)$ latches on the x data stream for each cell, as depicted by Fig. 3.11.

4.2 Complex Matrix Multiplication

Complex matrix multiplications are common in many signal processing applications such as beamforming. A complex multiply-accumulate operation

$$(x_r + jx_i) \cdot (w_r + jw_i) + (y_r + jy_i) = [(x_r \cdot w_r - x_i \cdot w_i) + y_r]$$
$$+ j[(x_r \cdot w_i + x_i \cdot w_r) + y_i]$$

involves four real multiplications and four real additions. They will be done in four cycles using the multiplier and adder of each cell. The real and imaginary parts of a complex number are processed in separate cycles; in particular, the real part of a result is computed two cycles earlier than its imaginary part. In the scheme of Fig. 3.12, the four multiplications, $x_r \cdot w_r$, $x_i \cdot w_i$, $x_r \cdot w_i$, and $x_i \cdot w_r$, occupy four consecutive stages of the multiplier. The resulting products enter directly into the adder to form $x_r \cdot w_r - x_i \cdot w_i$ and $x_r \cdot w_i + x_i \cdot w_r$, and these two additions occupy two stages of the adder. Interleaved with these two stages are stages for performing the other additions involving y_r and y_i. We append a three-stage shift register to the adder, so that two independent (complex) accumulations, that involve updating a total of four real numbers, can be simultaneously maintained inside the eight-stage pipeline.

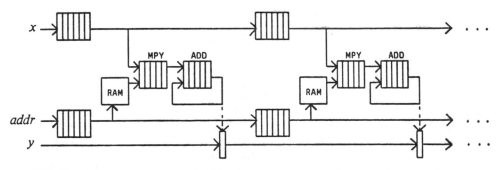

FIGURE 3.11 Two-level pipelined systolic array for multiple matrix multiplications.

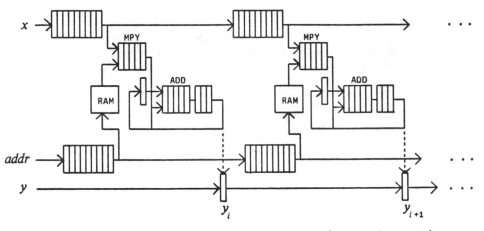

FIGURE 3.12 Two-level pipelined systolic array for complex matrix multiplication.

5 FAST FOURIER TRANSFORM WITH THE WARP PROCESSOR

The problem of computing an n-point discrete Fourier transform (DFT) is as follows: Given $x_0, x_1, \ldots, x_{n-1}$, compute $y_0, y_1, \ldots, y_{n-1}$ defined by

$$y_i = x_0 \omega^{i(n-1)} + x_1 \omega^{i(n-2)} + \cdots + x_{n-1}$$

where ω is a primitive nth root of unity.

Assume that n is a power of 2. The well-known fast Fourier transform (FFT) method solves an n-point DFT problem in $O(n \log n)$ operations, while the straightforward method requies $O(n^2)$ operations. The FFT involves $\log_2 n$ stages of $n/2$ butterfly operations, and data shufflings between any two consecutive stages. The so-called constant-geometry version of the FFT algorithm allows the same data shuffling to be used for all the stages [17]. This is depicted in Fig. 3.13 with n = 16. In the figure the butterfly operations are represented by circles, and number h by an edge indicates that the result associated with the edge must be multiplied by ω^h.

We show that the constant geometry version of the FFT can be implemented efficiently with the Warp array. In the systolic array, all the butterfly operations in the ith stage are carried out by cell i, and results are stored to the data RAM of cell i + 1. While the data RAM of cell i + 1 is being filled by the outputs of cell i, cell i + 1 can

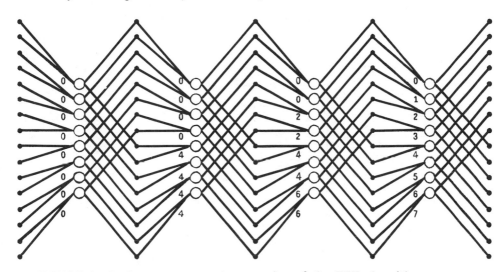

FIGURE 3.13 Constant-geometry version of the FFT algorithm.

work on the butterfly operations in the $(i + 1)$st stage of another FFT problem. In practical applications, there are often a large number of FFTs to be processed, or there are FFT problems being continuously generated. Thus it is possible that a new FFT problem can enter the first cell, as soon as the cell becomes free. In this way all the cells of the systolic array can be kept busy all the time.

We now describe how butterfly operations are executed by each cell. A butterfly operation,

$$(a_r + ja_i) \pm (b_r + jb_i) \cdot (w_r + jw_i) = [a_r \pm (b_r \cdot w_r - b_i \cdot w_i)]$$
$$+ j[a_i \pm (b_r \cdot w_i + b_i \cdot w_r)]$$

involves four real multiplications and six real additions. Thus, just to do the necessary additions, it will occupy six cycles of the adder of the cell. By techniques similar to those used in the design of Fig. 3.12, we can derive a two-level pipelined systolic design of Fig. 3.14, for which each cell can in fact process a new butterfly operation every six cycles. At any time, executions of up to four independent butterfly operations are interleaved at various stages of a cell. Note that inputs of the butterfly operation each are used twice and so are the intermediate results. As illustrated in Fig. 3.14, this can be efficiently supported by the pipeline registers linked to the input registers of the arithmetic units.

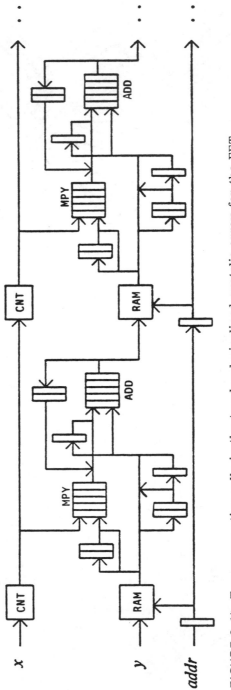

FIGURE 3.14 Two consecutive cells in the two-level pipelined systolic array for the FFT.

Inputs a_r, a_i, b_r, and b_i of the butterfly operations are obtained from the data RAM of each cell, and the outputs of the butterfly operations are stored in the data RAM of the next cell on the right. The constant geometry version of the FFT allows the use of the same address patterns for all the cells. Therefore, addresses for the RAM of each cell can be passed from cell to cell systolically along the *addr* stream.

The other inputs $w_r + jw_i$, needed for the butterfly operations are obtained from the x data stream. For illustration, consider the 16-point problem depicted in Fig. 3.13. The $w_r + jw_i$ needed by the eight butterfly operations at each stage are various powers of a primitive sixteenth root of unity ω. In particular, we use

$$\omega^0, \omega^0, \omega^0, \omega^0, \omega^0, \omega^0, \omega^0, \omega^0 \text{ for the first stage}$$

$$\omega^0, \omega^0, \omega^0, \omega^0, \omega^4, \omega^4, \omega^4, \omega^4 \text{ for the second stage}$$

$$\omega^0, \omega^0, \omega^2, \omega^2, \omega^4, \omega^4, \omega^6, \omega^6 \text{ for the third stage}$$

$$\omega^0, \omega^1, \omega^2, \omega^3, \omega^4, \omega^5, \omega^6, \omega^7 \text{ for the fourth stage.}$$

For each cell to obtain the proper ω^h at each cycle, we pump elements in $(\omega^0, \omega^1, \omega^2, \omega^3, \omega^4, \omega^5, \omega^6, \omega^7)$ sequentially into the leftmost cell of the systolic array, and move them from left to right systolically along the x data stream. Cell 1 buffers the first entry ω^0 and uses it for eight butterfly operations. Cell 2 buffers the first entry ω^0 and uses it for four butterfly operations, and after that it buffers the fifth entry ω^4 and uses it for four butterfly operations. Similarly, cell 3 buffers its inputs $\omega^0, \omega^2, \omega^4$, and ω^6, and uses each of them for two butterfly operations. Cell 4 just uses the entry on x data stream at every cycle. All these operations can be easily controlled, for example, by a counter at each cell, which is denoted by CNT in Fig. 3.14.

The prototype of the CMU Warp processor has 10 cells, each having a data RAM of 4K words. Thus by double-buffering the RAMs, the machine can independently compute 10 1024-point complex FFTs simultaneously. More precisely, at any given time all these FFTs are in different stages being carried out by different cells. Since the machine has a cycle time of 200 ns, a cell takes 1.2 μs to process a butterfly operation. This implies that we can process 1024-point complex FFTs at a rate of one FFT evey $(1/2)(1024) \times 1.2 \ \mu$s$(= 614.4 \ \mu$s). The I/O bandwidth requirement for the Warp array when performing FFTs is modest, in view of its high performance in throughput. Inputs $w_r + jw_i$ and *addr* to cell 1 are constants and thus can be supplied, for example, by a memory external to the host, as depicted in Fig. 3.15. The host delivers one set of inputs for a butterfly operation, which amounts to four words, every 1.2 μs. This is equiva-

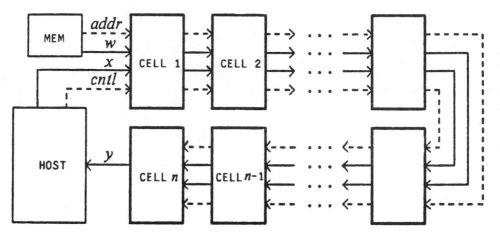

FIGURE 3.15 I/O interface of the Warp array when performing FFTs.

lent to a data rate of 13.4 megabytes per second (MB/s). At the same rate the host collects outputs coming our from cell n along the y data stream. Thus the total data bandwidth requirement between the host and the Warp need not excess 27 MB/s in order to make full utilization of the Warp array when performing FFTs. Note that the linear array of Fig. 3.15 is wrapped around so that both cell 1 and cell n are adjacent to the host. This layout arrangement allows an indefinite expansion of the array while maintaining the same I/O interface with the host.

6 CONCLUDING REMARKS

The Warp array is modular in the sense that the number of cells in the linear array can expand indefinitely to deal with problems of large sizes. For example, to handle two-dimensional convolutions for 5×5 kernels at the original rate of one output every 200 ns, we can extend the Warp array to 25 cells.

As discussed earlier, the array can also handle larger problems by time-multiplexing each cell to implement the functions of multiple cells. For example, a five-cell array can perform two-dimensional convolutions for 5×5 kernels at a rate of one output every micro-second instead of 200 ns. A pleasant side effect is that the I/O bandwidth requirement for the array to communicate with the outside world is reduced. Thus multiplexing cells is a useful technique to make full use of the Warp array when the communication speed with the outside world cannot keep up with the cell speed.

The Warp array can solve problems beyond those considered in this chapter. For example, it can perform polynomical evaluation and discrete Fourier transform at the peak speed [18]. In general, most of the so-called "local" operations in signal and image processing can be carried out efficiently by the Warp array.

Together with a boundary processor for performing fast divisions and square roots, the Warp array can efficiently implement most of the systolic arrays proposed in the literature for solving various types of linear systems. For example, it is known that a linear systolic array of Fig. 3.16 can solve triangular linear systems [2], and perform the QR-decomposition of a Hessenberg matrix [19] or singular-value decomposition [20], a key step in real-time adaptive signal processing [21,22]. The boundary processor is drawn in dashed lines in the figure.

Figure 3.17 shows that the Warp array augmented with the boundary processor can implement the systolic array of Fig. 3.16. Note that the y input of each cell now takes values from the y output of the cell to the right, so that the required bidirectional data flows can be implemented.

7 POSTSCRIPT: WARP CIRCA AUGUST 1986

Two copies of the wire-wrap prototype, built by Carnegie Mellon and its industrial partners, GE and Honeywell, have been operational since spring of 1986. These machines are being used for signal and vision processing and for scientific computing. Actual measurements have shown that for these applications, the new machines are typically several hundred times faster than the VAX 11/780. GE is under contract to build eight printed circuit board versions of the machine to support research in robot navigation and image analysis where computational demands can be extremely high. CMU is building a version of the boundary processor shown in Fig. 3.16. This boundary processor [23] will be attached to the printed circuit board version of the Warp machine that CMU will receive from GE in 1987.

While achieving a high computational throughput, Warp has a high degree of programmability. An optimizing compiler to support a high-level programming language has been developed [24]. To the application programmer, Warp is an array of simple sequential processors, communicating asynchronously. Based on the user's program for this abstract array, the compiler generates code for the host, interface unit, and Warp array automatically. The convolution, matrix multiplication, and FFT computations considered in this chapter are all running on Warp with near-maximum efficiency, using compiler-generated code.

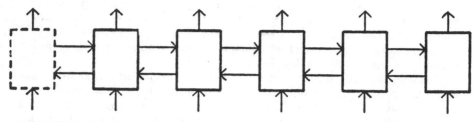

FIGURE 3.16 Bidirectional systolic array with a boundary processor.

Because of the availability of the compiler and the machine's
ability to support fine-grain parallelism, we have found that Warp is
suited for a broad set of applications, far beyond those described in
this chapter. For example, the machine is being used to find the
shortest paths for 450-node graphs; perform path planning for 512 ×
512 terrain images, and solve Poisson equations with 50,625 unknowns.
As a low-level vision engine, Warp is used routinely at CMU to perform the navigation for autonomous land vehicles.

Anticipating the future need for integrated Warp systems,
Carnegie Mellon and Intel have been developing a VLSI Warp chip,
called the iWarp chip, since April 1986. The resulting iWarp systems
are expected to represent an order of magnitude improvement in cost,
size and power consumption over the current Warp. Using tens of cells,
an iWarp array, configured in one-dimensional, two-dimensiona or
other interconnections, will be able to deliver over 1 billion floating-
point operations per second.

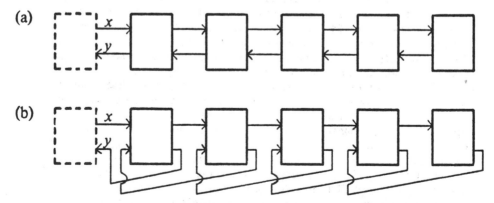

FIGURE 3.17 (a) Bidirectional systolic array, implemented by (b) the
Warp array.

ACKNOWLEDGMENT

A preliminary version of this chapter appeared in *Proceedings of the 7th International Conference on Pattern Recognition*, Montreal, Canada, July 1984, pp. 570–577, as the text of an invited talk. The research was supported in part by the Office of Naval Research under Contracts N00014-76-C-0370, NR 044-422 and N00014-80-C-0236, NR 048-659, in part by the Defense Advanced Research Projects Agency (DoD), ARPA Order No. 3597, monitored by the Air Force Avionics Laboratory under Contract F33615-81-K-1539, and in part by a Guggenheim Fellowship.

REFERENCES

1. H. T. Kung, Why systolic architectures? *Computer* (IEEE), *15*(1): 37–46 (Jan 1982).
2. H. T. Kung and C. E. Leiserson, Systolic arrays (for VLSI), *Sparse Matrix Proc. 1978* (I. S. Duff, and G. W. Stewart, eds.), Society for Industrial and Applied Mathematics, Philadelphia, PA. 256–282 (1979).
3. J. Avila and P. Kuekes, One-gigaflop VLSI systolic processor, *Proc. SPIE* 431: 159–165 (1983).
4. R. A. Evans, D. Wood, K. Wood, J. V. McCanny, J. G. McWhirter, and A. P. H. McCabe, A CMOS implementation of a systolic multi-bit convolver chip, *VLSI '83*, (F. Anceau and E. J. Aas, eds.,) North-Holland, Amsterdam 227–235 (1983).
5. H. T. Kung, On the implementation and use of systolic array processors, *Proc. IEEE International Conference on Computer Design: VLSI in Computers*, Port Chester, NY 370–373 (Nov. 1983).
6. J. J. Symanski, NOSC systolic processor testbed, *Technical Report NOSCTD 588*, Naval Ocean Systems Center (June 1983).
7. D.W. L. Yen, and A. V. Kulkarni, Systolic processing and an implementation for signal and image processing, *IEEE Trans. Comput.*, *C-31*: 1000–1009 (1982).
8. A. L. Fisher, H. T. Kung, and K. Sarocky, Experience with the CMU programmable systolic chip, *Proc. SPIE* 495: (Aug. 1984).
9. M. Annaratone, et al., Warp architecture and implementation, *Proc. 13th Annual International Symposium on Computer Architecture* 346–356 (June 1986).
10. H. T. Kung and O. Menzilcioglu, Warp: a programmable systolic array processor, *Proc. SPIE* 495: 130–136 (Aug. 1984).
11. A. L. Fisher and H. T. Kung, Synchronizing large VLSI processor arrays, *IEEE Trans. Comput.*, *C-34*: 734–740 (1985).

12. H. T. Kung and M. Lam, Wafer-scale integration and two-level
     pipelined implementations of systolic arrays, *J. Parallel Distribut.*
     Comput., *1*: 32–63 (1984). A preliminary version appeared in
     Proc. Conference on Advanced Research in VLSI, MIT, Cambridge,
     MA: 74–83 (Jan. 1984).

13. B. Y. Woo, L. Lin, J. Fandrianto, and E. Sun, A 32 bit IEEE
     floating-point arithmetic chip set, *Proc. 1983 International*
     Symposium on VLSI Technology, Systems and Applications
     219–222 (1983).

14. H. T. Kung and R. L. Picard, One-dimensional systolic arrays
     for multidimensional convolution and resampling, *VLSI for Pattern*
     Recognition and Image Processing, (K.-S. Fu, ed.), Spring-
     Verlag, New York 9–24 (1984). A preliminary version, "Hard-
     ware pipelines for multi-dimensional convolution and resampling,"
     appears in *Proc. 1981 IEEE Computer Society Workshop on*
     Computer Architecture for Pattern Analysis and Image Database
     Management, Hot Springs, VA: 237–278 (Nov. 1981).

15. H. T. Kung, L. M. Ruane, and D. W. L. Yen, Two-level pipe-
     lined systolic array for multidimensional convolution, *Image*
     Vision Comput., *1*: 30–36 (1983). An improved version appears
     as a CMU Computer Science Department technical report (Nov.
     1982).

16. H. T. Kung and S. Q. Yu, Integrating high-performance special-
     purpose devices into a system, *VLSI Architecture*, (B. Randel
     and P. C. Treleaven, eds.), Prentice-Hall, Englewood Cliffs, NJ:
     205–211 (1983). An earlier version also appears in *Proc. SPIE*
     341: 17–22 (May 1982).

17. L. R. Rabiner and B. Gold, *Theory and Application of Digital*
     Signal Processing, Prentice-Hall, Englewood Cliffs, NJ (1975).

18. H. T. Kung, Two-level pipelined systolic array for matrix
     multiplication, polynomial evaluation and discrete Fourier trans-
     form, *Proc. Workshop on Dynamical Behavior of Automata:*
     Theory and Applications, Academic Press, NY (Sept. 1983).

19. D. E. Heller and I. C. F. Ipsen, Systolic networks for orthogonal
     equivalence transformations and their applications, *Proc. Confer-*
     ence on Advanced Research in VLSI, MIT, Cambridge, MA:
     113–122 (Jan. 1982).

20. D. E. Schimmel and F. T. Luk, A new sytolic array for the
     singular value decomposition, *Proceedings of the 1986 Conference*
     on Advanced Research in VLSI, MIT, Cambridge, MA: 205–217
     (Apr. 1986).

21. B. A. Bowen and W. R. Brown, *VLSI Systems Design for Signal*
     Processing, Vol. 1; *Signal Processing and Signal Processors*,
     Prentice-Hall, Englewood Cliffs, NJ (1982).

22. R. A. Monzingo and T. W. Miller, *Introduction to Adaptive*
     Arrays, Wiley, New York (1980).

23.   M. Annaratone, et al. Extending the CMU warp machine with a boundary processor, *Proc. SPIE,* 564: (Aug. 1985).

24.   T. Gross and M. Lam, Compilation for a high-performance systolic array, *Proc. SIGPLAN 86 Symposium on Compiler Construction* 27—38 (June 1986).

4

Wavefront Array Processors

SUN-YUAN KUNG *University of Southern California, Signal & Image Processing Institute, Department of Electrical Engineering, Los Angeles, California*

1 INTRODUCTION

The basic discipline in a top-down design methodology, as depicted in Fig. 4.1, depends on a fundamental understanding of algorithm, architecture, and application. Note that the boundary between software and hardware has become increasingly vague under the new environment of very large scale integration (VLSI) system design. This enhances the already prevailing roles of the algorithm analyses and the mappings of algorithms to architectures. Therefore, a very broad spectrum of innovations will be required for obtaining highly parallel array processing: for examples new ideas on communication/computation trade-offs, parallelism extractions, array architectures, programming techniques, processor/structure primitives, and numerical performances of digital signal processing (DSP) algorithms.

In striving for a cohesive exploration of the overall implications of VLSI, a cross-disciplinary discussion on application, algorithm, and architecture is necessary. In fat, *integration* will be the keyword in VLSI. This means that innovations on a very broad spectrum of disciplines, including algorithm analyses, parallelism extractions, array architectures, programming techniques, functional primitives, structural primitives, and numerical performance of DSP algorithms will be needed.

1.1 VLSI Architectural Design Principles

VLSI architectures should exploit the potential of the VLSI technology and also take into account the cost of silicon area and input/output (I/O) pins; the layout constraint and the resultant interconnection

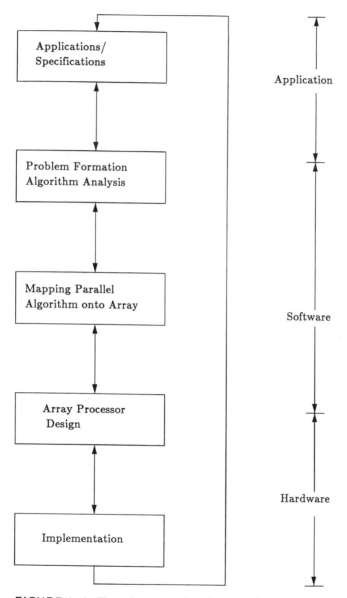

FIGURE 4.1 Top-down design integration.

costs in terms of area and time. When the delay time of the circuit
depends largely on the interconnection delay (instead of the logic gate
delay), minimal and local interconnections will become an essential
factor for an effective realization of the VLSI circuits. Consequently,

VLSI architecture design strategies stress modularity, regularity of data and control paths, local communication, massive parallelism, and multiple use of each input data item to minimize the I/O problem. Some design principles are summarized below.

1.2 Communication and Locality

In VLSI technology, computations per se are becoming very inexpensive and easily affordable. Therefore, the most critical factor in VLSI design is: communication. Architectures that balance communication and computation and circumvent communication bottlenecks with minimum hardware cost will eventually play a dominating role in VLSI systems.

"Principle of locality is seen at every level of VLSI design" [1]. In systolic arrays, both spatial locality and temporal locality are stressed [2]. The notion of locality can have two meanings in array processor designs: localized data transactions and localized control flow. In fact, most recursive signal processing algorithms permits both locality features, and they are fully exploited in the design of wavefront arrays, as we shall claborate in Sec. 2.

1.3 Regularity and Modularity

In VLSI, there is an emphasis on keeping the overall architecture as regular and modular as possible, thus reducing the overall complexity. For example, memory and processing power will be relatively cheap as a result of high regularity and modularity. Even in the communication or wiring, a careful algorithmic study may help create some form of regularity. This depends on special arrangements, realized in the course of topological mappings from algorithms to architectures.

1.4 Pipeline and Parallel Processing

Real-time DSP requires extensive concurrency by either pipeline processing or parallel processing. Moreover, in many DSP applications, throughput rate often represents the overriding factor dictating the system performance. To optimize throughput, a different design choice is often made than that of minimizing the total processing time (latency). Pipeline techniques fit naturally in our aim of improving throughput rate. Especially, for a majority of signal processing algorithms, suitable pipelining techniques are now well established.

For signal processing arrays, pipelining at all levels should be pursued. It may bring about an extra order of magnitude in performance with very little additional hardware. Although most of the current array processors stress only word-level pipelining, the new

trend is to exploit the potential of multiple-level pipelining (i.e.,
combined pipelining in all the bit-level, word-level, and array-level
granularities).

1.5 Global Clock Synchronization and Wafer-Scale Integration

For VLSI systems, the clocking scheme is critical [3–5]. In the globally
synchronous scheme, there is a global clock network which distributes
the clock signal over the entire array. For very large systems, the
clock skew incurred in global clock distribution is a nontrivial factor,
causing unnecessary slowdown in the clock rate. From a clock synchronization perspective. In fact, a detailed analysis [3] indicates that
clock skew may grow at a much higher than linear rate with the array
size. An immediate conclusion from this analysis is that, while for small
N, a globally synchronized array may be easier to implement, for large
values of N, an asynchronous system may become more favorable.
Moreover, complete synchrony of all the PEs in a large array also
implies a high power rate at the same instant, which is often accompanied with a even more serious heat-dissipation problem. This heat
problem is especially acute in wafer-scale-integration (WSI) systems.

1.6 Fault-Tolerant Designs

To enhance the yield and reliability of computing systems, array
processor architectures demand a special attention in compile-time and
run-time fault tolerances. Because of the communication constraints,
the fault-tolerance design features such as reconfiguration and rollback may become very involved.

1.7 Impact of Computer-Aided-Design Techniques

For special-purpose array processors, the system specifications required by the applications may change significantly if the development
cycle is too long; therefore, fast-turnaround implementation is critical.
Computer-aided-design (CAD) tools for all levels of array processor
design are essential. For VLSI implementation, the development of
simplified design rules and structured design methodology have already
allowed system designers to design their chips quickly. Silicon compiler technology is also becoming mature. It is therefore important
that a proper high-level (array processor) structured design and
description tools be developed that will be compatible with the existing
back-end (low-level) CAD tools.

2 VLSI ARRAY PROCESSORS

One way to satisfy the real-time requirement of digital signal processing is to use special-purpose array processors with extensive concurrency by either pipeline processing or parallel processing or both. As long as communication in VLSI remains restrictive, locally interconnected arrays will be of great importance. An increase of efficiency can be expected if the algorithm arranges for a balanced distribution of work load while observing the requirement of locality (i.e., short communication paths). The first such special-purpose VLSI architectures are systolic and wavefront arrays, which boast tremendously massive concurrency by utilizing combined pipeline and parallel processing.

A fundamental issue on mapping algorithms onto arrays is how to express parallel algorithms in a notation that is easy to understand by human beings and possible to compile into efficient VLSI array processors. Thus a powerful expression of array algorithms will be essential to the design of arrays. In this chapter we propose three primary ways of array algorithm expression: signal-flow-graph, systolic expressions, and wavefront expressions.

2.1 Signal Flow Graph

The signal-flow-graph (SFG) representations have been popularly used for signal-processing flow diagrams, such as FFT, digital filters, and many other domains of signal and system applications. For a reader from the signal processing community, it should be particularly enlightening to see how to apply SFG analysis to signal-processing architecture designs.

An SFG can be formally modeled by a directed graph $G = (V, E, D(E))$, where V is the set of nodes (vertices) and E is the set of edges, [2,6], as illsutrated in Fig. 4.2. Nodes model the computation and edges model the communication in a parallel algorithm. Each edge has a nonnegative weight, denoted as $D(e)$, which represents the number of delays (Ds) on the edge.

Directed loops of edges with zero delays are disallowed. It is assumed that the computations in nodes and communication between nodes takes zero time. A recursion is defined as the computation inside a SFG for a single set of input data. So once a data set is input to a SFG in a recursion, it is assumed to go through the nodes instantly, unless it is blocked by delays (Ds) on some edges. When the data are blocked by a delay, they will stay in the delay (D) until the next recursion. Therefore, the delays in the SFG have the function of separating successive recursions and keeping the state of the system. To start, all data in Ds are assigned according to the initial conditions of the algorithm. With these observations, it is not difficult

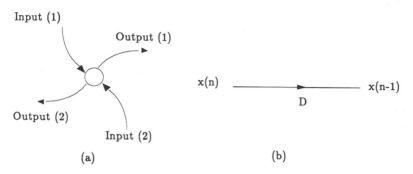

FIGURE 4.2 Examples of signal flow graph notation: (a) an operation node with (two) inputs and (two) outputs; (b) an edge as a delay operator. *Notation*: In general, a *node* is often denoted by a circle representing an arithmetic or logic function part *performed with zero delay*, such as multiply, add, etc. [see part (a)]. An *edge*, on the other hand, denotes either a function or a delay. Unless otherwise specified, for a large class of signal processing SFGs, the following conventions are adopted for convenience. When an edge is labeled with a capital letter D (or D', 2D, etc.), it represents a time-delay operator with delay time D (or D', 2D, etc.) [see part (b)]. A node is considered to be delay free unless otherwise specified. In fact, the SFG representation derives its power from the assumption that the computaions in the node are delay free, warranting simpler snap-shot descriptions than the systolic counterpart. Consequently, the undertaking of tracing the detailed space-time activities associated with pipelining is simplified.

to see that the SFG actually displays the activities in one recursion of the algorithm. This simplifies the complicated space-time activities associated with parallel processing.

The descriptions of array processing activities, in terms of the SFG representations, are often easy to comprehend. A typcial example used to illustrate a two-dimensional array operation is matrix multi-plication. The same matrix multiplication example will be used later to illustrate systolic and wavefront array processing.

*Example: Multiplication of a band matrix and
a full rectangular matrix*

Given two matrices $A = \{a_{ij}\}$ and $B = \{b_{ij}\}$, the problem is to compute $C = A \times b = \{c_{ij}\}$. Now we use a special but rather commonly encountered type of matrix multiplication problem, which involves a band matrix A, $N \times N$, with bandwidth P, and a rectangular matrix B, $N \times Q$:

AB =

$$\begin{array}{cc}
\times\times\times & \times\times\times\times \\
\times\times\times\times & \times\times\times\times \\
\times\times\times\times\times & \times\times\times\times \\
\times\times\times\times\times & \times\times\times\times \\
\times\times\times\times\times & \times\times\times\times \\
\times\times\times\times\times & \times\times\times\times \\
\cdots & \times\times\times\times
\end{array}$$

A band matrix A is one that has nonzero elements only on a finite "band" along the diagonal. This situation arises in many application domains, such as DFT (with low-time windowed data) and time-varying (multichannel) linear filtering. In most applications, $N \gg P$ and $N \gg Q$; therefore, it is very uneconomical to use $N \times N$ arrays for computing $C = A \times B$. Fortunately, with a special caution to the derivation of the SFG array, high-concurrency performance can be achieved with only a $P \times Q$ rectangular array (as opposed to an $N \times Q$ array). This is shown in Fig. 4.3.

The matrix A can be decomposed into columns A_i and matrix B into rows B_j, and therefore,

$$C = A_1 * B_1 + A_2 * B_2 + \cdots + A_N * B_N \tag{4.1}$$

where the product $A_i * B_i$ is termed the "outer product." The matrix multiplication can then be carried out in N recursions (each executing one outer product).

$$C^{(k)} = C^{(k-1)} + A_k * B_k \tag{4.2}$$

There will be N sets of wavefronts involved. More explicitly,

$$c_{i,j}^{(k)} = c_{i,j}^{(k-1)} + a_i^{(k)} b_j^{(k)} \tag{4.3}$$

$$a_i^{(k)} = a_{ik}$$

$$b_j^{(k)} = b_{kj}$$

for $k = 1, 2, \ldots, N$.

The notion of SFG array processing allows an extensive use of broadcasting, since a node or a zero-delay edge is considered to be delay free. This facilitates the mapping of algorithms onto an array.

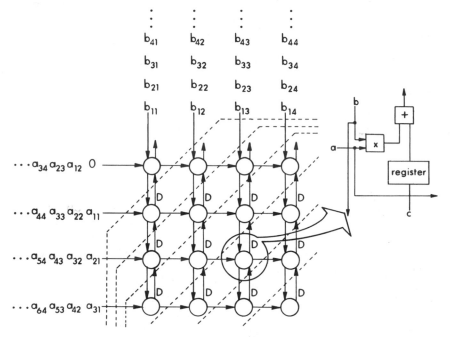

FIGURE 4.3 SFG Array for matrix multiplication. The left memory
module will store the matrix A along the band direction and the upper
module will store B in the ordinary column by column arrangement. A
straightforward SFG array design is to broadcast the columns A_k and
rows B_k [see Eqs. (4.2) and (4.3)] instantly along the square array
as shown in the figure. Multiply the two data meeting at node (i,j)
and add the product of $C_{ij}^{(k)}$, the data value currently residing in
a register in node (i,j). Finally, the newly updated result will be
sent upward to the partial sum register at node (i − 1). The upward-
shift arrangement compensates the loading of the input matrix A in a
skewed fashion. As succeding column and row input data arrive at
the nodes, all the outer products will be sequentially summed. The
final result (C) will be output from the I/O ports of the top-row PEs.
Although this design is not directly suitable for a VLSI circuit design
due to the use of global communication, it may be converted to a
systolic array as shown in Fig. 4.4, or a wavefront array as shown in
Fig. 4.5. A simple conversion strategy is discussed later.

The derivation of an SFG array for the matrix multiplication algorithm
is discussed in Fig. 4.3. Note that since the data are broadcast
throughout the entire array, the array operates just like a SIMD
machine. An SFG array can be considered a systolic array without
explicit pipelining, that is, the SFG may explicitly exhibit only the
parallel processing activities.

To make the system realizable in a locally interconnected array, proper delays on the edges need to be assigned. Moreover, the processing time required at each processing node will have to be reassigned. (This is done as opposed to the very idealistic zero-delay assumption adopted in SFG forms.) This implies a slowdown of input/output data rate—a price paid in necessity when adopting a locally interconnected network.

The abstraction provided by the SFG form is very powerful to use and the verification of the correctness can often be done with some extended notion of the Z-transform tool, as popularly used in DSP literature. In simplest abstraction, the SFG array often exhibits only the parallel processing part explicitly and leaves the pipelining part only implicitly expressed. The transformation of such an SFG description to a systolic array can often be accomplished automatically, for example, by means of a cut-set procedure for edge retiming. In other words, a simple way to compare a systolic array and its SFG form is as follows:

systolic array = SFG array + pipeline retiming

2.2 Systolic Array

Systolization of SFG array by a cut-set procedure [2].

A *cut set* in an SFG is a minimal set of edges that partitions the SFG into two parts. The systolization procedure is based on two simple rules:

1. Time scaling: All delays D may be scaled (i.e., $D \rightarrow \alpha D$) by a single positive integer α. Correspondingly the input and output rates also have to be scaled down by a factor α. The time-scaling factor (or, equivalently, the slowdown factor) α is determined by the slowest (i.e., maximum) loop delay in the SFG array.

2. Delay transfer: Given any cut set of the SFG, we can group the edges of the cut set into *in-bound edges* and *outbound edges*, depending on the directions assigned to the edges. Rule 2 allows advancing k time units on the in-bound edges. It is clear that, for a (time-invariant) SFG, the general system behavior is not affected because the effects of lags and advances cancel each other in the over-all timing. Note that the input/input and input—output timing relationships will also remain exactly the same only if they are located on the same side. Otherwise, they should be adjusted by a lag of +k time units or an advance of -k time units.

Example: With reference to Fig. 4.3, the dashed lines indicate a set of possible cuts. The systolication procedure is detailed as follows. According to the systolization rule (1), the slowdown factor α is

determined by the maximum loop delay in the SFG array. Refer to
Fig. 4.3; any loop consisting of one up-going and one down-going
edge yields a (maximum) delay of 2. This is why the final pipelined
systolic array has to bear a slowdown factor $\alpha = 2$. The pipelining rate
is 0.5 word per unit time for each channel. Apply rule 2 to the cut
sets shown in Fig. 4.3. The systolized SFG will have one delay assigned
to each edge and thus represent a localized network. Also based on
rule 2, the inputs from different columns of B and rows of A will
have to be adjusted by a certain number of delays before arriving at
the array. By counting the cut sets involved in Fig. 4.3, it is clear
that the first column of B needs no extra delay, the second column
needs one delay, the third needs two (i.e., attributing to the two
cut sets separating the third column input and the adjacent top-row
processor), and so on. Therefore, the B matrix will be skewed as
shown in Fig. 4.4. A similar arrangement can be applied to the input
matrix A.

Motivation for wavefront processing

The advantages of systolic arrays are that they are very amenable
to VLSI implementation and that they feature the very important
advantages of modularity, regularity, local interconnection, highly pipelined multiprocessing, and continous flow of data between the PEs.
They are especially suitable to a special class of computation-bound
algorithms and have a good number of digital signal processing applications. In fact, they are often derived as a direct mapping of computational algorithm onto a processor array.

The disadvantages of systolic arrays lie in the fact that the data
movements in a systolic array are controlled by global timing-reference
"beats": From a hardware perspective, global synchronization incurs
problems of clock skew and fault tolerance. The burden of having to
synchronize the entire computing network will be intolerable for ultra-
large-scale arrays. From a software perspective, in order to synchronize the activities in a systolic array, an exact number of additional
delays are required. A simple solution to these problems is to take
advantage of the control flow locality, as well as the data flow locality,
inherently existing in most DSP algorithms. This permits a data-driven,
self-timed approach to array processing. Conceptually, this approach
substitutes the requirement of correct "timing" by correct "sequencing."
This concept can be realized in data flow computers and wavefront
arrays.

Note that the interconnection and memory conflict problems remain very expensive in a general-purpose data flow multiprocessor.
Such problems can be greatly alleviated if modularity and locality are
incorporated into data flow multiprocessors. This motivates the concept
of wavefront array processors (WAPs).

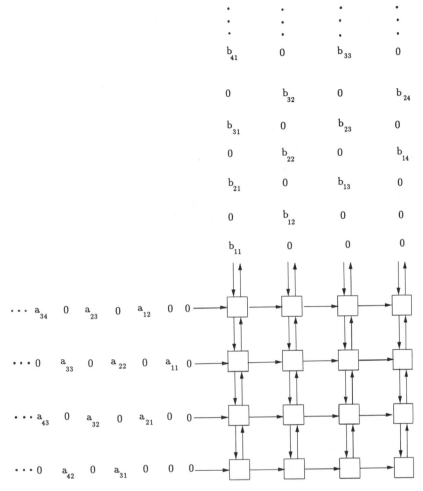

FIGURE 4.4 Systolic array for matrix multiplication. For this example, a two-dimensional square array forms a natural topology for the matrix multiplication problem. The figure specifically shows a 4 × 4 array of processing elements. In terms of one "snapshot" of the activities; the input data (from matrices A and B appearing at the upper left and upper right parts of the figure) are prearranged in an orderly sequence. The C data stay in the PEs and will be pumped out from one side of the array. (To become more convinced, the reader might want to "simulate" several consecutive snapshots of the data movements.) In general, the major characteristics of systolic arrays are [2]: (1) synchrony—the data are rhythmically computed (timed by a global clock) and pumped through the network; (2) regularity, modularity and spatial locality of interconnections; (3) temporal locality; and (4) effective pipelinability.

3 WAVEFRONT ARRAY PROCESSORS

3.1 Definition: Wavefront Array

A wavefront array is a computing network possessing the following features:

1. *Self-timed data-driven computation.* No global clock is needed, since the network is self-timed.
2. *Regularity, modularity, and local interconnection.* The array should consist of modular processing units with regular and (spatially) local interconnection. Moreover, the computing network may be extended indefinitely.
3. *Programmability in wavefront language or DFG (data flow graph).* The actual implementation of wavefront arrays can be either dedicated or programmable processors, but the latter are usually preferred. An important feature is that tracing the wavefronts or data flow graphs (DFGs) often provides a key to programming the wavefront arrays.
4. *Pipelineability with linear-rate speed-up.* A wavefront array should exhibit a *linear-rate speed-up*; that is, it should achieve an O(M) speed-up, in terms of processing rates, where M is the number of processor elements (PEs).

Note that the major difference distinguishing a wavefront array from a systolic array is the data-driven property and there is no global timing reference in the wavefront arrays. In the wavefront architecture, the information transfer is by mutual convenience between a PE and its immediate neighbors. Whenever the data are available, the transmitting PE informs the receiver of the fact, and the receiver accepts the data when it needs them. It then conveys to the sender the information that the data have been used. This scheme can be implemented by means of a simple handshaking protocol [3,7], which ensures that the computational wavefronts follow in an orderly manner, instead of crushing into other fronts. A good measure for the efficiency of the array is the following:

$$\text{speed-up factor} = \frac{T_s}{T_a}$$

where T_s is the processing time in a single processor and T_a is the processing time in the array processor.

In general, there are two approaches to deriving wavefront arrays:

1. In this section the approach to deriving wavefront arrays is to trace the computational wavefronts and pipeline the fronts on the processor array.

2. In the next section, however, we introduce a new approach based on converting an SFG array into a data flow graph (DFG) array and then into a wavefront array, by properly imposing several key elements in data flow computing.

3.2 Tracing the Computational Wavefronts

From an algorithmic analysis perspective, the notion of computational wavefronts offers a very simple way to appreciate the wavefront computing. This approach to deriving wavefront processing consists of three steps:

1. Decompose an algorithm into a sequence of recursions.
2. Map each of the recursions into a corresponding computational wavefront.
3. Successively pipeline the wavefronts through the processor array.

Illustration of wavefront array processing

The notion of a computational wavefront may be better illustrated by the example of the matrix multiplication algorithm. The topology of the matrix multiplication algorithm can be mapped naturally onto the square orthogonal $N \times N$ matrix array of the wavefront array processor (WAP), as in Fig. 4.5. The computing network serves as a (data) wave-propagating medium. To create a smooth data movement in a localized communication network, we propose a notion of the computational wavefronts. A wavefront in a processor array corresponds to a mathematical recursion in the algorithm. Successive pipelining of the wavefronts through the computational array will accomplish the computation of all recursions.

More elaborately, the computational wavefront for the first recursion in matrix multiplication is now examined. Suppose that the registers of all the processing elements (PEs) are initially set to zero:

$$c_{ij}^{(0)} = 0 \quad \text{for all (i,j)}$$

the entries of A are stored in the memory modules to the left (in columns), and those of B in the memory modules on the top (in rows). The process starts with PE(1,1):

$$c_{11}^{(1)} = c_{21}^{(0)} + 0 * b_{11} = 0 \tag{4.4}$$

is computed. The computational activity then propagates to the neighboring PEs (1,2) and (2,1), which will execute

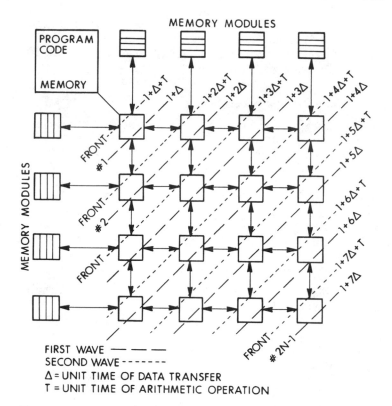

FIGURE 4.5 Wavefront processing for matrix multiplication. In this example, the wavefront array consists of $N \times N$ processing elements with regular and local interconnections. The figure shows the first 4×4 processing elements of the array. The computing network serves as a (data) wave propagating medium. Hence the hardware will have to support pipelining the computational wavefronts as fast as resource and data availability allow, which can often be accomplished simply by means of a handshaking protocol, such as that proposed in Ref. [7]. The (average) time interval (T) between two separate wavefronts is determined by the availability of the operands and operators. In this case there is a feedback loop, constituted of one down-going and one up-going edges. Therefore, the (average) time separating two consecutive fronts is $T = T_{MA} + 2\Delta$.

$$c_{12}^{(1)} = c_{22}^{(0)} + 0 * b_{12} = 0 \qquad (4.5)$$

and

$$c_{21}^{(1)} = c_{31}^{(0)} + a_{11} * b_{11} = a_{11} * b_{11} \qquad (4.6)$$

The next front of activity will be at PEs $(3,1)$, $(2,2)$, and $(1,3)$, thus creating a computation wavefront traveling down the processor array. This computational wavefront is similar to optical wavefronts, (they both obey Huygens' principle) since each processor acts as a secondary source and is responsible for the propagation of the wavefront. It may be noted that wave propagation implies localized data flow. Once the wavefront sweeps through all the cells, the first recursion is over. As the first wave propagates, we can execute an identical second recursion in parallel by pipelining a second wavefront immediately after the first one. For example, the $(1,1)$ processor will execute

$$c_{11}^{(2)} = c_{21}^{(1)} + a_{12} * b_{21} = a_{11} * b_{11} + a_{12} * b_{21} \qquad (4.7)$$

$$c_{ij}^{(k)} = c_{i+1,j}^{(k-1)} + a_{ik} * b_{kj} = a_{i1} * b_{1j} + a_{i2} * b_{2j} + \cdots + a_{ik} * b_{kj}$$

$$(4.8)$$

and so on. The pipelining is feasible because the wavefronts of two successive recursions will never intersect (Huygens' wavefront principle), as the processors executing the recursions at any given instant will be different, thus avoiding any contention problems.

In wavefront processing the (average) time interval (T) between two separate wavefronts is determined by the availability of the operands and operators. In the example above, a feedback loop is involved, which can be clearly identified in Fig. 4.5. For example, in PE$(1,1)$, the second front has to wait until the first front completes all the following steps: (1) propagates a data item downward (processing time Δ); (2) performs the arithmetic operations at PE$(2,1)$ (processing time T_{MA}); and (3) returns the result upwards to PE$(1,1)$ (processing time Δ). Once the result is returned to PE$(1,1)$, the second front can be activated immediately. The activations of all the later fronts follow exactly the same procedure; therefore, the (average) time separating two consecutive fronts is $T = T_{MA} + 2\Delta$.

Note that the successive pipelining of the wavefronts furnishes an additional dimension of concurrency. The separated roles of pipeline and parallel processing also become evident when we carefully inspect

how parallel processing "computational wavefronts" are pipelined
successively through the processor arrays. Generally speaking, parallel
processing activities always occur at the PEs on the same "front,"
while pipelining activities are often perpendicular to the fronts. With
reference to the wavefront processing example in Fig. 4.5, PEs on
the anti-diagonals execute in parallel, since eachof the PEs processes
information independently. On the other hand, pipeline processing is
enforced along the diagonal direction, that along which the computational wavefronts are piped.

Why the name "wavefront array"

As a justification for the name "wavefront array" we note that
the computational wavefronts are similar to electromagnetic wavefronts,
since each processor acts as a secondary source and is responsible
for propagation of the wavefront. The pipelining is feasible because
the wavefronts of two successive recursions will never intersect (by
Huygens' wavefront principle), thus avoiding any contention problems.
It is even possible to have wavefronts propagating in several different
fashions (e.g., in the extreme case of nonuniform clocking, the wavefronts are actually crooked). What is necessary and sufficient is that
the order of task sequencing be followed correctly. The correctness
of the sequencing of the tasks is ensured by the wavefront principle
[7].

The wavefront processing utilizes the localities of both data flow
and control flow existing inherently in many signal processing algorithms. Since there is no need to synchronize the entire array, a wavefront array is truly architecturally scalable. In other words, a simple
way to compare wavefront array and its systolic array counterpart is
as follows:

wavefront array = systolic array + self-timed data flow computing

3.3 Least-Square-Error Solver Using the Givens QR Method

The wavefront concept and the related architecture and language
designs may be better illustrated with an additional matrix algebraic
example. We shall use the example of least-square-error solver
(LSES), as they arise in many DSP applications, such as Wiener
filtering, image reconstruction, and sonar phased-array processing
problems. The LSES problem may be formulated mathematically as

$$Y = AX + V \qquad\qquad (4.9)$$

where Y is the vector of measurements, X the object vector, A the
projection matrix, and V the measurement noise vector. This amounts

to solving an overdetermined system, for which the least-square-error solution is popularly adopted:

$\hat{x}$: such that $\| Y - A\hat{x} \|_2$ is minimized

where $\| \cdot \|_2$ denotes the ℓ_2 norm of the vector.

A numerically attractive method of solving the least-square-error problem is the QR decomposition of matrix A. When the columns of A are linearly independent, A can be decomposed uniquely into the product of two matrices, $A = Q^T R$, where the Q^T is an orthonormal matrix and R is an upper triangular matrix [8]; T stands for the matrix transpose. The least-square solution for x can now be obtained by first applying

$$Q[A,Y] = \begin{bmatrix} R,c \\ O,d \end{bmatrix} \qquad (4.10)$$

and then solving a (triangular) linear system of equations:

$$Rx = c \qquad (4.11)$$

Givens rotation

The Givens algorithm enjoys a stable numerical behavior when used for performing QR decomposition. It is based on applying an orthogonal operator, $Q^{(q,p)}$, which performs a plane rotation of the matrix A in the (q,p) plane and annihilates the element $a_{q+1,p}$. When transforming a matrix into upper triangular form, the rotations are applied so as to annihilate the elements $a_{N,j}$, $a_{N-1,j}$, $\cdots$, $a_{j+1,j}$, $j = 1, \ldots, N - 1$ in that order. Thus, in the Givens algorithm, the subdiagonal elements of the first column are nullified first, then the elements of the second column, and so on, until an upper triangular form is eventually reached. The detailed upper triangularization procedure is described below.

$$QA = R$$

where R is an upper triangular matrix, and

$$Q = Q(N - 1) * Q(N - 2) * \cdots * Q(1), \qquad (4.12a)$$

and

$$Q(p) = Q^{(p,p)} * Q^{(p+1,p)} * Q^{(N-1,p)} \qquad (4.12b)$$

$Q(q,p)$ has the form

Columns: q q + 1

$$Q^{(q,p)} = \begin{bmatrix} 1 & & & & & & & \\ & 1 & & & & \phi & & \\ & & \ddots & & & & & \\ & & & \cos\Theta & \sin\Theta & & & \\ & & & -\sin\Theta & \cos\Theta & & & \\ & \phi & & & 1 & & & \\ & & & & & \ddots & & \\ & & & & & & 1 \end{bmatrix} \begin{matrix} \\ \\ \\ \text{qth row} \\ \text{(q + 1)th row} \\ \\ \\ \end{matrix} \qquad (4.12c)$$

Givens Generation

$$\Theta = \tan^{-1}\left[\frac{a_{q-1,p}}{a_{qp}}\right] \qquad (4.13)$$

where Θ is an abbreviation for the function $\Theta(q,p)$. The operation of creating $\cos\Theta$ and $\sin\Theta$ is referred to as "Givens generation" (GG). In fact, in a cordic implementation [9], this actually implies the computation of "cordic bits" for the rotation angle Θ (see Fig. 4.6).

Givens Rotation. The matrix product $A' = [Q^{(q,p)}] * A$ is then

$$a'_{q-1,k} = \cos\Theta * a_{q-1,k} + \sin\Theta * a_{qk} \qquad (4.14a)$$

$$a'_{q,k} = -\sin\Theta * a_{q-1,k} + \cos\Theta * a_{qk} \qquad (4.14b)$$

$$a'_{jk} = a_{jk} \quad j \neq q - 1, q$$

for all $k = 1, \ldots, N$. In the following, the operations in Eq. (4.13a,b) are referred to as Givens rotation (GR).

The effects of the GG and GR operations on the qth and (q + 1)th are displayed below.

Before GG and GR:

$$a_{q1} \quad a_{q2} \quad \cdots \quad a_{qN}$$

$$a_{q+1,1} a_{q+1,2} \cdots a_{q+1,N} \qquad (4.15a)$$

After GG and GR:

$$a'_{q,1} \quad a'_{q,2} \quad \cdots \quad a'_{q,N}$$

$$0 \quad a'_{q+1,2} \quad \cdots \quad a'_{q+1,N} \qquad\qquad (4.15b)$$

Two versions of mapping recursive
algorithms onto arrays

In view of matching the indexing of the algorithm and that of wavefront array, there two types of mapping from recursive algorithms onto arrays:

Version I. Match the PE space (PE row number) to the data elements
   (rows of matrix).
Version II. Match the PE space (PE row number) to the recursion
   number.

SFG and wavefront arrays for Givens QR
decomposition: version I

An SFG array for Givens QR decomposition is shown in Fig. 4.6. Before the Givens operation $Q^{(q,p)}$ is applied, the qth PE row, with a_{qk} originally residing within, first fetches $a_{q+1,k}$ from the $(q + 1)$th PE row via the up-going edges. The activities on of the SFG array in Fig. 4.6 follow the recursion described in Eqs. (4.11)−(4.15). The definition of the nodes are to support the operations above, and the two types of operation are shown in the figure.

As shown in Eqs. (4.14) and (4.15), the operation involves two distinct tasks: (1) GG: generation of the rotation parameters cos θ and sin θ, and (2) GR: modification of the elements. In (1) the rotation parameters cos θ and sin θ are generated at the $(q,1)$ PE, they are propagated through the entire qth PE row. In (2) the rotation parameters are used for the GR modification of the elements in the affected data through the rotations of Eqs. (4.14a) and (4.14b). This task is carried out by all the qth row PEs except the first PE.

Also note that the data of row $q + 1$ are transferred via the up-going edges to the row q for the rotation operation; see Eqs. (4.14a) and (4.14b). After each recursion, a new row of the R matrix is output.

Wavefront processing for $Q(1), Q(2)$,
$Q(3), \ldots$ operations

Note that $Q(1) = Q^{1,1} * Q^{2,1} \cdots * Q^{N-1,1}$. Its first operation is $Q^{N-1,1}$, that is, the row operations involved in annihilation of the elements of the first column. Now let us trace the fronts of the corresponding activities. For simplicity, let us suppose that the data $\{a_{j,k}, j,k = 1, 2, \ldots, N\}$ are now residing in their respective PE(j,k).

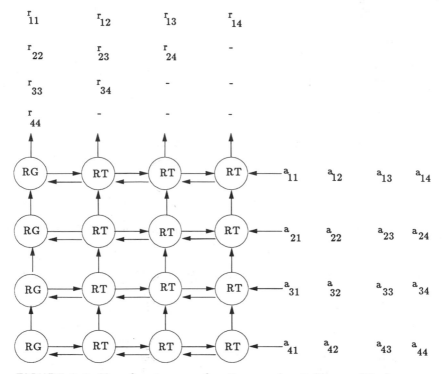

FIGURE 4.6 Wavefront array for the version I Givens QR decomposition. In general, the corresponding SFG array has the same form.

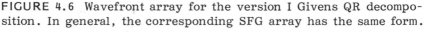

1. The wavefront starts at PE(N − 1, 1), fetching a_{N1} from below and performing the computation for generating the rotation parameters $C(N,1)$ and $S(N,1)$ which annihilate a_{N1}.

2. Upon completing this task, it will propagate the rotation parameters to PE(N − 1, 2), and then PE(N − 1, 3), and so on, and further trigger each of which to perform the rotation operations as in Eqs. (4.14a) and (4.14b). (Note that one of the operands is fetched from above, and the updated result will be returned to the PE above.) Almost simultaneously, PE(N − 2, 1) is ready to fetch a'(N − 1, 1) just created from below and generate the new rotation parameters corresponding to $Q^{N-2,1}$. Again, it will continue to trigger the successor PEs in a similar fashion. The processes for $Q^{N-3,1}$, $Q^{N-4,1}$, and so on, follow simularly. This process forms the first set of wavefronts as shown in Fig. 4.6.

3. Wavefront processing for $Q(2), Q(3), \ldots,$ operations. The first set of fronts will further trigger the activity of $Q(2)$, $Q(3)$, and so on, as discussed below. As soon as the elements a(N,2) and a

a(N − 1, 2) are updated (by operations $Q^{(N-1,1)}$) in PE(N − 1, 2) and transferred leftward to the PE(N − 1, 1), then the Q(N − 2) wavefront may be triggered therein. The Q(N − 2) wavefront, just like the Q(N − 1) wavefront, will propagate steadily along the column and row, generating the second set of wavefronts. Similarly, the PE(N − 1, 2) will produce the two numbers and propagate them leftward to PE(N − 1, 1) and initiate the wavefront for Q(N − 3), and so on. When all the N − 1 wavefronts are initiated and sweep across the array, the triangularization task is completed and the output from the top as shown in Fig. 4.6.

It is easy now to estimate the time required for processing the QR factorization. Note that for the least-square-error solver, the second computational wavefront need *not* wait until the first one has terminated its activity. In fact, the second wavefront can be initiated as soon as the matrix elements a_{N2} and $a_{N-1,2}$ are updated by the first wavefront. This occurs three task-time intervals after the generation of that first wavefront. Thereafter, the wavefronts for Q(N − 1), Q(N − 2), . . . can be pipelined with only three time intervals separating them from each other. The total processing time for the QR factorization would, therefore, be O(3N) on a triangular (i.e., half of N × N) array.

SFG and wavefront arrays for Givens QR
decomposition: version II

As shown in Fig. 4.7, the input array above the array is the data array of the matrix A. The final result on the matrix R will be output from the diagonal nodes.

Wavefront processing for Q(1), Q(2),
Q(3), . . . operations

Suppose that the first row PEs have fetched data {$a_{N,k}$, k = 1, 2, . . . ,N }.

1. Let us trace the fronts of activity relating to $Q^{N-1,1}$, that is, the row operations involved in annihilation of the elements of the first column. The wavefront starts at PE(1,1), fetching $a_{N-1,1}$ from above and performing the computation for generating the rotation parameters C(N,1) and S(N,1) which annihilate $a_{N,1}$.

2. Upon completing this task, it will propagate the rotation parameters to PE(N,2) and then PE(N,3), and so on, and further trigger each of which to perform the rotation operations as in Eqs. (4.14a) and (4.14b). (Note that one of the operands is fetched from above, and the updated result will be returned to the PE above.) Almost simultaneously, PE(1,1) is ready to fetch $a_{N-2,1}$ from above and [along with a'(N − 1, 1) created just now] generate the new rotation

parameters corresponding to $Q^{N-1,1}$. Again, it will continue to trigger the successor PEs in a similar fashion. The processes for $Q^{N-2,1}$, $Q^{N-3,1}$, and so on, follow similarly. In short, the computation activities are propagated sideways and they will further trigger the activity of PE(2,2), and so on, as discussed below.

3. As soon as the elements a(N,2) and a(N − 1, 2) are updated [by operations Q(N,1) and Q(N − 1, 1)] in PE(1,2) and "flown" downward to the PE(2,2), the Q(2) wavefront may be triggered therein. The Q(2) wavefront, just like the Q(1) wavefront, will propagate steadily along the second PE rows. Similarly, the PE(2,3) will produce the two numbers and propagate them downward to PE(3,3) and initiate the wavefront for Q(*,3), and so on. When all the N − 1 wavefronts are initiated and sweep across the array, the triangu-

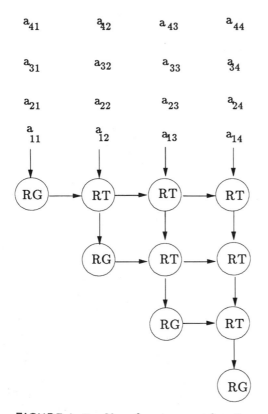

FIGURE 4.7. Wavefront array for the version II Givens QR decomposition. Again, the corresponding SFG array has the same form.

larization task is completed. In fact, the second wavefront can be initiated as soon as the matrix elements a_{N2} and $a_{N-1,2}$ are updated by the first wavefront.

Comparision and discussion

Let us compare the two versions of QR-decomposition array algorithms. Version I enjoys the following advantages: (1) there is no need for processor reprogramming (no change of processor functions), and (2) it is easily adaptable to the banded-matrix LU decomposition problem. On the other hand, version II also offers very attractive advantages: (1) there is no need of diagonal connections, and (2) there is a better utilization efficiency as it yields the same throughput rate with only 50% of PEs. Moreover, as we shall discuss in a moment, the systolized array of version II enjoys a better pipe-lining factor (α), equal to 1, compared to a value of 3 for version I, although they both result in the same latency. If CORDIC arithmetic units are adopted [10,11], the function variations from multiplication, square rooting, and rotation for the boundary PEs can easily be accommodated by adjusting control bits.

For solving a least square-error problem as depicted in Eq. (4.9), another important note should be made here

1. Using the version II design, we shall first cascade A and Y, and apply Q to both of them (i.e., perform the operation Q[A : Y]).
2. Matrix Q is N × N, R is a triangular M × M matrix, vectors c and × are M × 1, and d is (N − M) × 1.
3. In forming R from A, the Givens generator creates *cordic* bits for the rotation angle rather than C(q,p) and S(q,p), or $Q^{(q,p)}$. Applying the rotations on the row elements of A(respectively, Y) is equivalent to executing A' = $[Q^{(q,p)}]^T$ * A (respectively, Y = $[Q^{(q,p)}]^T$ * Y). Thus, the final QR procedure outcome in the processor array will be the triangular matrix, R and c. (In so doing, Q need not be retained.)
4. The vector d represents the residual, and is output from the bottom PE of the rightmost column.

3.4 Single-Wavefront Algorithms

There is a special class of wavefront processing algorithms which centers on the notion of single-wavefront algorithms. A single-wave-front algorithm is characterized by the fact that in any time of processing in a square array, there is at most one active wavefront in any column (or row) of the array. Although these problems may be solved by a square array, the process will be highly inefficient. In-stead, it is possible to map the square (virtual) array to a column-

reduced (or row-reduced) linear array. More precisely, one column
of PEs can be reduced to a single PE, with one column of memory
replacing one column of PEs. (Namely, the square processor array is
now replaced by a square memory array and a linear processor array.)
This can be accomplished without comprising much on the processing
speed, since at any time at most one PE in a column can be active. One
can envision square virtual wavefronts propagating on the (square)
virtual array just the same as before. Then mapping the (square)
virtual wavefronts to the linear real wavefronts is very straightforward.

Examples for this type of algorithm include the longest common
subsequence (LCS) problem, dynamic time warping in speech recognition, LU decomposition with partial pivoting for a numerically stable
solution of linear systems, and eigenvalue and singular-value decom-
positions in image processing applications [12].

4 DATA FLOW GRAPHS AND WAVEFRONT ARRAYS[†]

To facilitate the implementation and application of wavefront arrays,
an important subject is to formalize and systemize the design of such
arrays directly from algorithm descriptions. (An up-to-date review of
these contributions is given in Ref. 13.) As we discussed earlier,
signal flow graphs (SFGs) provide a popular description for recursive
parallel algorithms used in digital signal processing. In this section
we propose, further, a data flow graph (DFG) as a formal abstract
model for wavefront array processors. First, we address the issue of
transforming a signal flow graph into a data flow graph by an equivalence transformation from SFG to DFG. Then, we present a theorem
treating the timing analysis for all (cyclic or acyclic) DFG networks.
As a side product of the timing analysis theorem, the deadlock problem
associated with the DFG is also resolved naturally. Based on the same
analysis, it is also straightforward to predict the minimal number of
buffers required on all the edges of the DFG so as to achieve the best
possible throughput rate.

The SFG notation facilitates the understanding of recursive
algorithms in terms of the computaions required for each recursion
and provides a convenient tool for mapping algorithms onto parallel
processing arrays. To derive a data-driven wavefront realization of
an SFG, we introduced a theorem demonstrating the functional equivalence of each SFG to a particular data flow graph. We then introduce
the DFG model as an abstraction of a network of wavefront array
processors and explore some of its properties. In our treatment,

---

[†]This section is based largely on an earlier paper by Kung et al. [15].

the DFG may include cyclic graphs, and thus generalize earlier DFG timing analyses on acyclic graphs [14]. We propose several retiming theorems, which naturally lead to optimal queue assignment algorithms automatically transforming SFGs into optimal-throughput, cost-effective DFGs.

4.1 Data Flow Graphs

Data flow graphs (DFGs) are used extensively in the area of computer architecture, perhaps most commonly in data flow research [16]. Our purpose is to use a DFG as a formal abstraction of a network of wavefront array processing elements (PEs). To suit our needs, we use the following definition of a DFG.

A basic DFG is a directed graph $G = (V,E,D(E),Q(E))$, in which nodes in V model computation, and directed edges in E model asynchronous communication. Each edge e has a queue capacity, represented by a positive integer weight $Q(e)$. Each edge e is also associated with a nonnegative integer weight $D(e)$, representing the number of initial data tokens on the edge. The state of a DFG is represented by the distribution of tokens on its edges. Each edge may contain a nonnegative number of tokens that is less than or equal to its queue capacity. These tokens may be thought of as filling in a part of the edge queue. This leaves the remainder of the queue empty. Each empty queue slot is called a space. Therefore, each edge e is also associated with a nonnegative integer weight $S(e)$, representing the number of spaces on the edge. Obviously, the total edge queue capacity $Q(e)$ is equal to $D(e) + S(e)$, the sum of the initial tokens and spaces on an edge. Hence the state of a DFG may also be represented by the distribution of either tokens or spaces on its edges.

A node is enabled when all input edges contain a positive number of tokens and all output edges contain a positive number of spaces. The state of a DFG is altered by the firing of enabled nodes. The new state is determined by subtracting one token (and adding one space) from each input edge of the fired node, and adding one token (and subtracting one space) from each output edge.

DFGs have the following properties:

Persistence. Once a node is enabled, it remains enabled until it is fired.

Conservation. In any nondirected loop, the sum of the total number of tokens on edges in one direction and the total number of spaces on edges in the opposite direction is unchanged by state transition. A special case of this is the directed loop, in which the total number of tokens (and the total number of spaces) remains constant.

4.2 Equivalence Transformation Theorem

Equivalence relation between SFG and DFG

Theorem: Equivalence Transformation between SFGs and DFGs.
Barring deadlock situations, the computation of any SFG can be
equivalently executed by a self-timed, data-driven machine with a
topologically identical DFG. The number of initial tokens assigned on
each DFG edge is equal to the number of delays in the corresponding
SFG edge.

   Proof: What needs to be verified is that the global timing in the
SFG can be (comfortably) replaced by the corresponding sequencing
of the data tokens in the DFG. Note that the transfer of the data
tokens is now "timed" by the processing node. This ensures that the
relative "time" between data tokens received at the node is the same
as it was in the SFG, as far as that individual node is concerned. By
induction, this can be extended to show the correctness of the
sequencing in the entire network.

   For convenience, we shall term this transformation the *SFG/DFG
equivalence transformation*. This transformation helps to establish a
theoretical footing for the wavefront array as well as providing more
insights into the programming techniques. The transformation implies
that all regular SFGs can easily be converted into wavefront arrays,
making modularly designed wavefront processing elements very
attractive to use. Based on the equivalence relationship, the correctness of the DFG is assured.

   The major roles of Ds in SFGs is to locate the proper setting of
the initial conditions of the corresponding wavefront array. We stress
that the initial data token distribution plays a very important role in
assuring correct sequencing in a data-driven computing network. The
initial-state assignment is straightforward: For each delay in the SFG
there is an initial data token (regarded as initial values) assigned to
its corresponding DFG edge.

Example: Linear-phase filter design

   To illustrate the role of the initial states and the correctness of
task/data sequencing as guaranteed by the equivalence relationship,
let us discuss the SFG/DFG equivalence transformation via a linear-
phase filter example. Linear-phase filters have two key features:
(1) they have a symmetrical impulse response function [i.e., $h(n) =
h(N - 1 - n)$], and (2) they do not add phase distortion to the signal.
Figure 4.8a shows a SFG that takes advantage of the symmetry property and reduces the amount of multiplier hardware by one-half. By
the SFG/DFG equivalence transformation, the data flow graph is
derived as in Fig. 4.8b. Now note that each middle level edge has a

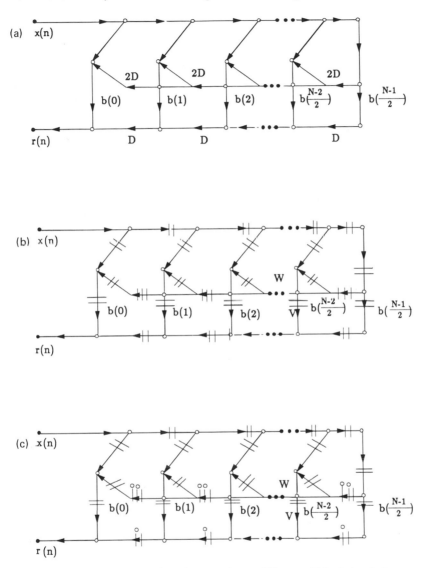

FIGURE 4.8. (a) SFG for linear-phase filters; (b) data-driven model (data-flow graph) for the linear-phase filter; (c) allocation and distribution of initial data token—important in ensuring a correct sequencing. DFG notations: An empty buffer is denoted as a bar on an edge, and a full buffer as a bar with a dot on it.

queue capacity of 3. To ensure the correct sequencing of data, the W data should propagate twice as slowly as the Y data. Note that the queues play the role of ensuring such a correct sequencing.

Let us now explain the initial conditions in the linear-phase filter example. Note that one initial zero-valued token is placed on each Y data edge; and two initial zero-valued tokens are placed on each W data edge. The first zero-valued token of the Y data edge, when requested by the Y summing node, will be passed to meet the V data token arriving from the upper node. When the operation is done, the way is cleared for sending the next Y data token from the right-hand PE. The situation is similar for the W summing node, but only one zero-valued token is "used" and the W data are still one token away from meeting X data in the summing node. They will have to wait until the Y data and the second "0" meet in the lower summing node. This explains why the propagation of W is slower than that of Y. (This is just what is needed to ensure a correct sequencing of data transfers.)

4.3 Formal Model of DFG and Effects of Queues

Once the correctness of the DFG is assured, the questions of pipe-lining speed (i.e., throughput rate) should be explored. To facilitate these analyses, we shall incorporate the notion of time in our DFG model. Each node is assigned a computation time (positive real number) and the node fires after it has been enabled for its computation time. Now a complete specification of a DFG consists of a five-tuple $G = (V, E, D(E), Q(E), T(V))$, where each node V is assigned a positive real number $T(V)$, representing the computation time of the node. $Q(E)$ represents the queue capacities of the edges in the DFG. The per-sistence property of DFG results in a deterministic behavior of state transitions over time.

Example: Effect of queues of throughput

Figure 4.9a shows an SFG that has two directed paths departing at one end and merging at the other. On the other hand, two cor-responding DFGs are shown in Fig. 4.9b and c. The operation times of the nodes are indicated by the numbers inside the nodes. With only one space on the lower branch in Fig. 4.9b, the DFG can accept input tokens only at times 1, 5, 9, ... However, in Fig. 4.9c, another equivalent DFG of the SFG with two spaces on the lower path can accept input tokens at times 1, 2, 5, 6, 9, 10, ... In Fig. 4.9d we show another DFG which does not correspond to the SFG in Fig. 4.9a, in which there are one space and one token on the lower path. This DFG can accept input only at times 1, 5, 9. .... It is clear from this example that the queue capacities (either empty or full) on the edges play an important role in determining the pipelining period of the DFG.

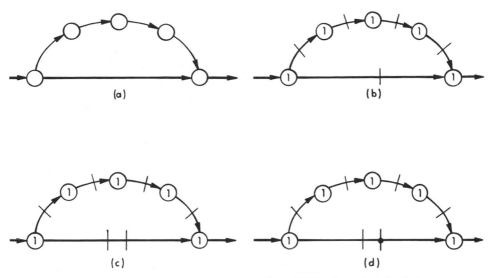

FIGURE 4.9 (a) SFG; (b) corresponding DFG of part (a); (c) corresponding DFG of part (a); (d) another DFG.

4.4 Timing Analysis of DFG

The foregoing example has demonstrated that the pipelining period depends heavily on the queue capacities of the DFG edges. In this section we provide a complete timing analysis for general (i.e., acyclic and cyclic) DFG networks. Our objectives can be stated as twofold: (1) Given a DFG, with defined initial tokens and spaces on the edges, is it deadlock free, and what is the average pipelining period? (2) Given a desired pipelining period, how do we to assign minimal queues on the edges of a DFG to achieve the maximal speed? These questions may be answered by a unified approach based on a duality concept of "token" and "space."

Duality of token and space

Figure 4.10a shows a directed loop with many spaces but only one token. Since the number of tokens in a directed loop is conserved, there will always be only one token in this loop. The token can traverse the loop repeatedly by a series of firings of the nodes in the loop. Each trip around the loop takes T_L time, which is the sum of all node operation times in the loop. The pipelining period is T_L. However, if we put one more token in the loop, then by the time one token traverses around the loop, the other one also finishes its trip around the loop. So the period is only half, $T_L/2$.

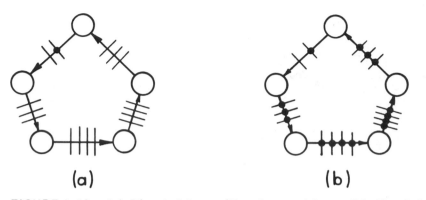

FIGURE 4.10. (a) Directed loop with only one token; (b) directed loop with only one space.

Suppose that we keep putting more tokens into the loop. We can observe that the period is decreasing for a while, and then it will increase again. Let us look at a contrasting case in Fig. 4.10b. Here we have many tokens but only one space. Since the firing of a node requires at least one space on all output edges, we see that only one node can fire at any time. So, effectively, this space will traverse (in reverse direction) around the loop in T_L time also. The pipelining period of this loop is again T_L. By the same analogy, if we put one more space in the loop, the period will become $T_L/2$. Recall that the number of spaces in a directed loop is conserved.

From this example, we see clearly that the token and space play a very similar role in firing a node. They are both the "resources" to be used by the nodes in order to fire. They display a duality relationship analogous to the duality of electron and holes in semiconductors. More precisely, a space in one direction plays the same role as a token in the reverse direciton. There are some observations on the roles of tokens and spaces in a DFG in Refs. 14 and 17, but they are not as extensive as our comments here.

Augmented flow graph

Based on this key observation, we can simplify the discussion by adopting an augmented graph of the original DFG, which is termed the *augmented flow graph* (AFG). The augmented graph is constructed by starting with the DFG and adding a reverse edge e' for every existing edge e. We leave the number of tokens on e unchanged. To e', we assign a number of tokens equal to the number of spaces on e. This conversion is shown in Fig. 4.11a. In effect, we treat spaces as tokens in the AFG.

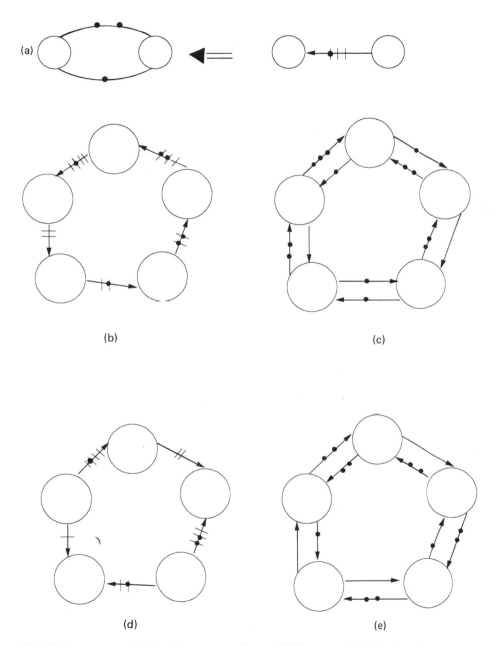

FIGURE 4.11. (a) Rule for converting a DFG to an AFG (note that tokens in AFG are denoted as dots on the edges; (b) directed loop DFG; (c) its AFG; (d) undirected loop DFG; (e) its AFG.

Firing rule of the AFG

Recall that in a DFG, a node can fire only when there is at least one token on all the input edges and one space on all output edges. However, in an AFG, tokens and dual tokens (i.e., spaces) are treated the same. Therefore, the firing rule of a node in an AFG is also modified: A node can fire when and only when there exists at least one token on every input edge. It can then be shown that the operation of this AFG is the same as the original DFG. In particular, the throughput analysis will remain the same.

As an example of this conversion, we show in Fig. 4.11b a DFG, which is a directed loop, and in Fig. 4.11c, its AFG. Note that because of augmenting all edges in the original loop, we obtain two directed loops. The inner loop contains only tokens and the outer loop contains only spaces. We also show in Fig. 4.11d a DFG, which is an undirected loop, and in Fig. 4.11e the AFG of Fig. 4.11d. Notice that in this case the two directed loops in Fig. 4.11e contain both spaces and tokens. The tokens and spaces here play exactly the same role in the timing analysis, which will be explained in the following theorem. Here, for notational convenience the DFG is represented by the five-tuple $(V, E, D(E), S(E), T(V))$, where $S(E) = Q(E) - D(E)$ denotes the spaces of the edges.

Main Theorem Given a DFG with preassigned initial data tokens and spaces, the pipelining period α of the DFG is

$$\alpha = \max \left[T_V, T_P, \frac{T_L}{D_{LC} + S_{LCC}}, \frac{T_L}{D_{LCC} + S_{LC}} \right], \text{ all L in DFG}$$

$$(4.16a)$$

where L is any undirected loop in the DFG, T_L the sum of all node operation times in loop L, D_{LC} the total number of tokens on edges in the clockwise direction in L, and S_{LC} the total number of spaces on edges in the clockwise direction in L. D_{LCC} and S_{LCC} are similarly defined for the counterclockwsie direction in L. T_V is the maximum of operation time of all nodes in the AFG. Finally, T_P is the maximum of any sum of the operation times of a pair of nodes, which are connected by an edge with only one buffer on it.

The main theorem has an equivalent but somewhat simpler statement in terms of the more convenient AFG notations.

Theorem (Rephrased Version) Given an AFG derived from a DFG with preassigned initial data tokens and spaces, the pipelining period α of the AFG (or DFG) is

$$\alpha = \max \left\{ T_V, \frac{T_L}{K_L} \text{ , for all L in the AFG} \right\} \qquad (4.16b)$$

where L is any directed loop in the AFG, T_L the sum of all node operation time in loop L, K_L the total number of tokens (including dual tokens) in L, and T_V the maximum of operation time of all nodes in the AFG.

Proof: Note that if L is a clockwise loop (respectively, a counterclockwise loop), then $K_L = D_{LC} + S_{LCC}$ (respectively, $K_L = S_{LC} + D_{LCC}$). Therefore, the two statements above are the same in content and differ only in notation. For convenience, our proof will be given only in terms of the AFG notation. For this, let us first establish several useful lemmas.

Lemma 1 The pipelining period of an AFG (or DFG) must be greater than or equal to every node operation time.

Proof: This is obvious, since no internal pipelining is assumed inside a node.

Lemma 2 Consider a directed loop L in the AFG. The pipelining period α_L of this loop, when it is operating independently of the remaining part of the AFG, is

$$\alpha_L = \frac{T_L}{K_L} \qquad (4.17)$$

Proof: To show this, first recall, from the conversion property of an AFG, that the number of data tokens in a directed loop L is constant. If we mark one token which begins its trip around the loop at one particular node, it is clear that there will be K_L tokens fired at that particular node during the period when this marked token completes its loop trip. This marked token will need T_L in time to return to the starting node, and this firing pattern will repeat at a period T_L, assuming that this loop operates independent of the rest of the AFG. So we obtain the average period for that loop L, which is T_L/K_L. Note also that all nodes in this loop must operate at the same pipelining period.

Lemma 3 All nodes in the AFG operate at a uniform period in the steady state.

Proof: Whereas all the nodes in a DFG are in general weakly connected only, the nodes in AFG are always strongly connected (i.e.,

there exists a directed path between any pair of nodes in the graph).
Consequently, all nodes in the AFG have to operate at a uniform period
in the steady state. Where this not the case, some edges of the AFG
would eventually have oversaturated tokens. This is not acceptable
for a deadlock-free network. Recall that there are only finite initial
tokens in the AFG, and all tokens are contained in some directed loops.
By the conservation of tokens in the directed loops, the number of
tokens in the AFG remains finite.

The proof of the main theorem has two parts: "necessity" and
"sufficiency."

Necessity: To prove this part, we need to show that the DFG cannot operate with a pipelining period of less than the α in Eq. (4.16).
Since the AFG is strongly connected, according to Lemma 3, all the
loops will operate at the same period. Obviously, the slowest loop will
prevail; therefore, the pipelining period for the total system will be
at least the maximum of all periods associated with the individual
directed loops. Thus the necessity part is proved.

Sufficiency: Here we want to prove that the DFG can operate at the
period of α in Eq. (4.16). Consider two directed loops A and B.
Assume that $\alpha_A = \alpha_B$, and loops A and B are coupled at some nodes.
Then the pipelining period of loops A and B together will be α_A,
except that for some "phase difference" between A and B when both
loops start computing. After a finite amount of time, both A and B
will be synchronized at the period of α_A.

Now assume that $\alpha_A > \alpha_B$. We first modify the loop B into B' by
adding extra operation time into nodes in loop B, which are not in
loop A, such that the new period for loop B' becomes α_A. This is
always possible, since the number of tokens in a directed loop is
constant, and we can increase the period by prolonging the total node
time around a loop. The two loops A and B' can now operate at the
period of α_A.

Comparing loops B and B', it is clear that any node time in B'
should be equal to or greater than the corresponding node time in
B. It is not hard to envision that the loop B can always operate at
the speed of loop B', since the nodes of B are as fast or faster than
their B' counterparts. Since loop B' can operate at the period of α_A,
it is straightforward to see that loop B can also operate at this speed.
So the resulting loops A and B can operate together at the period of
α_A.

By induction, the argument above can be extended to any number
of directed loops in the AFG. Namely, if the slowest period of all
directed loops in the AFG is α, all nodes in the AFG will eventually
operate at a period of α, and the time needed for the synchronization

of all loops will be finite, since there are only finite loops in the AFG. Thus the sufficiency part is proved.

A special case of this formula concerns the effect of putting only one buffer on an edge in a DFG. In Fig. 4.12a we show a simple DFG with two nodes and only one buffer on the edge between the two nodes. In Fig. 4.12b the corresponding AFG is shown. It is clear that the edge between the two nodes in the DFG induces a small loop in its AFG counterpart. This induces a special effect for the case when there is only one buffer on the edge. According to Eq. (4.16b), the pipelining period must be greater than or equal to $T(1) + T(2)$. This implies that these two nodes can only operate sequentially (i.e., only one can operate at any instant). This is the reason that the term T_p appears in Eq. (4.16a).

Another observation is that there are actually many directed loops in the AFG, but only some of them are simple directed loops (i.e., loops that visit nodes only once, except for the node where the loop begins and terminates). In Fig. 4.13 there are two simple directed loops, denoted as $(1,2,1)$, $(2,3,2)$. An example of a non-simple directed loop is $(1,2,3,2,1)$. Fortunately, we need to consider only the simple directed loops for the pipelining period computation. This claim is proved in the following lemma.

Lemma The pipelining period of a nonsimple directed loop L in an AFG is smaller than or equal to the maximum of the periods of all simple directed loops contain in L.

Proof: Without loss of generality, we can use Fig. 4.13 as the AFG. We want to show that

$$\max\left[\frac{T_1}{K_1}, \frac{T_2}{K_2}\right] \geq \frac{T_1 + T_2}{K_1 + K_2}$$

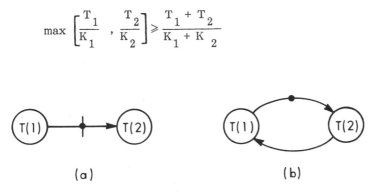

$$(a) \qquad\qquad (b)$$

FIGURE 4.12. (a) DFG with one buffer on the edge; (b) equivalent AFG of part (a).

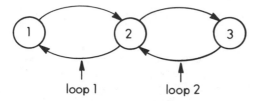

FIGURE 4.13. Example of AFG.

Assume that $T_1/K_1 \geq T_2/K_2$; it is easy to see that $T_1/K_1 \geq (T_1 + T_2)/ (K_1 + K_2)$.

Other cases can be proved similarly.

It can be deduced from this lemma that we can consider only the simple directed loops for the pipelining period computation. So in Eq. (4.1b) we can assume that L is a simple directed loop.

4.5 Deadlock Analysis

The issue of deadlock can be regarded as a special application of the above theorem. The result is stated in the following corollary:

Corollary The DFG is deadlock free if and only if there exists a finite solution α for Eq. (4.16).

Proof: Follows directly from the theorem. An example of deadlock is shown in Fig. 4.14.

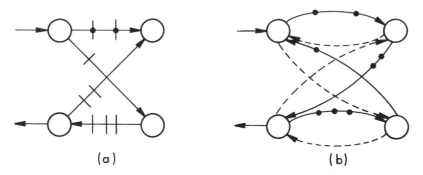

FIGURE 4.14. (a) Deadlocked DFG; (b) its corresponding AFG. Note that there are no tokens in the loop indicated by dashed edges. According to Eq. (4.16b), there exists no finite solution for α.

4.6 Optimizing Throughput of DFG via Queue Assignment

The theorem above gives the pipelining period of a DFG when the initial token assignments and number of queues on edges are given. The dual problem of determining the assignment of minimum queue capacities on the edges in a DFG to achieve the minimum (or optimal) pipelining period α^* is treated in the following theorem:

Theorem Given a DFG with initial token assignment; (1) the minimum (or optimal) pipelining period, and (2) the minimum queue capacities on the edges to achieve that pipelining rate can be determined as follows:

    1. The minimum pipelining period α^* is

$$\alpha^* = \left\{ \max\ T_V, T_P,\ \text{all}\ \frac{T_L}{D_L} \right\} \tag{4.18}$$

where L is a simple directed loop in the DFG, T_L the total node time in L, and D_L the total number of tokens in L. T_V and T_P are as defined earlier.

    2. The assignment of the minimum queue capacities on the edges of the DFG, in addition to that for an edge e: $Q(e)$ must be a positive integer equal to or greater than $D(e)$ and the queue capacities must be the minimum numbers satisfying the condition in Eq. (4.16a) in the main theorem, that is,

$$S_{LC} \geqslant \frac{T_L}{\alpha^*} - D_{LCC} \tag{4.19a}$$

$$S_{LCC} \geqslant \frac{T_L}{\alpha^*} - D_{LC} \tag{4.19b}$$

for any simple undirected loop L in the DFG, and

$$Q(e) \geqslant 2 \quad \text{if } T(\text{source}) + T(\text{sink}) > \alpha^* \tag{4.19c}$$

Note that α^*, T_L, D_{LC}, D_{LCC}, S_{LC}, and S_{LCC} are defined in the main theorem. $Q(e)$ is the queue capacity of an edge e, and $T(\text{source})$ and $T(\text{sink})$ are the operation times of the source and sink nodes of the edge e.

Proof: Note that the optimal pipelining period may be obtained by putting infinite spaces on all edges in the DFG. Theoretically, it is clearly the best one can do to maximize the pipelining rate. In this case, any undirected loop (excluding the directed loops) in the DFG will consist of at least an edge with infinite spaces. As a result, its individual pipelining period would be 0 [according to Eq. (4.17)], which can never adversely affect the pipelining period. Thus such undirected loops may be excluded for the purpose of applying Eq. (4.16) for determining α^*. Therefore, only those directed loops in the DFG remain to be considered. In this case, Eq. (4.16) becomes Eq. (4.18) and step (1) is thus verified. The validity of step (2) follows directly from the main theorem and Eq. (4.16).

In fact, the formula [Eq. (4.19)] in step (2) of this theorem is also valid for any given desired pipelining period which is greater than or equal to α^*. In general, there are nonunique solutions to this assignment. Among several possible approaches to deriving a feasible solution, we propose in the next subsection a simple cut-set procedure [2,18] to tackle this assignment problem.

4.7 Algorithms for Solving Queue Assignment

A direct approach to computing the optimal queue assignments would involve tracing all the possible loops—a very time consuming task. To avoid this, we propose a cut-set retiming procedure to assign queues on the edges in the DFG.

Procedure for transforming an SFG to a retimed DFG

A retimed DFG array is derived from the original SFG by the following procedure:

1. Set the initial α as the maximum of all node operation times. Scale all delays.
2. For each target edge (edge that does not have more or an equal number of delays as required by the node it incidents from), find a cut set containing the target edge such that after suitable delay transfer along the cut set, the target edge will have the exact amount of delay it needs, and at the same time, no new target edges are generated by this delay transfer.
3. If such a cut set cannot be found, a directed loop L that contains the target edge will be found. Rescale the α (i.e., $\alpha = T_L/D_L$).
4. Repeat step 3 until there are no more target edges. The final α is the optimal pipelining period α^*.

The cut-set procedure helps to determine the optimal pipelining period of the DFG, α^*, and also to assign to each edge e of the DFG an appropriate time delay, denoted as t(e), to achieve this optimal period.

After the cut-set procedure, the queue capacity Q(e) needed for the edge e in a DFG for optimal pipelining period α^* can then be computed as follows:

$$Q(e) = \max D(e), \left\lceil \frac{t(e) + T(sink)}{\alpha^*} \right\rceil \qquad (4.20)$$

where the "bracket operation" x denotes the smallest integer greater than or equal to x (the ceiling function). Also, t(e) and α^* are cut-set transfer timing and optimal pipeline period, respectively. D(e) is the number of initial tokens on edge e, and T(sink) is the operation time of the sink node corresponding to edge e.

Proof: We want to show that the queue capacities obtained from Eq. (4.20) satisfy Eq. (4.19). According to the rule used in the cut-set retiming procedure, t(e) should be greater than or equal to the operation time of its source node. Note that in our retimed graph the node operation time has to be absorbed by its output edges.) Now, consider any undirected loop L in the retimed DFG. For the purpose of illustration, an undirected loop and its retimed DFG are shown in Fig. 4.15. Let use assume first that there are no tokens in this loop. We want to assign queue capacities to the counterclockwise (CC) edge first. From the cut-set procedure, the total time assigned to the counterclockwise edges is the same as the total time assigned to the clockwise (C) edges. Note that the total time assigned to the C edges (or the CC edges) is equal to or greater than the sum of the node times of nodes A, B, D, since they are the source nodes of the C edges. Therefore, if we add the total time of the CC edges and the total operation times of the sink nodes (in this case, they are nodes C and E) of the CC edges together, as implied by Eq. (4.20), the sum is greater than or equal to the loop time T_L. So the queue assignment of the CC edges satisfies Eq. (4.19b), that is,

$$S_{LCC} \geq \frac{T_L}{\alpha^*} - D_{LC}$$

We can show that the queue assignment for the clockwise C edges satisfying Eq. (4.19a) follows basically the same argument except exchanging the roles of clockwise (C) edges and counterclockwise (CC) edges.

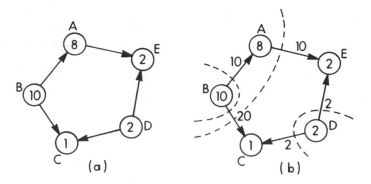

FIGURE 4.15. (a) Undirected loop in DFG; (b) retimed DFG.

Now let us consider the case when there are D_{LC} tokens on some
C edges in the undirected loop L. By the cut-set retiming procedure,
for each token on a C edge, a time delay α^* is assigned on that edge
initially. As a result, after the cut-set retiming, the total time of the
CC edges is $\geqslant$ (total source node times of the C edges) $- \alpha^*D_{LC}$. If
we sum up this total time and all the sink nodes time of the CC edges,
the sum is $\geqslant T_L - \alpha^*D_{LC}$. Therefore, Eq. (4.19b) is again satisfied.
Of course, the counterclockwise case can be proved similarly. Moreover, if the loop under consideration happens to be a directed loop,
it can simply be regarded as a special case, and there is no need for
any different treatment.

Next, we want to show that Eq. (4.19c) is also satisfied for all
edges. This is true since t(e) is always greater than or equal to its
source node time, T(source), and therefore t(e) + T(sink) $\geqslant$ T(source)
+ T(sink) of node e. It is then clear that Eq. (4.19c) is satisfied for
all edges in the retimed DFG.

Finally, we note that since the queue capacities have to be
integers, this quotient is rounded up by the bracket operation (i.e.,
the ceiling operation). It is obvious that the queue capacity Q(e)
must be $\geqslant$ D(e) in order to accommodate all the initial data tokens on
edge e. Therefore, the maximum of the two integers above is taken
for the queue assignment.

Discussion

The queue assignment above is not necessarily minimal for several
reasons. First, the cut-set retiming procedure leads to a nonunique
solution, and it requires a special retiming scheme to achieve the goal
of using the minimal total number of queues used in the graph. In
addition, the assignment has to bear the extra queue overhead due to
the quantization effects and the need for accommodating initial tokens

in Eq. (4.20). However, the real minimal queue problem is considerably more involved. It must consider all possible cut-set time transfers and choose the one that would minimize the overhead due to the quantization effects and the extra queues used for accommodating initial tokens.

Note that the discussion in this section basically extends the earlier timing analysis research on synchronous circuits [19]. To deal with generalized asynchronous data flow arrays, it requires a completely different DFG modeling, which abstracts an asynchronous circuit with initial conditions. Note that the initial conditions may affect the optimal pipelining period α.

Moreover, compared with the systolic array, the wavefront array represents an effective means to deal with multirate processing elements. In systolic arrays, a common procedure is to adopt a uniform clock unit based on the slowest node, not a good scheme in view of pipelining speed. It is also noted [15] that the DFG modeling, with queues accommodating initial tokens, has an important advantage. It allows us to avoid the tedious (and sometimes impossible) procedure of initial token reassignment as required in the systolic design.

Minimal queue algorithm

We propose to look into the following two approaches:

1. *Queue assignment algorithm via linear programming with edge retiming constraints.* After the cut-set procedure, optimal pipelining period α^* can be computed. It is also popular to use a special cut set around every node. Then the time transfer onto a edge, say e, can be represented by r(source) − r(sink). Here r(source) is time transferred into e via a cut around the source node, and r(sink) is time transferred away from a via a cut around the sink node. Then, according to Eq. (4.20),

$$Q(e) = \max\left[D(e), \left[\frac{t(e) + T(sink)}{\alpha^*}\right]\right]$$

and the suboptimal queue assignment can be reformulated in terms of a linear programming formulation:

$$\min \Sigma \; [r(source) - r(sink)] \tag{4.21a}$$

subject to the constraints that for all edges e,

$$r(source) - r(sink) \geqslant T(source) - \alpha^*D(e) \tag{4.21b}$$

where r(source) and r(sink) are time transfer on the source and sink nodes of the edge e, and T(source) is the source node time for edge e.

2. *Integer programming with (fundamental) loop constraints.* The assignment of the minimum queue capacities on the edges of the DFG requires that in addition to that for an edge e, $Q(e)$ must be a positive integer equal to or greater than $D(e)$ and the queue capacities must be the minimum numbers satisfying the condition in Eqs. (4.19a)– (4.19c), that is,

$$\min \Sigma Q(e) \qquad\qquad (4.22a)$$

under the linear constraints

$$S_{LC} \geqslant \frac{T_L}{\alpha^*} - D_{LCC} \qquad\qquad (4.22b)$$

$$S_{LCC} \geqslant \frac{T_L}{\alpha^*} - D_{LC} \qquad\qquad (4.22c)$$

for any simple undirected loop L in the DFG, and

$$Q(e) \geqslant 2 \quad \text{if } T(\text{source}) + T(\text{sink}) > \alpha^*$$

The questions to be addressed are how to (1) choose the non-unique fundamental loops among all simple loops, (2) develop a simple technique for the computation, and (3) compare the node and loop techniques in terms of their computational efficiency. These are currently under our investigation.

4.8 Examples of Timing Analysis and Optimization

To illustrate the use of the proceeding theorems, let us consider two examples.

Lattice filter example

First, an SFG of a lattice filter is shown in Fig. 4.16a, and one equivalent DFG of the SFG is shown in Fig. 4.16b. The initial data tokens, queue capacities, and node operation times are also displayed.
   We first apply the main theorem to determine the pipelining period of the DFG Fig. 4.16b. The augmented flow graph (AFG) is shown in Fig. 4.16c, and the loop that has the maximum period α is shown in Fig. 4.16d. To form this loop, we obtain $\alpha = 11$.

---

†This gives an optimal solution modulo the extra queues due to quantizations and initial token queues.

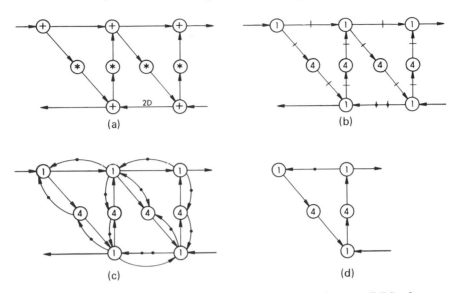

FIGURE 4.16. (a) Lattice filter SFG; (b) one equivalent DFG of part (a); (c) AFG of part (b); (d) slowest loop in the AFG.

To assign minimum queues on the edges in order to achieve optimal pipelining period, we first derive the retimed DFG of the lattice filter by the cut-set procedure. The retimed DFG of the lattice filter is shown in Fig. 4.17a. The optimal pipelining period α^* obtained by the cut-set procedure is 5.5. Applying Eq. (4.20) to all edges in the DFG, the assignment of minimum queues is as shown in Fig. 4.17b.

IIR filter example

An IIR digital filter, which is commonly used in digital signal processing, serves as our second example. In Fig. 4.18a a digital network of this filter is shown, in Fig. 4.18b we show the SFG representation of this filter, and in Fig. 4.18c one equivalent DFG of this filter is shown.

Assume that any node time in the DFG in Fig. 4.18c is equal to T. We can compute the pipelining period α by using Eq. (4.16a). The resulting $\alpha = 2T$. To obtain the minimal queue assignment for optimal pipelining period, the retimed DFG is shown in Fig. 4.19a. The optimal pipelining period α^* is T. From Eq. (4.20) we can compute the queue assignment, which is shown in Fig. 4.19b.

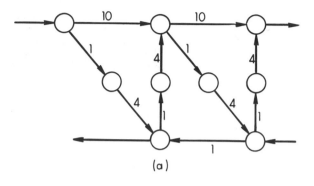

(a)

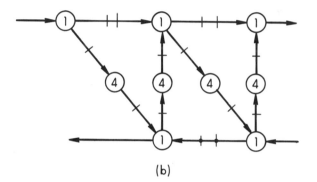

(b)

FIGURE 4.17.   (a) Retimed DFG; (b) queue assignment for optimal period.

Summary

We have presented a systematic approach to converting parallel algorithm described by a signal flow graph to a functionally equivalent data flow graph, which sets the stage for its wavefront array implementation. Based on the notion of token and space (and their intriguing duality property), pipeline timing analysis and optimal queue design of generalized data flow arrays are investigated. It is expected that data flow, wavefront, and fault-tolerant architectures will play a central role in future VLSI or wafer-scale computing systems. It is our belief that for these applications, our analyses and algorithms will prove to be very useful.

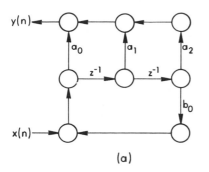

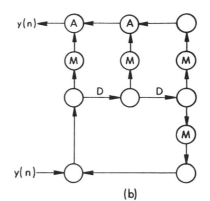

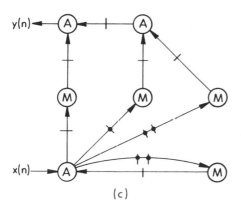

FIGURE 4.18. (a) Digital network of a IIR filter; (b) SFG of the filter; (c) one equivalent DFG.

5 SOFTWARE TECHNIQUE, HARDWARE IMPLE-MENTATION, AND APPLICATION DOMAIN OF WAVEFRONT ARRAYS

5.1 Programming Languages and Software for Array Processors

The actual implementation of wavefront arrays can be either dedicated or programmable processors. Programmable arrays are preferred, due to the high cost of hardware implementation and the increasing varieties of application demands. Therefore, it is equally important to develop a complete set of software packages for most wavefront-type processing. For that, a formal algorithmic notation and programming language will be indispensable.

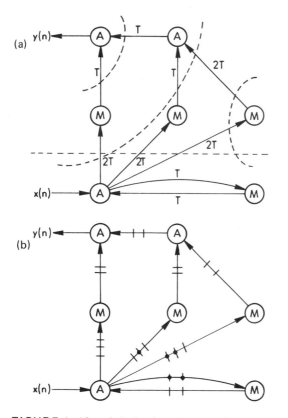

FIGURE 4.19. (a) Retimed DFG of the IIR filter; (b) queue assignment for optimal pipelining period.

General guidelines for algorithmic notations for array processors
are problem orientation, executability, and semantic simplicity. More
important, adequate language criteria must take into account the
characteristics and constraints of the arrays. Examples for appropriate
array processing notations, which incorporate the language criteria
for systolic/wavefront array processors, are CRYSTAL [20], data
space notation [21], and the wavefront language (MDFL) [7].

Wavefront Language

The wavefront notion greatly reduces the complexity in the
description of parallel algorithms. The mechanism provided for this
description is a special-purpose, wavefront-oriented language termed
matrix data flow language (MDFL) [7]. The wavefront language is
tailored toward the description of computational wavefronts and the

corresponding data flow in a large class of algorithms (which exhibit the recursitivity and locality mentioned earlier). This notation is in many ways very similar to the data space notation, which is based on the notion of applicative state transition systems described in [21,22]. Among other commonalities shared by the two notations is, in particular, that they are both based on the data flow principle [23].

The wavefront notion can facilitate the description of parallel and pipelined algorithms and drastically reduce the complexity of parallel programming. There are two appraoches to wavefront programming:

1. Rather than requiring a program for each processor in the array, a wavefront language (e.g., MDFL) allows the programmer to address an entire front of processors. To translate the global MDFL language into instructions for the PEs, a preprocessor is needed. For a wavefront array the design of such a preprocessor is relatively easy since we do not have to consider the timing problems associated with the synchronous systolic array.

2. The second approach to programming the wavefront array is based on the SFG/DFG descriptions. Many one- or two-dimensional digital filters are given initially in an SFG form. Therefore, it is desirable to have a simple mechanism to convert its SFG or DFG representation into an array processing programming codes.

Example: Wavefront programming for
matrix multiplication

Let us first take a look at the band-matrix and full-matrix multiplication $C = A \times B$. The matrix multiplication can be carried out in N (outer product) recursions (see Eqs. (4.1)–(4.3)). A general configuration of computational wavefronts traveling down a processor array is illustrated in Fig. 4.5. By the DFG/SFG equivalence, the corresponding DFG will be just as shown in Fig. 4.3. Based on the DFG, an example of a MDFL program for the corresponding array processing of the matrix multiplication is given in Fig. 4.20.

Note that initially matrix A is stored (row by row) in the left memory module (MM). Matrix B is in the top MM and is stored column by column. The final result will be in the C registers of the PEs. This example illustrates the typical simplicity of the MDFL programming language.

Wavefront programming example on
linear-phase filtering

To demonstrate the simplicity of programming based on the DFG representation, an MDFL program implementing the DFG for the linear-phase filter (see Fig. 4.8b) is shown in Fig. 4.21. The simple mapping between the DFG and the programming codes suggests a potentially

Begin

Set Count N;

Repeat;

While Wavefront In Array Do

Begin

Fetch C, Down;

Fetch A, Left;

Fetch B, Up;

Flow A, Right;

Flow B, Down;

(*comment: Now compute C: = C + A × B*)

Mult A, B, D;

Add C, D, C;

Flow C, Up;

End;

Decrement Count;

Until Terminated;

Endprogram.

FIGURE 4.20. MDFL program for matrix multiplication.

significant impact of the SFG/DFG equivalence transformation to both the hardware and software developments of array processors.

5.2 Hardware Implementation Aspects

A programmable wavefront array must be able to execute the entire range of instructions of the MDFL repertoire (i.e., it will be able to carry out any of the algorithms applicable to the computational wavefront notion).* Each PE, therefore, includes the full complement of hardware necessary to support all the MDFL commands. However, within the above-mentioned class of algorithms there is a finer division

---

*Note that the separators in the DFG are implemented simply by adding three lines of (internal register transfer) code to the program, as opposed to adding a separate buffer register external to the PE.

Begin

Repeat;

While Wavefront In Array Do

Begin

Fetch X, Left;

Flow X, Right;

Transfer W2 to W1;

Flow W1, Left;

Fetch W2, Right;

(comment: now compute V: = (W1 + X) × h(k))

Add W1, X, U;

Mult U, h(k), V;

Fetch Y, Right;

(comment: now compute Y: = Y + V)

Add Y, V, Y;

Flow Y, Left;

End;

Decrement Count;

Until Terminated;

Endprogram.

FIGURE 4.21. MDFL program for linear-phase filter.

of the applications, which may lead to significant simplification in architectural structures and great saving in hardware implementation. When the scope of applications of the wavefront arrays is limited, it is clearly beneficial, to streamline the PE design to the need at hand by deleting unnecessary (general-purpose) hardware and enhancing certain hardware components of the PE. The benefits involved include decreased PE area (i.e., increased number of PEs that can be assembled on a chip) and an increase in the speed and throughput rate of the PEs and the wavefront arrays.

It is very obvious that the structure of the control unit (CU), the ALU, and the program memory will be application dependent. In the next section we provide the configuration options available, then look at a possible basis for the final selections.

PE-level hardware design

In this subsection we give an overview of the architecture of a processing element (PE) to be used as a basic module in a programmable array processor. A typical example will be the design of a PE to be used in a wavefront array. The basic wavefront array is either a square array of N × N PEs, a linear array of 1 × N PEs, or a bilinear array of 2 × N PEs. The PEs are orthogonally connected and are identical. The hardware of the PE is designed to support the features of the matrix data flow language (MDFL) introduced previously. Given the current state of the process technology, with a minimum feature size of 2 μm or less, we estimate the area of the chip taken by a PE to be 6 × 6 mm^2.

Architectural outline

The processor element (PE) that we have designed is a special-purpose microprocessor. The functional block diagram of the PE is shown in Fig. 4.22. The main functional blocks are the data path, program memory, I/O control units, and instruction decoder.

Our design objective is to limit the complexity of the data path, preferring a regular and easy layout design. We have adopted a 32-bit-wide data path for fixed-point computations. Moreover, the ALU in the PE is designed to support the operations that are of major importance for signal processing applications, such as multiplication and rotation. To speed up the throughput of the PE, we used a two-level pipelining scheme.

The PE can perform data transfers in four directions simultaneously. The transfer of data is controlled by an I/O controller, one for each of the four directions, which handle the two-way handshaking functions.

Instruction set

The instruction set of the PE was selected to optimize the performance of the wavefront array as a whole. To reduce the complexity of the control unit in a manner similar to that used in the RISC design [24], we wanted each instruction to take exactly one clock cycle. This implies that complex instructions should be decomposed into sequences of simpler (primitive) instructions. An example is the multiplication instruction, which is decomposed into three instructions: one for initializing the processor registers with the correct data, one that does the main multiplication step, and the last one, which transfers the result back to the register file. The instruction set is divided into arithmetic instructions, register transfer instructions, conditional and unconditional jump instructions, and program loading instructions.

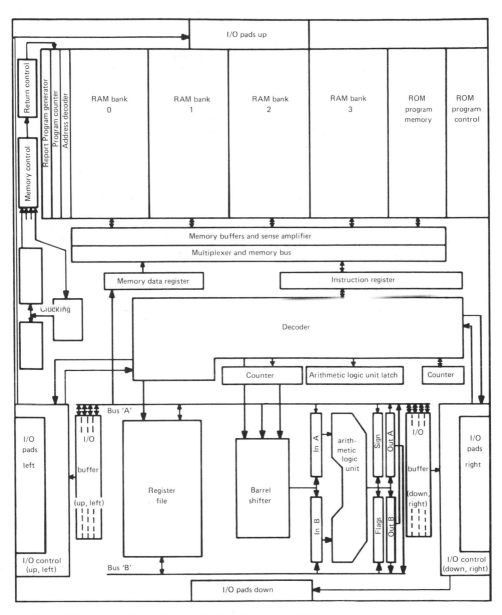

FIGURE 4.22. Functional block diagram of PE.

Design specification and verification

The PE described in this section is currently being specified and verified using the ISPS language. The ISPS language allows not only the specification of the design of a single PE, but also the simulation of an entire wavefront array. More important, it facilitates the verification of the correctness and suitability of the architecture of the PE before designing the lower levels (logic, circuit, layout) of the PE. It will also be an important tool for the development of the host interface, the memory units, and so on.

Array system-level implementational considerations

From an image processing point of view, the excessive supervisory overhead incurred in general-purpose supercomputers often severely hampers the processing rates. On the other hand, VLSI has made direct array implementation of the parallel algorithm possible. To achieve a throughput rate adequate for real-time image processing, the only effective alternative appears to be massively concurrent processing. Evidently, for special applications, a dedicated array processor (stand-alone or peripheral) will be more cost-effective and perform much faster. For example, concurrent array processors will speedily execute (compute-bound) functional primitives such as FFT, digital filtering, correlation, and matrix multiplication/inversion and handle other possible computation bottleneck problems.

An array processor is in general used as an attached processor, enhancing the computing capability of the host machine. An array processor system should provide

More computing power, with help from multiple devices
More flexibility, to cope with the partitioning or fault-tolerance
    problems
More reliability and predictability

Therefore, we propose an array processor system consisting of the following major components:

1. Host computer
2. Interface system, including buffer memory and control unit
3. Connection networks (for PE-to-PE and PE-to-memory connections)
4. Processor array, comprising a number of processor elements with local memory

A possible overall system configuration is depicted in Fig. 4.23, where the design considerations for the four other major components are further elaborated. In general, in an overall array processing system, one seeks to maximize the following performance indicators: computing power using multiple devices; communication support, to enhance the performance; flexibility, to cope with the partitioning problems; reliability, to cope with the fault-tolerance problem; and practicality and cost-effectiveness.

Host Computer. The host computer should provide batch data storage, management, and formatting; determine the schedule programs that control the interface system and connection network; and generate and load object codes to the PEs. A very challenging task to the system designer is to identify a suitable host machine for interfacing with high-speed array processor units.

Interface System. The interface system, connected to the host via the host bus, has the functions of down-loading and up-loading data. Based on the schedule program, the controller monitors the interface system and array processor. The interface system should also furnish an adequate hardware support for many common data management operations. A very challenging task for the system designer is the management of blocks of data. Another task is to make sure that the memory (buffer) unit is able to balance the low bandwidth of system I/O and the high bandwidth of array processors.

Connection Network. Connection networks provide a set of mappings between processors and memory modules to accommodate certain common global communication needs. Incorporating certain structured inter-connections may significantly enhance the speed performance of the processor arrays.

In array processing, data are often fetched and stored in parallel memory modules. However, the fetched data must be realigned appropriately before they can be sent to individual PEs for processing. When some global interconnection algorithms are to be executed effectively or certain array reconfigurations or partitioning are to be supported, such alignments also become necessary. They are implemented by the routing funcitons of the interconnection network, such as the wraparound connection (e.g., spiral or torus) or perfect shuffle, or hypercube network. It is shown in Ref. 25 that the shuffle-exchange network can realize arbitrary permutations of N data in $3 \log_2 N$ passes. However, using such a network requires a major effort for generating the somewhat complicated control codes. A crossbar connection network, when affordable, represents another convenient choice, which offers maximum flexibility at the expense of extremely high hardware costs.

In short, the performance for VLSI arrays depends critically on the communication cost for data transactions. The degree of achievable parallel processing also depends heavily on the effectiveness of matching communication needs as called for by the parallel algorithm with interconnectivity as offered by the real array architecture. It is therefore desirable to have a formal way of characterizing the communication requirement and evaluating its cost.

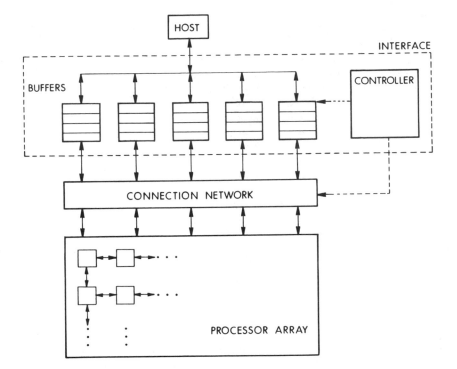

FIGURE 4.23. Example of array processing system configuration.

Processor Arrays. For simplicity, only one processor array is depicted in Fig. 4.23. However, the concept of networking several processor arrays has now attracted a good deal of attention. For example, when a problem is decomposible into several subproblems, to be executed one after the other, it will be useful to have each sub-problem executed in its own processor array, while utilizing the network to facilitate the data pipelining between the arrays. This suggests a pipelining scheme in the array level, which may speed-up the processing by one additional order of magnitude.

Constructing a wavefront array based on transputer chips

Array computer design is better based on commerically available chips, at least for constructing experimental prototypes. These chips must provide computing power, communication capabilities, and interface mechanisms. A transputer is a state-of-the-art chip tailored to meet such requirements. Its 32-bit wordlength provides enough precision for many signal processing applications. Implementation of the wavefront array processor using these chips and additional VLSI/LSI devices appears to be very feasible and cost-effective.

The transputer is designed to execute the Occam language directly. Occam can be used to program a network of transputers— forming a wavefront array—each executing an Occam process. A link between two transputers implements an Occam channel in each direction between two processes. All aspects of the Occam model are implemented within the transputer hardware, so there is no need for scheduling and communications software.

Implementation Example. In RSRE, United Kingdom [26], a wavefront processor for the least-squares minimization algorithm has been implemented on an array of microprocessors programmed in Occam. As the transputer was not yet available, transputer emulator boards have been built to enable algorithms to be evaluated and performance assessments made. Each emulator consists of a circuit board, containing an 8-MHz Intel 8086 processor with transputer-like communication channels. Four ports operate synchronously (i.e., handshaking) at 200K baud with system software ensuring that the channel facilities appear the same as transputer silicon. A fifth channel is programmed to operate as an asynchronous RS232 link to enable programs to be loaded and for data storage, and 8K bytes of EPROM for the bootstrap loader and monitor. Additionally, each emulator has an Intel 8254 time chip so that Occam timing facilities can be implemented in hardware. The ratio of CPU execution time to I/O time is similar to that of the transputer, with operation at approximately one-fifth of transputer speeds.

DSP-oriented array processor chips

The key characteristics for digital signal processing (DSP) applications are adequate word length, fast multiply/accumulate, high-speed RAM, fast coefficient table addressing, a new sample fetching mechanism. The transputer offers 32-bit fixed-point word length, a precision adequate for most (but not for all) signal processing applications. As to the computing speed, arrays of transputers can be used to provide concurrent computing performance very close to what is needed in many signal processing applications. Take the FFT as an example. To cover the full audio spectrum up to 100 kHz, one will need 6, 8, and 10 transputers for 64-, 256-, and 1024-point FFT computations, respectively. For future real-time signal processing requirements a revised transputer has to trade on-chip RAM for a fast multiplier, fast shifter, and if necessary, floating-point microcode. At any rate, an optimized DSP transputer chip is likely to closely resemble a wavefront processor.

A potential weak point of transputers for DSP applications is that DSP often requires fast and/or versatile arithmetic units (e.g., multiplier or cordic processor), high-speed table addressing, and other features [27]. Therefore, an upgraded and optimized DSP transputer chip will be of great interest to the DSP community. For DSP, a revised transputer chip should trade on-chip RAM for a fast multiplier, fast shifter, and for specific applications, floating-point microcode. The current handshake links between PEs have to be retained as they will be essential for the construciton of large-scale array processors. On the other hand, the use of internal parallelism (supported by the current transputer model) is somewhat unclear and depends greatly on the class of applications for which it is intended. The other commerical VLSI chips worthy of exploration for array processor implementation are NEC's data flow chip [28], NCR's GAPP [29], and 64-bit floating-point arithmetic chips [30], among others. To enhance the transputer's arithmetic speeds, especially for multiplication, one should consider combining the transputer chip and a floating-point arithmetic chip. The possibility of multilevel parallelism and pipelining (including both external and internal PE levels) are also worth further exploration.

5.3 Application Domain of Wavefront Array Processors

The power and flexibility of the systolic/wavefront-type arrays and MDFL programming are best demonstrated by the broad range of signal processing and scientific computation algorithms suitable for systolic/wavefront arrays [2]. Such algorithms can be roughly classified into three groups.

1. *Basic matrix operations*: matrix multiplication, band-matrix multiplication, matrix-vector multiplication, LU decomposition, LU decomposition with localized pivoting, Givens algorithm, back substitution, null-space solution, matrix inversion, eigenvalue decomposition, and singular-value decomposition.
2. *Special signal processing algorithms*: Toeplitz system solver, one- and two-dimensional linear convolution, circular convolution, ARMA and AR recursive filtering, linear-phase filtering, lattice filtering, DFT, and two-dimensional correlation (image matching).
3. *Other algorithms*: PDE (partial difference equation) solution, sorting, transitive closure, and shortest-path problems.

These signal processing algorithms share the critical attributes of regularity, recursiveness, and local communication, which are effectively exploited in systolic and wavefront array processors. These arrays maximize the strength of VLSI in terms of intensive and pipelined computing and yet circumvent its main limitation on communication. The application domain of such array processors covers a very broad range, including digital filtering, spectrum estimation, adaptive array processing, image/vision processing, and seismic and tomographic signal processing.

Furthermore, if the communication constraint is relaxed, our technique for converting an SFG for a given application into a wavefront array will also work with other global-type algorithms, such as the FFT algorithm, the Kalman filter network, or other nonregularly interconnected arrays. The only additional requirement lies in routing the physical connections between PEs. This is the topic of the following section.

6 COMPARISON OF PARALLEL PROCESSORS FOR DIGITAL SIGNAL PROCESSING

6.1 Comparing Pipeline (Systolic/Wavefront) Arrays with Other Architectures

In this section we compare the systolic/wavefront array processors with the other array or multiprocessors discussed above. A totally objective comparison is almost impossible, because very complex trade-offs are involved among numerous criteria, such as programmability, software simplicity, modularity, synchronization, and interprocessor communication.

Data flow multiprocessor

The data flow multiprocessor [30] and the wavefront array share the property of self-timed data-driven computing. The basic principle of wavefront architecture is that each PE waits for the arrival of a primary wavefront, then executes the required computation and acts as a secondary source of new wavefronts. This is equivalent to the key concept in data flow machines: arrival of data fires each PE, which subsequently sends relevant data to the next PE. Hence the WAP can be regarded as an array of homogeneous data flow processing elements. The "wait" data feature, provided by handshaking, allows for globally asynchronous operation of processors (i.e., there is no need for global synchronization).

For both data flow and wavefront processing concepts, scheduling and synchronization are built in at the hardware level and are distributed, not centralized. This allows for efficient, fine-grain parallelism. Being language-based architectures, they support the mapping of application algorithms directly onto the multiprocessor in a way that achieves high performance.

Due to its general-purpose nature, a data flow machine often involves a great amount of data and resource management and needs a powerful supervising system. By the same token, the construction of a working data flow machine has not demonstrated convincing effectiveness. It is therefore extremely advantageous to focus on a special (yet major) class of applicational algorithms, and achieve a much greater efficiency. This, in fact, leads to the notion of wavefront processing.

The inherent advantages of restricting applications to a special class are the WAP's programming ease, functional modularity, and consequent reduction in design costs. However, again, a trade-off exists between the general-purpose nature of data flow multiprocessors and the simplicity of the WAP. A research effort is being conducted into how to compromise optimally between specialized applications (of the WAP) and the general-purpose approach (of conventional data flow multiprocessors).

The wavefront array and data flow multiprocessors are both globally asynchronous. This is in contrast to the global synchronization adopted in both systolic and ILLIAC IV arrays. The adoption of global asynchrony has been a controversial issue and has already attracted a great deal of attention. Nevertheless, it is bound to have a major impact on the design of future VLSI systems.

In a sense, the wavefront array processor possesses all the advantages of the systolic array processors, such as extensive pipelining and multiprocessing, regularity, and modularity. More significantly, it shares the asynchronous waiting capability of data-flow machines and therefore can accommodate the critical problem of timing uncertainities, which is bound to be a cause for concern in future VLSI systems.

SIMD machine

A SIMD (single-instruction-multiple-data) computer is often implemented as an array of arithmetic processors. The Illiac IV system has been a typical example for the study of SIMD computers. Just like the systolic array, the ILLIAC IV processors are all synchronized. Centralized control and global synchronization are possible for today's LSI technology, since it is a relatively simple task to add a few delay buffers to the control signals to compensate for clock skew in LSI systems. In VLSI, however, due to large random clock skew and expensive communication, the adoption of the SIMD scheme in future multiprocessors has to be reviewed carefully. Consequently, asynchronous distributed control and localized communication (for both data and control) are more attractive in VLSI.

In contrast to the ILLIAC IV, the WAP employs local instruction storage, data flow—based control, local communication and needs no global synchronization, all of which make the WAP very appealing for

VLSI. Owing to the drastically simpler structure of the wavefront array, its range of applications is naturally much more limited than that of the ILLIAC IV. Nevertheless, the wavefront array is applicable, with inherent locality and recursivity, to a large number of algorithms.

6.2 Comparison of Systolic and Wavefront Arrays

The two types of arrays share the important common feature of using a large number of modular and locally interconnected processors for massive pipelined and parallel processing. They are, however, very different in hardware design (e.g., clock and buffer arrangements) and in software requirements especially on using SFG/DFG as a programming tool. As a result, the wavefront design offers some additional useful advantages on, for example, maximizing pipelining efficiency, architectural extendability, facilitating a simpler wavefront language, and flexibility to cope with time uncertainties.

Maximum pipelining

A major thrust of the wavefront array derives from its maximizing the pipelinability by exploring the data driven nature inherent in many parallel algorithms. This becomes specially useful in the case of uncertain processing times used in individual PEs. As reported in Ref. 26, wavefront pipelining may yield a significant speed-up compared with pure systolic pipelining.

In their case study, a recursive least-sqaures minimization may be carried out using a triangular wavefront array, and it is easy to modify the array to operate as a systolic proceesor. In the least-squares minimization example used in the simulation study [26], the improvement is a factor of almost 2.

Synchronization and architectural extendibility

The clocking scheme is a critical factor for large-scale array systems, and global synchronization often incurs severe hardware design burdens in terms of clock skew and heat dissipation. The data-driven model allows a self-timed design since a global timing reference is no longer required. As a result, it requires no global timing and local synchronization will suffice; thus it is termed a *locally synchronized* (as opposed to globally synchronized) *system*. In short, the asynchronous model in the wavefront arrays incurs a fixed-time-delay overhead due to the handshaking processes.The synchronization time delay in the systolic arrays is due primarily to the clock skew, which changes dramatically with the size of the array. Therefore, the wavefront array enjoys a better extendibility of the array size.

Programming simplicity

The notion of computational wavefronts also facilitates the program-mability of array processors. By tracing the wavefronts, the descrip-tion of the space-time activities in the array may be significantly simplified. Programming a wavefront array is also made easy based on the DFG descriptions of many one- or two-dimensional computing structures. The previously mentioned parallel processing language, Occam, is very suitable for programming wavefront arrays.

Fault Tolerance of Array Processors

To enhance the reliability of computing systems, real-time signal processing architectures demand special attention in run-time fault tolerance. However two-dimensional systolic arrays are in general not feasible for run-time fault tolerance design since a global stoppage of PEs is required when any failure occurs. It is known that certain fault-tolerance issues (rollback, suspension of execution, etc.) are simpler to handle in data flow architecture than in other multiprocessors [16]. Since wavefront arrays incorporate the data-driven feature into the arrays, they pose similar advantages in dealing with time uncer-tainties in a fault-tolerance environment. For example, once a fault is detected, further propagation of the wavefront can be automatically suspended, due to its data-driven nature. Systolic arrays, in com-parison, would require global broadcasting of an "interrupt" signal and a corresponding complicated rollback scheme.

As an example, suppose that, say, PE(2,2) in the array fails in run time; how does the failure affect the other PEs in the array? Because of the distributive computing nature, the failure information can be immediately available only to the neighboring PEs. This shows a major barrier (or impossibility) to implementation of a run-time fault-tolerant systolic array. On the other hand, wavefront processing appears to offer a promising solution. Wavefront arrays also seem to be a good candidate for wafer-scale integration, where the actual communication paths will be fluctuating due to the necessary rerouting around faulty PEs.

In summary, to choose between a systolic and a wavefront array, there are several important factors, such as (global) synchrony, programmability, hardware complexity, extendibility, reliability, fault tolerance, and testability. The final choice between the two array processors hinges on the specific applications intended. In general, a systolic array is useful when the PEs are simple primitive modules, since the handshaking hardware in a wavefront array would represent a nonnegligible overhead for such applications. On the other hand, a wavefront array is more favored when the PEs involve more complex modules (such as multiply-and-add and lattice or rotation operations), or when a robust and reliable computing environment (such as fault tolerance) is essential.

We have presented a systematic approach to converting a parallel algorithm described by a signal flow graph to a functionally equivalent data flow graph, which sets the stage for its wavefront array implementation. Based on the notion of token and space (and their intriguing duality property), pipeline timing analysis and optimal queue design of generalized data flow arrays are investigated. It is expected that data flow, wavefront, and fault-tolerant architectures will play a central role in future VLSI or wafer-scale computing systems. It is our belief that for these applications, the theorems and the timing analyses proposed in this chapter will prove to be very useful.

7 CONCLUSION

The emergence of new VLSI technology, together with modern CAD tools and other hardware and software advances in computer technology, virtually assure a revolutionary information processing era in the near future. The research and development community will soon face the impact of modern VLSI technologies via vertically integrated design techniques and facilities. VLSI technology, starting as a device research area, provides opportunities and constraints that will open up new areas of research in computer architecture. From a scientific research perspective, a close interaction between VLSI and array architecture research areas will be essential. In this chapter we have identified several novel architectures that maximize the strengths of VLSI in terms of intensive and pipelined computing, yet circumvent its main limitations of reliability and communication. In the author's opinion, research and development in array processors not only benefit from the revolutionary VLSI technology, but will play a central role in shaping the course of algorithmic, architectural, and applicational trends in future supercomputer technology.

ACKNOWLEDGMENTS

This research was supported in part by the National Science Foundation under Grant ECS-82-13358 and by the Innovative Science and Technology Office of the Strategic Defense Initiative Organization and was administered through the Office of Naval Research under Contracts N00014-85-K-0469 and N00014-85-K-0599.

REFERENCES

1. C. Seitz, Concurrent VLSI architectures, *IEEE Trans. Comput.* *C-33*: 1247—1265 (1984).

2. S. Y. Kung, On supercomputing with systolic/wavefront array processors, *Proc. IEEE, 72,* 867–884 (1984).

3. S. Y. Kung and R. J. Gal-Ezer, Synchronous vs. asynchronous computation in VLSI array processors, *Proc. SPIE* 341: 53–65 (1982).

4. M. Franklin and D. Wann, Asynchronous and clocked control structures for VLSI based interconnection networks. *9th Annual Symp. Computer Architecture,* Austin, TX (Apr. 1982).

5. A. T. Fisher, H. T. Kung, L. M. Monier, H. Walker, and Y. Dohi, Design of the PSC: a programmable systolic chip, in *3rd CalTech Conference on VLSI* (R. Bryant, ed.), Computer Science Press, Rockville, MD 287–302 (1983).

6. S. J. Mason, Feedback theory—some properties of signal flow graphs, *Proc. IRE, 41:* 920–926, (1953).

7. S. Y. Kung, K. S. Arun, R. J. Gal-Ezer, and D. V. Bhaskar Rao, Wavefront array processor: language, architecture, and applications, *IEEE Trans. Comput. C-31:* 1054–1066 (1982).

8. G. W. Stewart, *Introduction to Matrix Computations,* Academic Press, New York (1973).

9. S. Y. Kung, and J. Annevelink, VLSI design for massively parallel signal processors, *Microprocess. Microsys.,* 7: 461–468 (1983).

10. H. M. Ahmed, Alternative arithmetic unit architectures for VLSI digital signal processors, *VLSI and Modern Signal Processing* (S. Y. Kung, H. J. Whitehouse, and T. Kailath, eds.), Prentice Hall, Englewood Cliffs, NJ 277–303 (1985).

11. P. Dewilde, E. Deprettere, and R. Nouta, Parallel and pipelined implementation of signal processing algorithms, *VLSI and Modern Signal Processing* (S. Y. Kung, H. J. Whitehouse, and T. Kailath, eds.), Prentice-Hall, Englewood Cliffs, NJ, 257–276 (1985).

12. S. Y. Kung and R. J. Gal-Ezer, Eigenvalue, singular value and least square solvers via the wavefront array processor, *Algorithmically Specialized Computer Organizations* (L. Snyder et al., eds.), Academic Press, New York (1983).

13. J. A. B. Fortes, K. S. Fu, and B. W. Wah, Systematic approaches to the design of algorithmic specified systolic arrays, *Proc. IEEE International Conference Acoust. Speech Signal Processing,* Tampa, FL. 300–303 (Mar. 1985).

14. J. D. Brock and L. B. Montz, Translation and optimization of data flow programs. *International Conf. on Parallel Processing 1979,* 46–54 (1979).

15. S. Y. Kung, S. C. Lo, and P. S. Lewis, Timing analysis and optimization of VLSI data flow arrays, *Proc. 13th International Symposium on Computer Architecture.* Tokyo (June 1986). Submitted for publication.

16. J. B. Dennis, Data flow supercomputers. *Computer* (IEEE), 13: 48–56 (Nov. 1980).

17. J. B. Dennis and D. P. Misunas, A preliminary architecture for a basic data-flow processor, *The 2nd Annual Sumposium on Computer Architecture,* 126–132 (Jan. 1975).

18. S. Y. Kung, S. C. Lo, and J. Annevelink, Temporal localization and systolization of signal flow graph (SFG) computing networks, *Proc. SPIE,* (Aug. 1984).

19. C. E. Leiserson, F. M. Rose, and J. B. Saxe, Optimizing synchronous circuitry by retiming, *Proc. Caltech VLSI Conference,* Pasadena, CA (1983).

20. M. Chen and C. Mead, Concurrent algorithms as space-time recursion equations, *VLSI and Modern Signal Processing* (S. Y. Kung, H. J. Whitehouse, and T. Kailath, eds.), Prentice-Hall, Englewood Cliffs, NJ (1984).

21. A. B. Cremers and T. N. Hibbard, The semantic definition of programming languages in terms of their data spaces. *Inf.-Fachber.,* 1: 1–11 (1976).

22. J. Backus, Can programming be liberated from the von Neumann style? A functional style and its algebra of programs, *Commun. ACM 21:* 613–641 (1978).

23. A. B. Cremers and S. Y. Kung, On programming VLSI concurrent array processors, *Proc. IEEE Workshop on Languages for Automation,* Chicago, pp. 205–210 (1983). Also in *Integration, 2* (1984).

24. D. A. Patterson and C. H. Sequin, A VLSI RISC, *Computer* (IEEE), *14:* (Sept. 1981).

25. C. L. Wu, and T. Y. Feng, On a class of multistage interconnection networks, *IEEE Trans. Comput., C-29:* 694–702 (1980).

26. D. S. Broomhead, et al., A practical comparison of the systolic and wavefront array processing architectures. *Proc. IEEE International Conference Acoust. Speech Signal Processing,* Tampa,

FL, 296–299 (Nov. 1985). Also in the *Proc. IEEE Workshop on VLSI Signal Processing*, Los Angeles, (Nov. 1984).

27. P. Wilson, Thirty-two bit micro supports multiprocessing, *Comput. Des.*,: 143–150 (June 1984).

28. M. Chase, A pipelined data flow architecture for digital signal processing the NEC μPD7281, *Proc. IEEE Workshop on VLSI Signal Processing*, Los Angeles (Nov. 1984).

29. R. H. Davis and D. Thomas, Systolic array chip matches the pace of high-speed processing, *Electron. Des.*, 207–218 (Oct. 1984).

30. F. Ware, et al., Fast 64-bit chip set gangs up for double-precision floating-point work, *Electronics*, 99–103 (July 1984).

5

Space Time Transformation
of Cellular Algorithms

PETER R. CAPPELLO *Computer Science Department, University of California, Santa Barbara, California*

1 INTRODUCTION

Our desire for faster computation seems insatiable. Computer architecture has evolved along several paths, all of which ultimately are guided by economics. In the *sequential process model*, known today as the von Neumann model, both data and instructions are stored in memory. A bottleneck of this architecture is that caused by the memory access circuits; all activity must pass through this portion of the machine. The sequential nature of program execution also limits speed. John von Neumann, a mathematician, did not create the so-called "von Neumann bottleneck." It was created by the economics of pre-integrated-circuit technology and is being destroyed by the economics of very large scale integration (VLSI). Electronic circuits are nearing their asymptote in switching speed. Speed-ups in uniprocessors are getting smaller, therefore, while the costs of attaining the speed-ups are getting larger. Concurrent process architecture may prove to be a path of less economic resistance. VLSI technology is making feasible the physical construciton of cellular automata, a highly concurrent architectural class. Interestingly, cellular automata were first, investigated by John von Neumann [1]. Technology and economics are just now catching up.

Two aspects of cellular automata are to be noted. The first has to do with recurrence equations. Cellular automata, a generalization of Turing machines, can compute any computable function. Uniform recurrence equations, in particular, however, map in a natural way onto cellular automata. The relationship between cellular automata and uniform recurrence equations have been either studied or used by the authors of all of the following: Refs. 2–7. Indeed, all the computations discussed in this chapter can be described by uniform recurrence equations.

The second aspect of cellular automata worth noting is that they include an important architectural subclass: systolic arrays. H. T. Kung and Leiserson first investigated systolic arrays because they possess two attributes that make their use especially attractive in VLSI technology. The two properties may be interpreted economically.

The first property concerns design cost. VLSI reduces system costs. It reduces the cost of system fabrication, automating a greater portion of system integration. It also reduces the cost of system operation, in both energy and time. It does not directly reduce the cost of system design.† Systolic arrays do. The number of cells in a systolic array is related to the size of the problem that it is intended to solve (e.g., the size of the matrix in a matrix-vector product). The number of cell types, though, is independent of the size of the problem: Increasing the size of the system being constructed does not, in principle, increase the cost of component design. The second property concerns communication cost. A cellular automaton's cells reside at the integer coordinates of an n-dimensional space (e.g., the vertices of an n-dimensional cube). Cells communicate directly with "neighbors" in the n-dimensional space. The cells of a systolic array, however, reside in a one- or two-dimensional space, and a cell's neighbors must be physically separated by a distance that is independent of the number of cells. That is, as the size of the array is increased, the distances between neighboring cells remains constant.

Systolic arrays are thus distinguished by characteristics that reflect real-world costs of design and communication. Wavefront arrays (as described in this volume) may be thought of as systolic arrays that take the trend of decentralized computation one step further. In systolic arrays, intercell synchronization is conducted centrally via global clocks. In wavefront arrays, intercell synchronization is decentralized. It is accomplished by substituting a local handshake for a global clock.

In this chapter we illustrate a space-time representation of cellular automata. The computations used to illustrate the space-time representation are of the systolic variety‡; their low dimension and homogeneous character make them easy to visualize. Systolic arrays have been formally represented and manipulated in a variety of ways. Fortes et al. [8] survey and classify this work. A space-time representation, or variations of it, have been studied and reported in Refs. 3, 4, 7, 9, and 10—15. The space-time representation presented here is

---

†It reduces system design cost indirectly, of course, via the bootstrap effect: VLSI leads to more powerful CAD hardware, which leads to better VLSI chip design, which leads to even better CAD hardware,....
‡ The space-time transforms apply equally well to the wavefront paradigm, however.

intended to be a framework for unifying array design. It permits different systolic array designs for the same computation to be related in a way that is intuitive but formal. This is done via transformations. The representation can enhance a designer's intuition. It is easy to "see," for example, how to transform an array design that uses broadcasting into one that is pipelined (defined below).

The model of computation used in this chapter is the synchronous model of VLSI [16, 17]. Two time measures are of interest:

Definition The *functional latency* (or time [18]) of a circuit is the amount of time separating the appearance of the first input bit on some port, from the appearance of the last output bit on some port, for one computation of the function, f, denoted T_f. This corresponds to the usual use of the term "speed of operation." A "100-nanosecond" multiplier, for example, means that 100 ns elapse between the appearance of the first input bit of the multiplicands and the last output bit of the product.

Definition The *functional period* of a circuit that calculates function f is the time interval separating corresponding bits of successive inputs (or outputs), denoted P_f. *Period* is the reciprocal of throughput rate.

Definition A circuit is *pipelined* with respect to function f when $P_f < T_f$. That is, the computation of one instance of f starts even before the computation of the previous instance has been completed. Pipelined circuits, in this respect, resemble assembly lines.

Definition A circuit is *completely pipelined* with respect to function f when P_f is independent of the size of f. The size of a completely pipelined circuit does depend on the size of f, however. For example, if a circuit design for an adder is completely pipelined, the period of such an adder for 8-bit words is the same as the period of such an adder for 64-bit words. The size of the 8-bit adder would, of course, be smaller than the size of the 64-bit adder.

The chapter is structured as follows. In Sec. 2 the space-time representation is introduced and applied to the computation of matrix-vector product, convolution, and matrix product. The transformations used are linear. In Sec. 3 two nonlinear transforms are illustrated: They address a problem in processor sharing and the partitioning of large problems onto smaller arrays. These sections are intended to illustrate techniques whose applicability is much wider than the scope of the illustrations themselves. The chapter concludes with a brief discussion of research topics that are related to space-time representation and transformation.

2 LINEAR TRANSFORMS

2.1 Matrix-Vector Product

In this section a space-time representation of systolic arrays is introduced by way of example. Consider the computation of the matrix-vector product $y \leftarrow A \cdot x$. It can be written as follows: $y_i \leftarrow \Sigma a_{ij} \cdot x_j$. To make the example more concrete, the expressions are written explicitly for a 3×3 matrix.

$$y_1 \leftarrow a_{11} \cdot x_1 + a_{12} \cdot x_2 + a_{13} \cdot x_3 \qquad (5.1)$$

$$y_2 \leftarrow a_{21} \cdot x_1 + a_{22} \cdot x_2 + a_{23} \cdot x_3$$

$$y_3 \leftarrow a_{31} \cdot x_1 + a_{32} \cdot x_2 + a_{33} \cdot x_3$$

We may think of this set of expressions as constituting an *algorithm* for transforming x into y. That is, the expressions indicate that each output y_i can be obtained by computing certain specified products and adding them together. Considerable freedom remains as to how this algorithm can be implemented. For example, the notation above does not dictate a particular association for the additions; any will do. The following recurrence relation fixes a particular association:

$$y_{i0} \equiv 0 \qquad (5.2)$$

$$y_{ij} \leftarrow y_{ij-1} + a_{ij} \cdot x_j$$

$$y_i \equiv y_{in}$$

where n is the size of the vector x.

Now, to make this algorithm more specific, an index representing time is adjoined to y:

$$y_{ijt} \leftarrow y_{i,j-1,t} + a_{ij} \cdot x_j \qquad \text{for } t = 0 \qquad (5.3)$$

Time is taken to be discrete and is measured in cycles. In this example, the time index, t, is set to zero. Thus as formulated presently, this entire computation occurs in one cycle: cycle$_0$. (The meaning of this time index is explained further shortly). The algorithm is not yet completely specified; no particular method has been specified for performing addition and multiplication. One must have some primitive notions.

Definition A computation is *primitive* if it is assumed that it can be done in constant area and with constant latency. The algorithm for carrying out a primitive computation is not specified.

To give this recurrence relation a space-time interpretation, the symbol "←" is associated with a location and a primitive computation.

Definition The *primitive computation* represented by ← is specified on its right-hand side.

In this case the computation is an inner-product step.

Definition The *location* represented by ← is the set of index values of the left-hand side interpreted as coordinates.

Figure 5.1 illustrates the example. To interpret the figures properly, note that the meaning of a recurrence relation is unaffected by adding a constant to all occurrences of an index (e.g., $y_{ij} \leftarrow y_{ij-1}$ represents the same computation as $y_{i-1j-2} \leftarrow y_{i-1j-3}$). Equivalently, in the space-time representation a computation is unaffected by translation. The reader is cautioned that axes in the figures are intended merely to associate dimensions with indices: For ease of viewing, the space-time representations are translated to the nonnegative orthant.

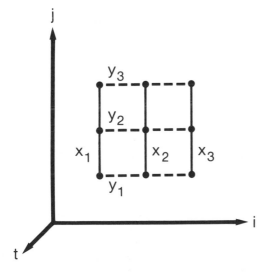

FIGURE 5.1 Spacetime representation of the canonical linear transform design. Contours are labeled, both input distribution and output accumulation.

Figure 5.1 is interpreted as follows. At location $(1,1,0)$ the computation $y_{110} \leftarrow y_{100} + a_{11} \cdot x_1$ occurs. This means that the value of x_1 and a_{11} must "be" at location $(1,1,0)$ [i.e., at spatial coordinates $(1,1)$ at time coordinate 0]. The solid lines indicate the path of a particular value of x. We refer to them as x-value contours. The dashed lines represent summation paths. These lines, or contours, are intended to make interpretation of the figures more intuitive. Movement of data, both input data distribution and output accumulation paths, can also be determined unambiguously from the recurrences themselves by using an ordering rule. In this chapter the ordering rule is simply the lexicographic order of the spatial indices within time. So, for example, if an output value is accumulated at more than one location with the same time index value, it is accumulated at these locations according to the lexicographic order of their spatial coordinates. We will call Eq. (5.3) the *canonical design* of the algorithm denoted by Eq. (5.2) and denote its space-time representation by Γ. This representation of the computation illustrates an important aspect of the representation of time. If two computations, c_{ijt} and d_{kls} are located at distinct points in time, with $t < s$, then c occurs before d. Again, however, if $t = s$, c and d take place during the same cycle but not necessarily "simultaneously." In fact, there is no notion of "simultaneity" in our interpretation of time. Interpreting Fig. 5.1, we see that the computation occurs in nine distinct locations in space and at one location in time.

Together, a space-time representation and an ordering rule indicate what data move, where they move to, and when they move. The number of physically distinct computational elements is simply the number of computational locations whose spatial coordinates are distinct. The communication topology of these elements emerges as well when we project out the time dimension. Thus the space-time representation associates a particular schedule of computation (an algorithm) with a particular network of computational elements. The association of an algorithm with a computational structure is termed a *design*. In what follows various linear transformations of space-time are applied to the canonical design. The transformations, then, relate distinct designs in a formal way.

The canonical design is not well suited to implementation because it is not pipelined. Kung and Leiserson [6] present a systolic design for matrix-vector product that is better. Their design, denoted Λ, is illustrated in Fig. 5.2. The communication structure is simply a linearly connected array of cells. Each cell performs an inner-product-step computation. Briefly, the design works as follows. X values move to the right through the array, while output values are accumulated as they move to the left. The matrix elements move down through the array as indicated by Fig. 5.2. More detail is given in Ref. 6.

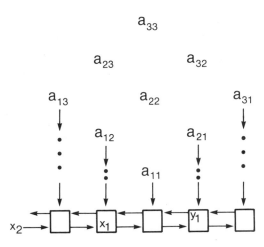

FIGURE 5.2 Conventional representation of the Kung and Leiserson linear transform design.

This design now is shown to be a linear transformation of the canonical design. We first apply R, a rotation-like transformation. We then "interchange" a space dimension with the time dimension. That is, we select a dimension that represents space and interpret it as time. Since there can be only one time idmension, we must now interpret the old one as space. A permutation transformation, T, is used to effect this semantic interchange. The result is depicted in Fig. 5.3. It is a space-time representation of the Kung-Leiserson design. To see this, one needs to interpret the representation. Since we interchanged one of the two space dimensions with the time dimensions, the resulting design has only one nontrivial spatial dimension, (Recall that the initial time coordinate values of the computation were all 0. Those coordinate values now represent a spatial dimension. That spatial dimension is unused: It is trivial.) The communication structure thus is a linearly connected array. The x values move down in space over time; the outputs move up in space over time. The transform coefficients, x values, and y values move through the linear array with the same schedule as in the Kung-Leiserson design. We have derived the following formal representation of Λ:

$$\Lambda = T \cdot R(\Gamma) \quad \text{where } R = \begin{bmatrix} 1 & 1 & 0 \\ 1 & -1 & 0 \\ 0 & 0 & 1 \end{bmatrix} \quad \text{and} \quad T = \begin{bmatrix} 0 & 0 & 1 \\ 0 & 1 & 0 \\ 1 & 0 & 0 \end{bmatrix}$$

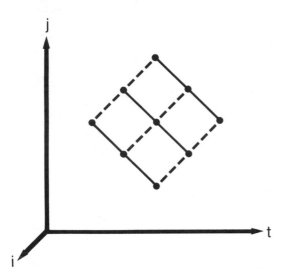

FIGURE 5.3 Spacetime representation of the Kung and Leiserson linear transform design.

Notice that Λ uses $2n - 1$ cycles and $2n - 1$ inner-product-step cells. Figure 5.4 motivates a notion of efficiency. The bounding rectangle represents the volume of space-time that is available during the computation $(2n - 1)^2$. Only n^2 of the space-time is used. Approximately half of the available space-time is unused because cells are idle every other cycle. Half of the remainder is unused because of the time needed for pipe filling and pipe draining. That is, approximately 25% of the available space-time is used.

Accordingly, let us define the asymptotic efficiency of design D, denoted $\varepsilon(D)$.

$$\varepsilon(D) = \lim_{n \to \infty} \frac{\text{space-time used by D}}{\text{space-time available to D}}$$

where n is the size parameter of D's computation.

$$\varepsilon(\Lambda) = \lim_{n \to \infty} \frac{n^2}{(2n - 1)^2} \to \frac{1}{4}$$

This efficiency can be improved if we interleave pairs of matrix-vector product computations, as is commonly advocated. Figure 5.5 is a space-time diagram for such a computation pair. Let Λ_2 denote this design

space

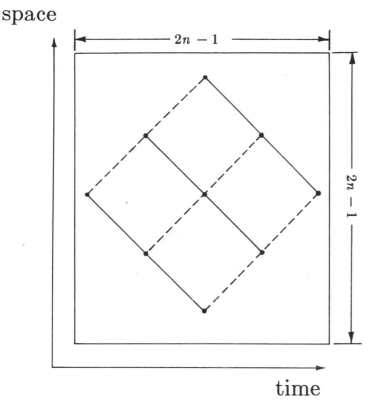

time

FIGURE 5.4 The rectangle encloses the portion of spacetime that is available for use during the Λ design: $(2n - 1)^2$. Only n^2 of the spacetime is used.

for matrix-vector product pairs. Notice from the figure that the intercellular communication rate of Λ_2 is twice that of Λ.

$$\epsilon(\Lambda_2) = \lim_{n \to \infty} \frac{2n^2}{2n(2n - 1)} \to \frac{1}{2}$$

A systolic design, Ψ, is now presented that uses $2n - 1$ cycles but uses only n inner-product-step cells. Thus, without resorting to interleaving, $\epsilon(\Psi) = 1/2$. This is illustrated in Fig. 5.6. We first apply a transformation, S, which skews the canonical representation, Γ. The time-space interchange transformation, T, is then applied as before. We interpret the result as follows. Due to the time-space interchange, the array again extends only in one dimension in space.

space

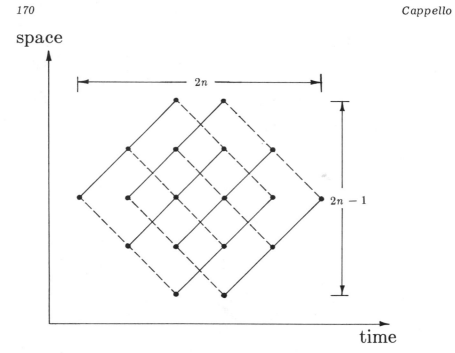

FIGURE 5.5 Two Λ designs are superimposed, offset 1 unit in time.

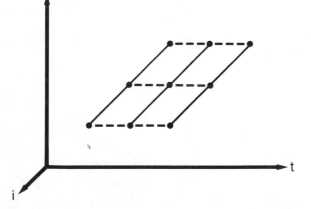

FIGURE 5.6 Spacetime representation of the design Ψ.

But now its image in space is a linearly connected array of n cells, not 2n − 1. X vectors, whose components as skewed in time, are piped through the array while the transformed vector's components are accumulated in distinct processors, also over time. Ψ is related to the canonical design by the transformations T and S:

$$\Psi = T \cdot S(\Gamma) \quad \text{where } S = \begin{bmatrix} 1 & 1 & 0 \\ 0 & 1 & 0 \\ 0 & 0 & 1 \end{bmatrix}$$

This formal specification of the design illustrates the utility of the space-time approach. With Ψ, there is no fill-up and drainage period between successive computations. New x vectors and transform coefficients can follow on the "heels" of the preceding ones. This motivates a notion of asymptotic stream efficiency. Let D be a computational design in space-time. Let D^i denote i instances of D, placed as closely as possible one after another in time. Figure 5.7 depicts Λ^3 and Ψ^3. Define the *asymptotic stream efficiency*, denoted $\sigma(D)$, as

$$\sigma(D) = \lim_{i \to \infty} \frac{\text{space-time used by } D^i}{\text{space-time available to } D^i}$$

For example, $\sigma(\Lambda) = 1/4$; $\sigma(\Lambda_2) = 1/2$; $\sigma(\Psi) = 1$.

Neither the Λ design nor the Ψ design is completely pipelined. A new design that is completely pipelined is described next. It is asymptotically optimal with respect to the complexity measure area × period^2 (AP^2). Period rather than latency (delay) is a good measure in applications where high throughput (rather than short latency) is of interest.

Vuillemin [18] has shown that

1. Fixed matrix-vector product and convolution are transitive functions.
2. Any circuit computing a transitive function at data rate D must have wire area $A_W \geq a_W D^2$ for some technology-dependent constant a_W.

Vuillemin's lower bound for fixed matrix-vector product is not valid for every matrix. The identity matrix, for example, clearly requires wire area only linear in the data rate. The bound, however, is an existence bound: It says that even when we fix the matrix, there exist some matrices which require a circuit of area $\Omega(D^2)$.† Many important transformations, such as the discrete Fourier transform

---

†That is, the area is asymptotically bounded from below by D^2.

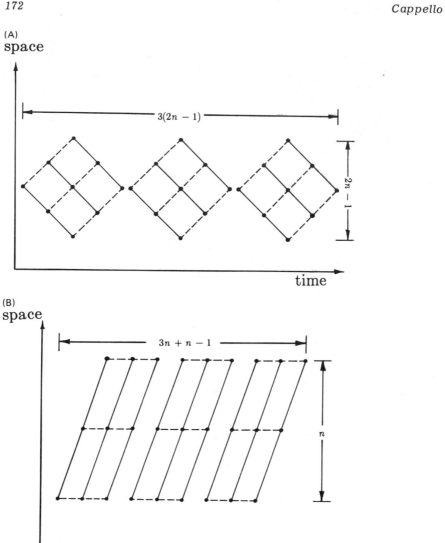

FIGURE 5.7 (A) Λ^3: 3 Λ designs placed consecutively in time. (B) Ψ^3: 3 Ψ designs placed consecutively in time.

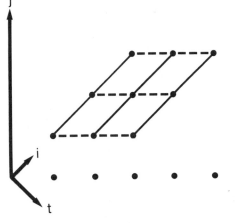

FIGURE 5.8 Spacetime representation of an AP^2 optimal design Δ. Dots below are the projection of the computation onto the i—time sub-space.

(DFT), however, are among those for which the quadratic bound holds [16]. Note that since the period $P = n/D$, where n is the number of input bits and D is the rate at which they are read in, we have for transitive functions (such as the DFT) that $AP^2(n) = \Omega(n^2)$.

A systolic array that has $AP^2(n) = O(n^2)$† is now shown. The canonical design is first skewed as before. But now, instead of inter-changing the (trivial) time dimension with a space dimension yielding a linear array of cells, we pipeline the computation, producing a two-dimensional mesh of cells. A rotation-like transform, N, accomplishes this (see Fig. 5.8). The resulting design is denoted by Δ, where

$$\Delta = N \cdot S(\Gamma) \qquad \text{where } N = \begin{bmatrix} 1 & 0 & 0 \\ 0 & 1 & 0 \\ 1 & 0 & 0 \end{bmatrix}$$

As one can see from Fig. 5.8, inputs are piped through the mesh. Input vectors have their components skewed in the time dimension. Output vector components are similarly skewed. This systolic array is completely pipelined. Since the matrix is fixed, we can assume that the

---

†That is, the AP product is asymptotically bounded from above by n^2.

a_{ij}'s are encoded into their proper cell. Thus $AP_\Delta{}^2(n) = O(n^2)$. Because Δ is completely pipelined, it is wasteful to use it for a single computation:

$$\varepsilon(\Delta) = \lim_{n \to \infty} \frac{n^2}{n^2(2n-1)} \to 0$$

Of course, it is well suited to streams of matrix-vector products: $\sigma(\Delta) = 1$. Rather than continue our examination of matrix-vector product designs, we shall move on to another computation: convolution.

2.2 Convolution

Consider the computation of convolution: $y_i \leftarrow \Sigma_j x_j \cdot w_{i-j}$. This one operation can represent an FIR filter, a discrete Fourier transform [19], or multiplication when x and w are single bits so that " $\cdot$ " is interpreted as a bit product (i.e., a logical AND operation) and a carry propagation is included. Also, Kung and Leiserson [6] have noted that convolution describes string pattern matching when " $\cdot$ " is interpreted as string compare and "+" is interpreted as Boolean and. For the purposes of this chapter, convolution can represent any computation where X, Y, and W are (not necessarily distinct) sets, " $\cdot$ " is a map from $X \times W$ to Y, and (Y, +) is a monoid. In a recent article, H. T. Kung [20] enumerated seven known designs for convolution. Their space-time representations are now shown to be related by linear transformations. Then two new designs are presented. In this way, much of the work on convolution designs is unified. Finally, the Kung designs are transformed to AP^2 asymptotically optimal designs. A space-time representation of convolution is established first. It is very similar to the space-time representation of matrix-vector product. Writing out an example convolution for x and w signals of length 3, we obtain

$$y_0 \leftarrow x_0 \cdot w_0$$

$$y_1 \leftarrow x_0 \cdot w_1 + x_1 \cdot w_0$$

$$y_2 \leftarrow x_0 \cdot w_2 + x_1 \cdot w_1 + x_2 \cdot w_0 \qquad\qquad (5.4)$$

$$y_3 \leftarrow x_1 \cdot w_2 + x_2 \cdot w_1$$

$$y_4 \leftarrow x_2 \cdot w_2$$

These are reformulated as a recurrence relation

$$y_{io} \equiv 0$$

$$y_{ij} \leftarrow y_{i,j-1} + x_j \cdot w_{i-j}$$

$$y_i \equiv y_{in} \tag{5.5}$$

where n is the signal size. We let " $\leftarrow$ " represent an inner-product-step computation located in space-time as before. To do this we again adjoin a time coordinate (see Fig. 5.9):

$$y_{ijt} \leftarrow y_{i,j-1,t} + x_j \cdot w_{i-j} \quad \text{for } t = 0 \tag{5.6}$$

This is the canonical design of the algorithm expressed by Eq. (5.5) and is denoted by Γ, as before.

We now proceed to derive some known designs by transforming Γ to them. Table 5.1 summarizes the seven designs in Kung's article. B1 is a design especially close to the canonical design. By appyling the time-space interchange transformation T to Γ, B1 is derived (illustrated in Fig. 5.10): B1 = T(Γ). We interpret that figure as follows. The same x value appears at processors that are spatially but not temporally distinct: x values are broadcast to their processors. Similarly, w values (weights) appear at processors that are temporally but not spatially distinct: w values "stay." y values (results) appear at processors that are distinct in both space and time: They "move."

Table 5.1 Convolution Designs Enumerated by Kung [20]

Design name	Results (y)	Input (y)	Weights (w)
B1	Move	Broadcast	Stay
B2	Stay	Broadcast	Move
F	Fan-in	Move	Stay
R1	Stay	Move in opposite directions	
R2	Stay	Move in same direction at different speeds	
W1	Move in opposite directions		Stay
W2	Move is same direction at different speeds		Stay

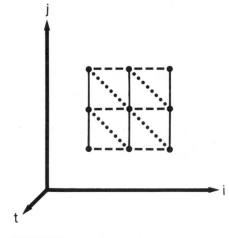

FIGURE 5.9 Spacetime representation of the canonical convolution design. In all convolution figures, x contours are represented by solid lines, w contours are represented by dashed lines, y accumulation paths are represented by dotted lines.

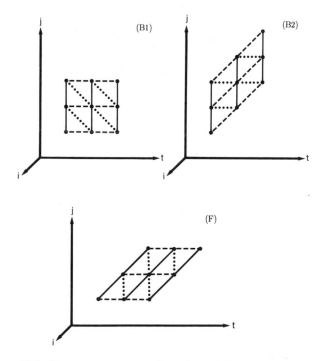

FIGURE 5.10 Spacetime representations of the designs enumerated by Kung.

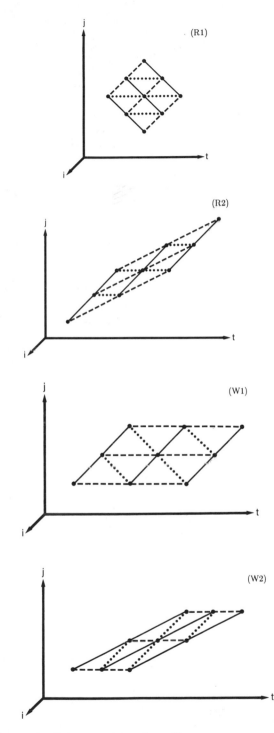

FIGURE 5.10 (Continued)

Table 5.2   Space-time Design Definitions for the Serial Convolvers and Their Parallel AP^2-Optimal Counterparts

	Geometric Design Definitions	
Design name	Transform from Γ	AP^2 optimal counterpart
B1	T	N
B2	$T \cdot S_1$	$N \cdot S_1$
F	$T \cdot S$	$N \cdot S$
R1	$T \cdot R$	$N \cdot R$
R2	$T \cdot S_2$	$N \cdot S_2$
W1	$T \cdot S_4$	$N \cdot S_4$
W2	$T \cdot S_3$	$N \cdot S_3$
A1	$T \cdot S_1(R2) = T \cdot S_1 \cdot S_2$	$N \cdot S_1 \cdot S_2$
A2	$T \cdot S_5(W1) = T \cdot S_5 \cdot S_4$	$N \cdot S_5 \cdot S_4$

Other derivations are summarized in Table 5.2. The transforms used are as follows (T, R, S and N are as in Sec. 1):

$$S_1 = \begin{bmatrix} 1 & 0 & 0 \\ 1 & 1 & 0 \\ 0 & 0 & 1 \end{bmatrix} \quad S_2 = \begin{bmatrix} 2 & 1 & 0 \\ 1 & 1 & 0 \\ 0 & 0 & 1 \end{bmatrix} \quad S_3 = \begin{bmatrix} 1 & 2 & 0 \\ 0 & 1 & 0 \\ 0 & 0 & 1 \end{bmatrix} \quad S_4 = \begin{bmatrix} 2 & 1 & 0 \\ 0 & 1 & 0 \\ 0 & 0 & 1 \end{bmatrix}$$

$$S_5 = \begin{bmatrix} 2 & 0 & 0 \\ -1 & 2 & 0 \\ 0 & 0 & 1 \end{bmatrix}$$

Figure 5.10 illustrates the space-time representations of the designs. H. T. Kung [20] refers to designs B1, B2, and F as being semisystolic. This is because, in these designs, either input data are broadcast or output data are "fanned in." The space-time representation of such systolic arrays have data contours that are perpendicular to the time axis. (There are, in fact, many ways to accomplish design

F's "fan-in" because the addition of products is associative. In particular, an add tree can be used to fan-in the products. Such a design, however, is not a linear transform of the canonical design.)

The algorithm indicated by Eq. (5.5) admits other designs. A1 and A2 are two such designs. In design A1, the input, weights, and results all move through the array in the same direction but at three different speeds. This design is depicted in Fig. 5.11. In design A2, depicted in Fig. 5.12, input moves through the array in one direction; weights and results, the other. The weights and results must move at different speeds, so that they come into contact with one another at some point in space-time. As in the other designs, the physical network is a linear array of processors.

Some AP^2 asymptotically optimal designs are presented now. Like their word-serial counterparts, these designs have different data movement characteristics. Such qualities are important in practice; a design (for, say, integer multiplication) may have data movement constraints deriving from the larger application in which it is embedded. For each of the linear systolic arrays for convolution, there is a corresponding AP^2 optimal design. As with the matrix-vector product computation, the linear array convolution designs are transformed to AP^2 optimal designs by pipelining them. In fact, the same transform is used. The T transform in Table 5.2 is replaced by the N transform. Consider design R2, for example. Its AP^2 optimal counterpart is illustrated in Fig. 5.13. Spatially it is a hex-connected mesh of processors. X signal components, skewed in time, move along their contours (the solid lines), w signal components, skewed in time, move along their contours (the dashed lines), while output signal components are accumulated along the dashed lines. An input and output signal can be accepted every cycle of the processor assemblage: The array is completely pipelined. Its period thus is O(1) and its area is $O(n^2)$: $AP^2_A(n) = O(n^2)$. As noted in Sec. 2.1, Vuillemin [18] shows that convolution is a transitive function: $A(n) = \Omega(D^2)$. Thus $AP^2(n) = \Omega(n^2)$, and these new designs are asymptotically optimal. (For the same reasons that H. T. Kung refers to designs B1, B2, and F as being semisystolic, their counterpart designs might be referred to as semipipelined. Even under the synchronous model of VLSI, design F's semipipelined counterpart is not AP^2 optimal). The word-serial designs displayed in the left column of Table 5.2 all have $P(n) = O(n)$ and $A(n) = O(n)$: none is AP^2 optimal.

Again, using a completely pipelined design for a single computation is wasteful. The asymptotic efficiency of these completely pipelined is zero. But all the asymptotic stream efficiencies are 1. Finally, we note that all the designs presented in this section—the ones using linear arrays and the ones using hexagonal meshes—have $AP(n) = O(n^2)$. Put another way, designs implementing the same algorithm all have the same switching energy, E_{sw} (see Ref. 21 for a definition

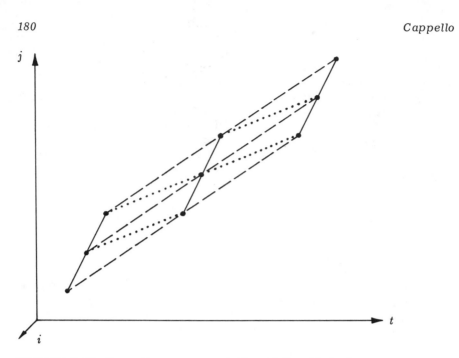

FIGURE 5.11 Spacetime representation of the A1 design.

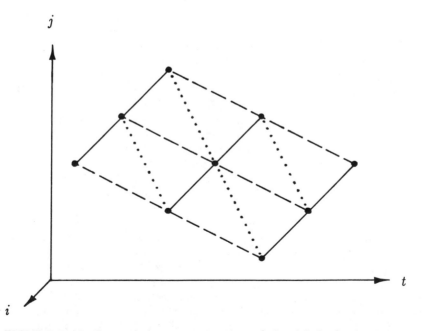

FIGURE 5.12 Spacetime representation of the A2 design.

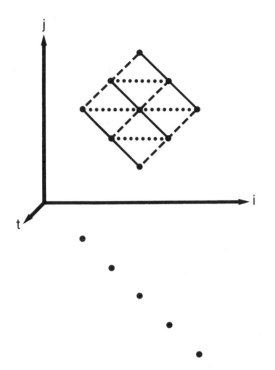

FIGURE 5.13 Spacetime representation of design R 2's AP^2-optimal counterpart. Dots below are the projection of the computation onto the i—time subspace.

of this quantity); this energy is just distributed differently in spacetime. Clearly, one-to-one transformations of space-time (such as nonsingular linear transformations) conserve E_{sw}.

Related work that can be represented quite naturally in the spacetime framework includes Danielsson's [13] serial/parallel convolvers, and Swartzlander's [22] quasi-serial multipliers.

2.3 Matrix Product

Given two matrices A and B, their product is denoted C $\leftarrow$ A $\cdot$ B. A more algorithmic description of matrix product is

$$c_{ij} \leftarrow \sum_k a_{ik} \cdot b_{kj} \quad \text{for } 1 \leqslant i, j, k \leqslant n \qquad (5.7)$$

where A, B, and C are n × n matrices. Equation (5.7), for n = 2, is

$$c_{11} \leftarrow a_{11} \cdot b_{11} + a_{12} \cdot b_{21}$$

$$c_{12} \leftarrow a_{11} \cdot b_{12} + a_{12} \cdot b_{22}$$

$$c_{21} \leftarrow a_{21} \cdot b_{11} + a_{22} \cdot b_{21} \qquad (5.8)$$

$$c_{22} \leftarrow a_{21} \cdot b_{12} + a_{22} \cdot b_{22}$$

Again we reformulate using a recurrence relation:

$$c_{ijo} \equiv 0$$

$$c_{ijk} \leftarrow c_{i,j,k-1} + a_{ik} \cdot b_{kj} \qquad (5.9)$$

$$c_{ij} \equiv c_{ijn}$$

In order to let " $\leftarrow$ " represent an inner-product-step computation located in space-time, a fourth coordinate modeling time, is adjoined:

$$c_{ijkt} \leftarrow c_{i,j,k-1,t} + a_{ik} \cdot b_{kj} \qquad (5.10)$$

This is the canonical design of the algorithm expressed by Eq. (5.9), and its space-time representation is denoted by Γ (see Fig. 5.14).

The matrix product design of Chern and Murata [23], which we denote by Φ, is especially close to the canonical design. Φ is obtained from Γ by applying a time-space permutation, T, which interchanges one spatial coordinate with the time coordinate:

$$\Phi = T(\Gamma) \qquad \text{where } T = \begin{bmatrix} 1 & 0 & 0 & 0 \\ 0 & 1 & 0 & 0 \\ 0 & 0 & 0 & 1 \\ 0 & 0 & 1 & 0 \end{bmatrix}$$

This design is depicted in Fig. 5.15. That figure is interpreted as follows. Projecting out time, we are left with an $n \times n$ square mesh of processors. At each cycle:

1. One column of matrix A is broadcast into the mesh of processors, one A column component into each column of processors.
2. One row of matrix B is broadcast into the mesh of processors, one B row component into each row of processors.
3. Processor(i,j) accumulates $a_{it} \cdot b_{tj}$ at cycle t.

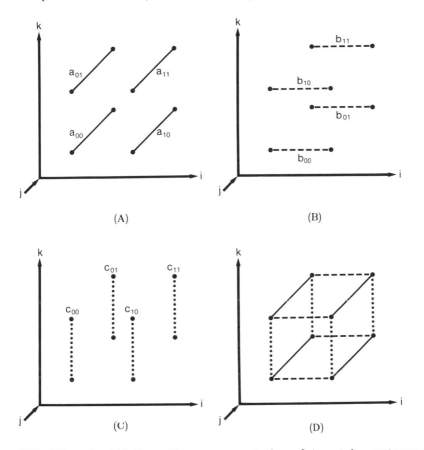

FIGURE 5.14 (A) Spacetime representation of A matrix contours in the canonical matrix product design, (B) B matrix contours, (C) C matrix accumulation in the canonical matrix product design. (D) Putting parts (A), (B), and (C) together, we get the spacetime representation of the canonical matrix product design. Time, the fourth dimension, is not shown.

After n cycles, processor(i,j) has accumulated all of product matrix element c_{ij}. A more detailed description of this design is given in Ref. 23. Note that $\varepsilon(\Phi) = \sigma(\Phi) = 1$.

Ullman [24] presents a matrix product design for a square mesh which does not use broadcasting. This design, denoted T, is perhaps best thought of as juxtaposing n of the Kung and Leiserson linear transform designs, each producing one column of the matrix product.

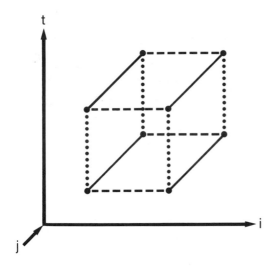

FIGURE 5.15 Spacetime represnetation of Φ.

Figure 5.16 depicts two juxtaposed Kung-Leiserson linear transform designs. That is, an $(n \times n) \times (n \times n)$ matrix product design, for $n = 2$. Notice that the price here for eliminating broadcasting is that $(2n - 1) \cdot n$ processors and $2n - 1 + n - 1 = 3n - 2$ cycles are used. Thus $\varepsilon(T) = 1/6$. This design is obtained from Γ in much the same way as the Kung-Leiserson linear transform design is obtained from its canonical design. We (1) rotate the canonical design, as in the Kung-Leiserson linear transform design, (2) interchange a spatial coordinate with the temporal coordinate, and (3) skew the linear transform designs in time so that each one starts one cycle after its predecessor:

$$T = S \cdot T \cdot R(\Gamma) \quad \text{where } R = \begin{bmatrix} 1 & 0 & 0 & 0 \\ 0 & 1 & 1 & 0 \\ 0 & 1 & -1 & 0 \\ 0 & 0 & 0 & 1 \end{bmatrix}$$

$$T = \begin{bmatrix} 1 & 0 & 0 & 0 \\ 0 & 0 & 0 & 1 \\ 0 & 0 & 1 & 0 \\ 0 & 1 & 0 & 0 \end{bmatrix} \quad S = \begin{bmatrix} 1 & 0 & 0 & 0 \\ 0 & 1 & 0 & 0 \\ 0 & 0 & 1 & 0 \\ 1 & 0 & 0 & 1 \end{bmatrix}$$

Other linear transform designs can be used as a basis for a matrix product design. For example, the Ψ linear transform design of Sec. 2

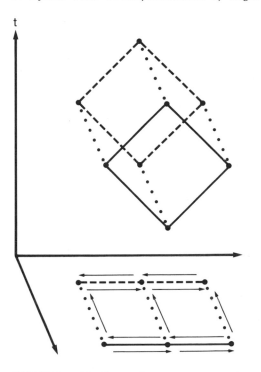

FIGURE 5.16 Spacetime representation of T. The solid diamond represents the linear transform for the first product column. The dashed diamond represents the linear transform for the second product column. The dotted lines reprent b coefficient contours. Below this cube-like computation is its projection onto the spatial subspace.

can be juxtaposed to obtain a new matrix product design. This design is denoted by Ξ and depicted in Fig. 5.17. This design, which also does not use broadcasting, uses only n^2 processors and $3n - 2$ cycles: $\varepsilon(\Xi) = 2\varepsilon(T) = 1/3$. Of course, there is a dual design that uses $(3n - 2) \cdot n$ processors and n cycles.

$$\Xi = S \cdot T \cdot S_1(\Gamma) \qquad \text{where } S_1 = \begin{bmatrix} 1 & 0 & 0 & 0 \\ 0 & 1 & 1 & 0 \\ 0 & 0 & 1 & 0 \\ 0 & 0 & 0 & 1 \end{bmatrix}$$

We now turn our attention to band matrix product. An important special case of matrix product, it is illustrated conventionally by the

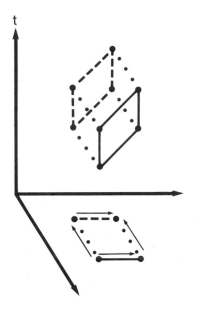

FIGURE 5.17 Spacetime representation of Ξ. The solid diamond
represents the linear transform for the first product column. The
dashed diamond represents the linear transform for the second product
column. The dotted lines represent b coefficient contours. Below this
cube-like computation is its projection onto the spatial subspace.

matrix expression in Fig. 5.18, whereas Fig. 5.19 illustrates a space-
time representation of the computation.
Figure 5.20 displays a summary (and conventional representation) of
the systolic design to do this computation that was devised by Kung
and Leiserson. Reference 6 provides more detail. Their design, which
we denote by Λ, is based on the same algorithm as the canonical
design: Eq. (5.9). We now present a space-time representation of the
Kung-Leiserson design. To obtain it one can take the staircase-like
structure of the canonical design and situate it vertically using two

$$
C = \begin{bmatrix} a_{11} & a_{12} & 0 \\ a_{21} & a_{22} & a_{23} \\ 0 & a_{32} & a_{33} \end{bmatrix} \cdot \begin{bmatrix} b_{11} & b_{12} & 0 \\ b_{21} & b_{22} & b_{23} \\ 0 & b_{32} & b_{33} \end{bmatrix}
$$

FIGURE 5.18 Band matrix product.

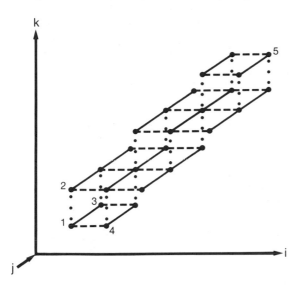

FIGURE 5.19 Spacetime representation of the band matrix product of Fig. 5.18. To obtain a geometric representation of the Leiserson and Kung design: 1) rotate the canonical representation such that a line connecting computations labeled 1 and 5 is parallel to the k axis, and computations labeled 2, 3, and 4 all have the same k coordinate value; 2) interchange the k and t dimensions (i.e., interpret the k axis as time). See Fig. 5.21.

rotation-like transforms. The Λ design, illustrated in Fig. 5.21, emerges when the vertical space dimension is interchagned with the (previously unshown) time dimension (i.e., interpret the vertical dimension as time). The reader is invited to verify this informally. The K matrix is a transformation that combines the three transformations just described:

$$\Lambda = K(\Gamma) \qquad \text{where } K = \begin{bmatrix} 1 & 0 & -1 & 1 \\ 0 & 0 & 1 & 1 \\ -1 & 0 & 0 & 1 \\ 1 & 1 & 0 & 0 \end{bmatrix}$$

In Λ, approximately one-third of the cells are active on any given cycle. Weiser and Davis present a design that improves Λ in this respect [25]. Their design, which we denote by Ψ, is depicted conventionally in fig. 5.22. Like Λ, it uses a hex-connected array. In

Ψ, however, the A band matrix flows through a row at a time, and the B band matrix, a column at a time, producing the C product band matrix a column at a time. (There is, of course, a dual design producing a row at a time.) Ψ, obtained from Γ by a linear transformation, is

$$\Psi = D(\Gamma) \qquad \text{where } D = \begin{bmatrix} 1 & -1 & 0 & 0 \\ 0 & 0 & 0 & 1 \\ -1 & 0 & 1 & 0 \\ 1 & -1 & 1 & 0 \end{bmatrix}$$

Its throughput rate is three times that of Λ. Ψ is represented in space-time by Fig. 5.23.

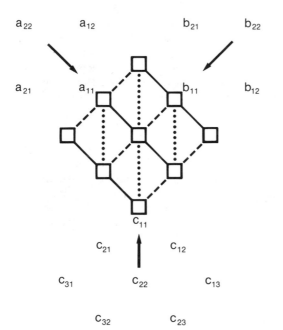

FIGURE 5.20 Conventional representation of the Leiserson and Kung design for band matrix product.

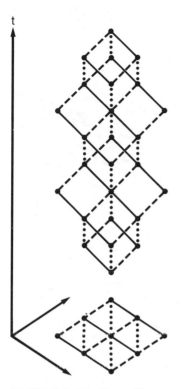

FIGURE 5.21 Spacetime representation of the Leiserson and Kung design for band matrix product.

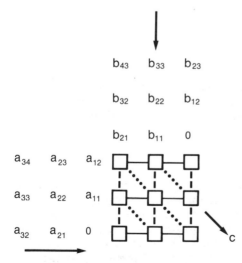

FIGURE 5.22 Conventional representation of the Weiser and Davis design for band matrix product.

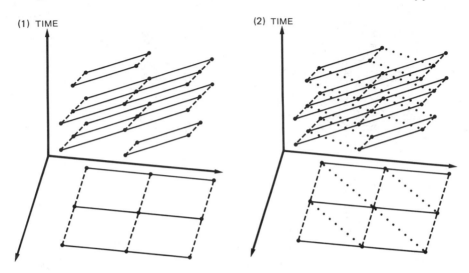

FIGURE 5.23 With respect to data communication, part (1) shows only the input distribution of the Weiser-Davis design. In part (1), the 2-dimensional mesh below is the projection of this subdesign onto the spatial subspace. Part (2) depicts both input distribution and output accumulation. The hexagonally-connected mesh below is the projection of this design onto the spatial subspace.

3 NONLINEAR TRANSFORMS

Two nonlinear transformations are presented. The first, although nonlinear, preserves topology. It is used to formally represent processor sharing in bidirectional linear arrays. The second is used to map large systolic computations onto smaller systolic arrays.

3.1 Cell-Sharing Transformation

In Sec. 2.1 we introduced Kung-Leiserson's design, Λ, for matrix-vector product. The reader may recall that in their design, each cell is active on every other cycle (see Fig. 5.3). The systolic design Ψ is twice as efficient but alters the data flow characteristics. For example, Λ uses a bidirectional data flow, whereas in Ψ the y components are accumulated in place; they "stay." Depending on the context of the larger application, one may require bidirectional data flow. Kung and Leiserson claim "yes": "Observe also that at any given efficiency of Λ without interleaving or eliminating bidirectional data flow. Kung and Leiserson claim "yes": Observe also that at any given time alternating processors are idle. Indeed, by coalescing pairs of adjacent processors, it is possible to use w/2 processors in the network for a general band matrix with band width w" [6].

By inspecting the space-time diagram we can see what needs to be done to achieve this cell sharing. A nonlinear transformation then can be used to describe cell sharing in a precise and formal way. Figure 5.24 illustrates the space-time diagram for a 2 × 2 matrix-vector product. The second part of that figure represents a nonlinear transformation that eliminates the idle cycle, using one less cell. Figure 5.25 illustrates this pairwise cell coalescing on a 4 × 4 matrix-vector product computation. With this design, denoted Λ', the number of cells is reduced from

$$2n - 1 \quad \text{to} \quad \left\lceil \frac{2n - 1}{2} \right\rceil = n$$

doubling the asymptotic efficiency of the design: $\varepsilon(\Lambda') = 2\varepsilon(\Lambda)$. The transformation needed, our first nonlinear transformation, is

$$\Lambda' = \lceil C \rceil (\Lambda) = C \cdot T \cdot R(\Gamma)$$

where

$$\lceil C \rceil = \begin{bmatrix} 1 & 0 & 0 \\ 0 & 1/2 & 0 \\ 0 & 0 & 1 \end{bmatrix}$$

and

$$C x = \begin{bmatrix} \lceil y_1 \rceil \\ \lceil y_2 \rceil \\ \vdots \\ \lceil y_n \rceil \end{bmatrix} \quad \text{for } y = Cx$$

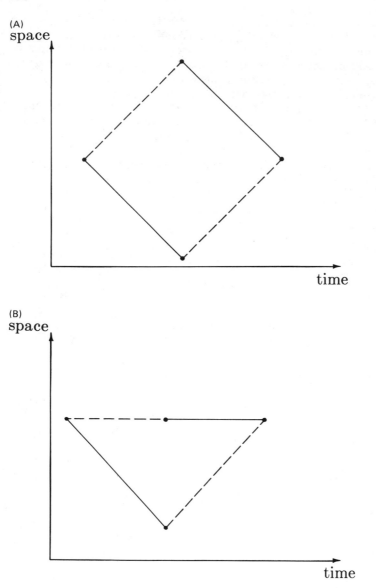

FIGURE 5.24

(A)

space

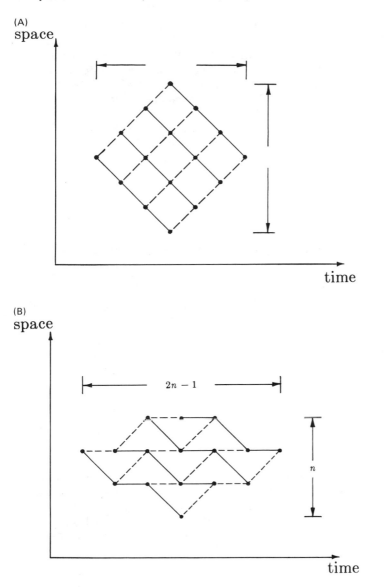

time

(B)

space

time

FIGURE 5.25

That is, we take the component-wise ceiling of the linear transform.

It perhaps should be stressed that this cell-sharing transformation can be used generally on linear arrays that employ bidirectional data flow. First let us reconsider Ullman's matrix product design, T. Recalling that T is just the juxtaposition of n Λ designs, we can construct a new Ullman design, T' with a commensurately improved efficiency: $\epsilon(T') = 2\epsilon(T) = 1/3$.

Figure 5.26 illustrates the cell-sharing transformation applied to convolution. As a final example, consider the problem of solving an upper triangular linear system, such as the eight-dimensional example given below.

$$u_{11}x_1 + u_{12}x_2 + u_{13}x_3 + u_{14}x_4 + u_{15}x_5 + u_{16}x_6 + u_{17}x_7 + u_{18}x_8 = 0$$

$$u_{22}x_2 + u_{23}x_3 + u_{24}x_4 + u_{25}x_5 + u_{26}x_6 + u_{27}x_7 + u_{28}x_8 = 0$$

$$u_{33}x_3 + u_{34}x_4 + u_{35}x_5 + u_{36}x_6 + u_{37}x_7 + u_{38}x_8 = 0$$

$$u_{44}x_4 + u_{45}x_5 + u_{46}x_6 + u_{47}x_7 + u_{48}x_8 = 0$$

$$u_{55}x_5 + u_{56}x_6 + u_{57}x_7 + u_{58}x_8 = 0$$

$$u_{66}x_6 + u_{67}x_7 + u_{68}x_8 = 0$$

$$u_{77}x_7 + u_{78}x_8 = 0$$

$$u_{88}x_8 = 0$$

The spatial arrangement of the expressions is intended to suggest an cellular organization for their computation. The value of a particular x component is computed incrementally from right to left. Once computed, the x-component value propagates up its column. Figure 5.27 depicts a two-dimensional mesh that solves for x. Square cell$_{ij}$ contains the value u_{ij}. Its input/output equations follow.

$$x_j^{out} \leftarrow x_j^{in}$$

$$s^{out} \leftarrow s^{in} - u_{ij}x_j$$

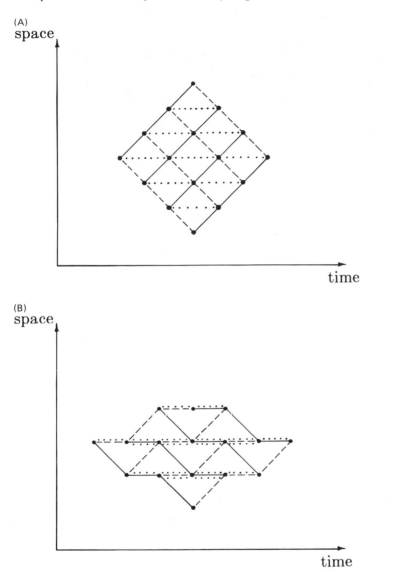

FIGURE 5.26 (A) Spacetime diagram of the R2 convolution design.
(B) The R2 design after application of the coalescing transformations.
Each cell now needs 2 registers; it is responsible for the accumulation
of 2 result values (two dotted lines). Their computation is interleaved
in time.

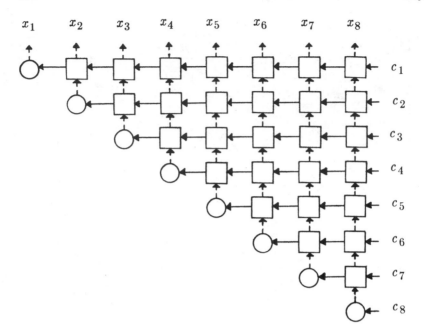

FIGURE 5.27 A cellular organization for solving an upper triangular linear system. In the figure, n = 8.

Round cell$_{ii}$ contains the value u_{ii}. Its input/output equation is simply

$$x^{out} \leftarrow \frac{s^{in}}{u_{ii}}$$

It is not difficult to embed this mesh in space-time and rotate it so that the division (circle) cells are vertically aligned, parallel to the time axis. The result is the familiar linear array for solving the system that is proposed by Kung and Leiserson [6]. Its space-time representation is given in Fig. 5.28. Other designs clearly are possible. Figure 5.29, however, shows this design after the nonlinear cell-sharing transformation has been applied. It reduces the number of cells from n to n/2 while preserving bidirectional data flow.

This cell-sharing transformation also can be used profitably on the Heller and Ipsen [26] QR factorization design. A similar transform can be used to remove the idle time from the Kung-Leiserson matrix-product design, while preserving its data flow characteristics. The ideas in this section were developed by the author and Ilse Ipsen. Lisper also discusses nonlinear transforms.

time

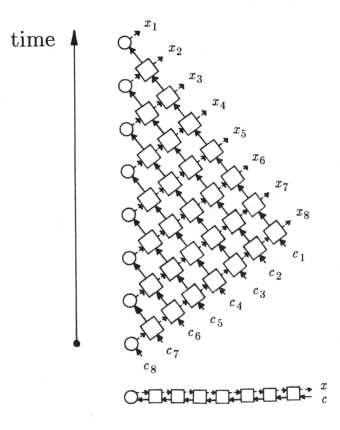

space

FIGURE 5.28 A spacetime representation of the Kung and Leiserson design for solving an upper triangular linear system. Beneath the spacetime mesh is a linear array obtained by projecting the computation onto the spatial subspace.

3.2 Working with Arrays of Fixed Size

Recall Heller's tripartite corollary [27] to Murphy's law:

No matter what special-purpose device is available, there is a problem too large for it.

The problem will manifest itself only after the device is acquired and can no longer be modified.

The problem cannot be ignored.

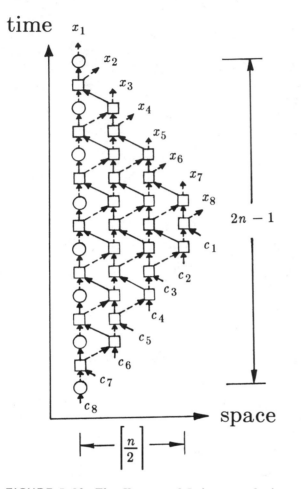

FIGURE 5.29 The Kung and Leiserson design after the coalescing transform has been applied. Notice that the round (division) cell now must perform a square cell function on alternate cycles.

In this section a transformation is introduced that is intended to treat the problem of mapping a large computation onto a smaller array. We will work with the computation of matrix-vector product. Although simple, it is adequate to convey the ideas. Suppose that we have a 9×9 matrix-vector product but that our linear array has only three cells. We start with the Γ design for matrix-vector product depicted in Fig. 5.30. It is "stretched" along the two horizontal dashed lines so that there are three distinct sections that can be

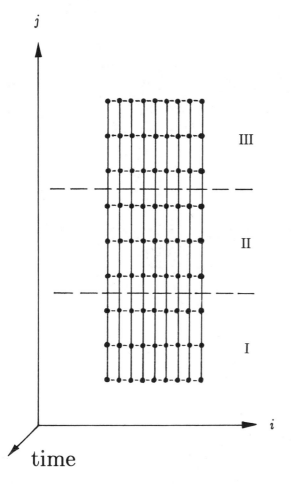

j

III

II

I

i

time

FIGURE 5.30 A 9 × 9 Γ design for computing matrix-vector product.

placed along side one another in space. The resulting design is
depicted in Fig. 5.31a. This design then is skewed in time, via the
linear transformations S and T of Sec. 2.1, resulting in the design
shown in Fig. 5.31b, denoted Σ.

$$\Sigma = T \cdot S \cdot C(\Gamma) \qquad \text{where } C: \begin{bmatrix} i \\ j \\ t \end{bmatrix} \rightarrow \begin{bmatrix} \left\lfloor \dfrac{j}{k} \right\rfloor \cdot n + i \\ j \bmod k \\ t \end{bmatrix}$$

(A)

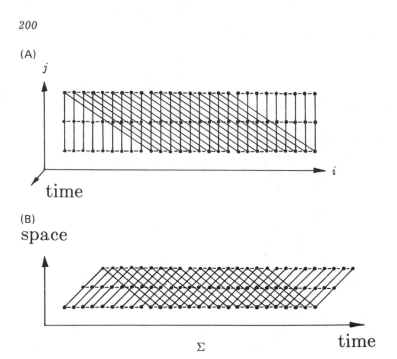

(B)

space

FIGURE 5.31 (A) The 9 × 9 Γ design after 3 sections have been placed alongside one another. x-data contours are "stretched" by this transformation. (B) This design is then skewed in time. The result, denoted Σ, is like the matrix-vector product Ψ design, but it works on a ring of k cells. The x-data circulate n/k = 3 times through the ring of cells.

where k is the number of cells. In our example, k = 3. The x-data contours are according to the lexicographic rule. Notice that x data emerge from the "top" cell and wrap around down to the "bottom" cell. For these cells to be neighbors in space, the linear array of cells needs to be ring connected. The x data circulate through the ring. The circulation is not uniform in time, however. The wraparound takes place over n − k + 1 cycles. Thus n − k words of memory are required. There are some interesting alternatives. Two are discussed that change the design's communication topology. One design cuts the "stretched" x-data contours. As a result it requires that we repeat the input of x values every n cycles. The result, denoted Π, is depicted in Fig. 5.32.

   Another alternative again uses memory. After the first n cycles, the lower cell has seen all the x values and can remember them. If they are placed in the cell's memory, repeated input is unnecessary. The number of cells required by these designs is k; the time, $(n^2/k) + k - 1$.

This raises the question of memory versus communication in VLSI circuits. Data communication in space-time is of three types (see Fig. 5.33):

1. Communication that is normal to the temporal subspace (i.e., instantaneous broadcasting)
2. Communication that is normal to the spatial subspace (i.e., memory)
3. Communication that is normal to neither the temporal nor the spatial subspace (which, like type 1, requires physical ports)

Of these three space-time communication types, VLSI technology most favors the second: memory [28]. Throughout this chapter, those designs that propagate some data normal to the spatial subspace (i.e., through time only) enjoy an efficiency greater than those that communicate all data through both space and time. The best data commuincation characteristics depend ultimately on the application context. Nonetheless, it is generally desirable to transform designs so that some space-time communication is normal to the spatial subspace.

In the case under discussion, the use of memory would generally be favored over repeated input (which relies on external memory anyway). Taking this idea one step further, we may install memory in each cell so that subsequent x-data communication is unnecessary, as depicted in Fig. 5.34.

The transformation discussed in this section is only one way of mapping large computaions onto smaller arrays. Many nonlinear transformations may prove useful in treating this practical problem. The space-time representation facilitates the exploration of such treatments.

space

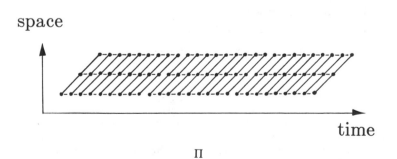

time

П

FIGURE 5.32 The Σ design after the stretched x-data contours have been severed. In this design, denoted П, x-input is repeated 3 times.

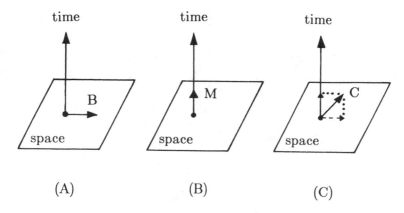

FIGURE 5.33 The 3 types of communication in spacetime (A) information propagation that is perpendicular to the temporal subspace,
(B) information propagation that is perpendicular to the spatial
subspace, (C) information propagation that is perpendicular to neither
the temporal subspace nor the spatial subspace.

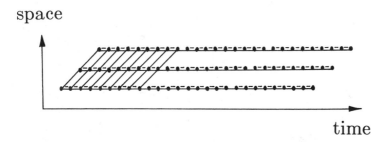

FIGURE 5.34 Spacetime diagram of memory intensive design. Spatial
communication occurs only during the first n cycles. From that point
on, communication is through time only. The solid horizontal lines
indicate x-data flow through time. Note that each horizontal line
represents n x-data contours. Each cell therefore needs a memory
capacity of n words (as well as its y-component register).

4 CONCLUSION

Although only two nonlinear transforms were discussed, many others may be useful. Folding a two-dimensional surface sometimes can be used to achieve proximity of directly communicating cells. For example, suppose that a square mesh of cells contained the elements of a matrix. Then folding this mesh along the main diagonal would facilitate matrix transposition, as has been noted by Muraga. Guibas et al. [29] have given a square mesh systolic design for computing the transitive closure of a relation. If this is properly represented in space-time (i.e., directly communicating cells are neighbors in space-time), the spatial projection is really a two-dimensional torus. Another application concerns the problem of mapping large computations onto smaller systolic arrays. Space-time folding is one way of accomplishing this. A large computation also can be partitioned into neighborhoods, each of which is mapped to a single cell of a small cellular array. Such non-linear mappings require the cells to have memory (at least proportional to the size of the neighborhood). But, as discussed in Sec. 3.2, memory (i.e., communication in time only) is perhaps the communication of choice in VLSI technology. Indeed, this fact has an architectural consequence that transcends systolic communication structures. Applications that can use local memory exclusively will perform better than those that must resort to large shared memories. We can expect that general-purpose multiprocessors will be equipped with large local memories enabling such applications to reduce communication. The use of local memory as an alternative to large arrays is also discussed in Refs. 30 and 31.

The technique presented in this chapter can be applied to a wide variety of systolic computations, including the systolic designs given in Ref. 6 for LU decomposition and the solving of triangular linear system. These designs involve arrays of more than one kind of processing element, an important generalization. Space-time representations can be applied in an even more general setting, however. They are suited to represent any n-dimensional cellular automaton [1] in an (n + 1)-dimensional space. A wealth of computation designs can thus be explored using transformations of space-time. Bit-level systolic algorithms, where the primitive computation involves only a few bits, also have received much attention (see, e.g., Refs. 22 and 32–42). Kuhn [43] uses linear transformations of space-time to create elegant time-redundant fault-tolerant systolic designs. Another direction of research concerns optimization. Li and Wah [44] and Delsme and Ipsen [5] have worked on the problem of synthesizing linear transformations that optimize the systolic array according to some complexity measure.

In this chapter we have presented a technique for placing cellular computations in a space-time framework and have used this framework to relate them formally. Some new designs have been given for the computations discussed, illustrating the ease with which computational designs can be explored using space-time transformations. Design properties such as broadcasting and pipelining can be defined formally, and their presence or absence in a particular design can be ascertained readily. A design's communication topology can be disclosed similarly by projecting onto the spatial subspace of the representation.

VLSI has precipitated a generalization of the algorithm designer's task: A particular algorithm may be implemented via a wealth of different communication structures, each with different properties. In the context of VLSI it has become necessary to distinguish between an algorithm and an implementing structure. In this chapter a design is the association of an algorithm (or schedule) with a particular communication structure. The main value of a space-time representation is its ability to facilitate visualizing and mainpulating cellular designs in an intuitive but formal way.

ACKNOWLEDGMENTS

This work was partially supported by National Science Foundation under Grant ECS-8307955, and the Office of Naval Research under Contracts N00014-84-K-0664 and N00014-85-K-0553.

REFERENCES

1. A. W. Burks (ed.), *Essays on Cellular Automata*, University of Illinois Press, Urbana, Ill. (1970).

2. R. M. Karp, R. E. Miller, and S. Winograd, The organization of computations for uniform recurrence equations, *J. ACM.*, *14*: 563–590 (1967).

3. M. C. Chen, and C. Mead, Concurrent algorithms as space-time recursion equations, *Proc. USC Workshop on VLSI and Modern Signal Processing,* Los Angeles, 31–52 (Nov. 1982).

4. P. Quinton, Automatic synthesis of systolic arrays from uniform recurrent equations, *Proc. 11th Annual Symposium on Computer Architecture,* 208–214 (1984).

5. J.-M. Delosme and I. C. F. Ipsen, Efficient systolic arrays for the solution of Toeplitz systems: an illustration of a methodology for the construction of systolic architectures in VLSI, *Proc. 2nd International Symposium on VLSI Technology, Systems and Applications,* Taipei, Taiwan, 268–273 (1985).

6. H. T. Kung and C. E. Leiserson, Algorithms for VLSI processor arrays, *Introduction to VLSI Systems* (C. Mead and L. Conway, eds.), Addison-Wesley, Reading, Mass., Sec. 8.3 (1980).

7. P. R. Cappello and K. Steiglitz, Unifying VLSI array design with linear transformations of space-time, *Advances in Computing Research*, Vol. 2, *VLSI Theory* JAI Press, Greenwich, Conn., 23–65 (1984).

8. J. A. B. Fortes, K. S. Fu, and B. W. Wah, Systematic approaches to the design of algorithmically specified systolic arrays, *Proc. International Conference on Acoustics, Speech, and Signal Processing*, Tampa, Fla., 300–303 (Mar. 1985).

9. P. R. Cappello, VLSI architectures for digital signal processing, Ph. D. dissertation, Princeton University (1982).

10. D. I. Moldovan, On the design of algorithms for VLSI systolic arrays, *Proc. IEEE, 71*: 113–120 (1983).

11. P. R. Cappello and K. Steiglitz, Unifying VLSI array design with geometric transformations, *Proc. International Conference on Parallel Processing*, Bellaire, MI. pp. 448–457 (Aug. 1983).

12. W. L. Miranker and A. Winkler, Spacetime representations of computational structures, *Computing, 32*: 93–114 (1984).

13. P. E. Danielsson, Serial/parallel convolvers, *IEEE Trans. Comput. C-33*: 652–667 (1984).

14. J.-M. Jover and T. Kailath, Design framework for systolic-type arrays, *Proc. International Conference Acoust. Speech Signal Processing*, San Diego, CA. (Mar. 1984).

15. M. C. Chen, Space-time algorithms: semantics and methodology, Ph. D. thesis, California Institute of Technology (1983).

16. C. D. Thomson, Area-time complexity for VLSI, *Proc. 11th Annual Symposium on the Theory of Computing* (Apr. 1979).

17. G. Bilardi, M. Pracchi, and F. P. Preparata, A critique and an appraisal of VLSI models of computation, *VLSI Systems and Computations* (H. T. Kung, R. Sproull, and G. Steele, eds.), Computer Science Press, Rockville, MD., 81–88 (1981).

18. J. Vuillemin, A combinatorial limit to the computing power of VLSI circuits, *IEEE Trans. Comput., C-32*: 294–300 (1983).

19. Lawrence R. Rabiner and Bernard Gold, *Theory and Application of Digital Signal Processing*, Prentice-Hall, Englewood Cliffs, NJ (1975).

20. H. T. Kung, Why systolic architectures? *Computer* (IEEE) *15*: 37–45 (Jan. 1982).

21. C. Mead and Lynn Conway, *Introduction to VLSI Systems*, Addison-Wesley, Reading, MA (1980).

22. E. E. Swartzlander, Jr., The quasi-serial multiplier, *IEEE Trans. Comput.*, *C-22*: 317–321 (1973).

23. Ming-Yang Chern and Tadao Murata, Efficient matrix multiplication on a concurrent data-loading array processor, *Proc. IEEE International Conference on Parallel Processing*, Bellaire, MI 90–94 (Aug. 1983).

24. J. D. Ullman, *Computational Aspects of VLSI*, Computer Science Press, Rockville, Md. (1984).

25. U. Weiser and A. Davis, A wavefront notation tool for VLSI array design, *VLSI Systems and Computations* (H. T. Kung, R. Sproull, and Guy Steele, eds.), Computer Science Press, Rockville, MD., 226–234 (1981).

26. D. E. Heller and I. Ipsen, Systolic networks for orthogonal equivalence transformations and their applications, *Proc. Conference on Advanced Research in VLSI*, Cambridge, MA 113–122 (Jan. 1982).

27. D. E. Heller, Partitioning big matrices for small systolic arrays, *VLSI and Modern Signal Processing*(S. Y. Kung, H. J. Whitehouse, and T. Kailath, eds.), Prentice-Hall, Englewood Cliffs, 185–199 (1985).

28. H. Garcia-Molina, R. J. Lipton, and J. Valdes, A massive memory machine, *IEEE Trans. Comput.*, *C-33*: 391–399 (1984).

29. L. J. Guibas, H. T. Kung, and C. D. Thompson, Direct VLSI implementation of combinatorial algorithms, *Proc. Caltech Conference on VLSI*, 509–525 (1979).

30. A. L. Fisher, Memory and modularity in systolic array implementations, *Proc. International Conference on Parallel Processing*, St. Charles, IL 99–101 (Aug. 1985).

31. P. R. Cappello, A mesh automation for solving dense linear systems, *Proc. International Conference on Parallel Processing*, St. Charles, IL 418–425 (Aug. 1985).

32. K. Hwang, *Computer Arithmetic*, Wiley, New York (1979).

33. F. C. Hennie, *Finite-State Models for Logical Machines*, Wiley, New York (1968).

34. P. R. Cappello and K. Steiglitz, A note on free accumulation in VLSI filter architectures, *IEEE Trans. Circuits Syst.*, *CAS-32*: 291–295 (1985).

35. P. R. Cappello and K. Steiglitz, Completely pipelined architectures for signal processing, *Proc. International Conference on Circuits and Computers*, New York (Sept. 1982).

36. P. R. Cappello and K. Steiglitz, Bit-level fixed-flow architectures for signal processing, *Proceedings of the International Conference on Circuits and Computers*, New York (Sept. 1982).

37. P. R. Cappello and K. Steiglitz, Digital signal processing applications of systolic algorithms, *VLSI Systems and Computations* (H. T. Kung, R. Sproul, and G. Steele, eds.), Computer Science Press, Rockville, MD, pp. 245–254 (1981).

38. R. A. Evans, D. Wood, K. Wood, J. V. McCanny, J. G. McWhirter, and A. P. H. McCabe, A CMOS implementation of a systolic multi-bit convolver chip, in *VLSI '83*, (F. Anceau and E. J. Ans, eds.), North-Holland, Amsterdam 227–235 (1983).

39. P. B. Denyer and D. J. Myers, Carry-save arrays for VLSI signal processing, *VLSI '81*, (J. P. Gray, ed.), Academic Press, New York, 151–160 (1981).

40. J. V. McCanny, J. G. McWhirter, J. B. G. Roberts, D. J. Day, T. L. Thorp, Bit-level systolic arrays, *Proc. 15th Asilomar Conference on Circuits, Systems, and Computers*, Monterey, CA (Nov. 1981).

41. J. V. McCanny and J. G. McWhirter, Bit-level systolic array circuit for matrix-vector multiplication, *IEE Proc.*, *130*, pt. G: 125–130 (1983).

42. J. V. McCanny, K. W. Wood, J. G. McWhirter, and C. J. Oliver, The relationship between word and bit level systolic arrays as applied to matrix X matrix multiplication, *Proc. SPIE* (Aug. 1983). *Symposium*, Real Time Signal Processing VI, San Diego, (Aug. 1983).

43. R. H. Kuhn, Yield enhancement by fault-tolerant systolic arrays, *VLSI and Modern Signal Processing* (S. Y. Kung. H. J. Whitehouse, and T. Kailath, eds.), Prentice-Hall, Englewood Cliffs, N.J., 178–184 (1985).

44. G. J. Li and B. W. Wah, The design of optimal algorithms, *IEEE Trans. Comput.*, *C-34*: 66–77 (1985).

6

Architecture Design for Regular Iterative Algorithms

SAILESH K. RAO *AT&T Bell Laboratories, Holmdel, New Jersey*

THOMAS KAILATH *Information Systems Laboratory, Stanford University, Stanford, California*

1 INTRODUCTION

This chapter is about regular iterative algorithms (RIAs), their properties, and their implementation on processor arrays. The importance and significance of RIAs rest on the following facts, to be partially substantiated here:

1. RIAs are sufficiently well structured that the detection and exploitation of instruction-level concurrency in RIAs is not a horrendously difficult task. In fact, RIAs are very well matched to mesh-connected and related architectures, and it is not unrealistic to conceive of a compiler that maps RIAs into such architectures as the massively parallel processor (MPP) [1].
2. The algorithm executed by every systolic array can be identified to belong to a certain subclass of RIAs. Conversely, every member of this subclass of RIAs can be mapped onto a systolic array architecture. This mapping can be accomplished in many different ways, although the resulting systolic arrays all execute the same algorithm. Thus a unified framework can be devised, using which many of the existing systolic arrays (and many new ones) can be derived simply by changing some parameters in the synthesis procedure.
3. Examples of RIAs are scattered throughout the literature. Thus systolic array designs can be realized by isolating and identifying RIAs and then applying the synthesis procedure developed in Sec. 3.

RIAs arise naturally in the discretized numerical solution of differential and integral equations. The use of RIAs in the solution of linear algebraic problems can also be demonstrated. In addition, many digital signal processing problems can be solved by means of RIAs. A few examples of such problems are:

1. Digital filtering in its various forms: adaptive, linear, time-invariant, time-varying, multi-input multioutput, multidimensional, and so on
2. Discrete fourier transformation (DFT), discrete cosine transformation (DCT), and other transform algorithms, in both their one-dimensional and multidimensional versions
3. Factorization, inversion, multiplication, and so on, of matrices
4. Convolution, correlation, prediction, modeling, and so on, in several variations

The application of RIAs to various graph-theoretic problems, such as the transitive-closure problem, the connected-components problem, and various graph decomposition problems can also be demonstrated. In a more general setting, one would expect RIAs to function in a role similar to that of do and while loops in standard sequential programming languages: namely, as powerful constructs that are components of an algorithm and that can be mapped onto an universal architecture to extract a high degree of instruction-level concurrency.

1.1 Overview

A formal definition of regular iterative algorithms is given in Sec. 2. For the present, it is best to explore the various features of an RIA by means of a simple example. This example is an RIA for matrix multiplication, obtained by reformulating the standard algorithm

For i = 1 to N_1 do

For j = 1 to N_2 do

For k = 1 to N_3 do

$$c_{ij} = c_{ij} + a_{ik} b_{kj}$$

and writing it in the form

For i =·1 to N_1 do

For j = 1 to N_2 do

For k = 1 to N_3 do

$a(i, j + 1, k) = a(i,j,k)$

$b(i + 1, j,k) = b(i,j,k)$

$c(i,j,k) = c(i,j, k - 1) + a(i,j,k)b(i,j,k)$

Note that the RIA in the example above is a *single-assignment algorithm;* that is, a variable is assigned a unique value in a particular instance of the algorithm. A variable itself is identified by a variable name ($\{x_1, \ldots x_V\}$ in general, and a, b, c, in the example) and by an index vector [$I = (i,j,k)^T$, in the example]. The range of the index vector I is referred to as the index space, which is a lattice of integer points enclosed within a region in the S-dimensional Euclidean space. For the example above, S = 3 and the index space, shown in Fig. 6.1, is a rectangular parallolepiped with sides of length N_1, N_2, and N_3.

In a single-assignment algorithm, a variable, say x, is said to be directly dependent on another, say y, if and only if the value of x is computed using the value of y. In an RIA, these direct dependences are regular throughout the algorithm; that is, x(I) is directly dependent on y(I − D) regardless of the particular value of I. In view of this regularity, it is possible to express the direct dependences concisely by means of the *reduced dependence graph* (RDG), shown in Fig. 6.2 for the example above; this RDG consists of three nodes, one for each of the variable names a, b, and c in the algorithm, and it has vector-weighted arcs that represent the direct dependences.

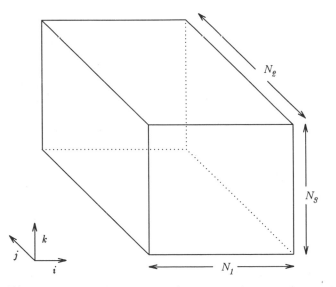

FIGURE 6.1 Index space of the RIA for matrix multiplication.

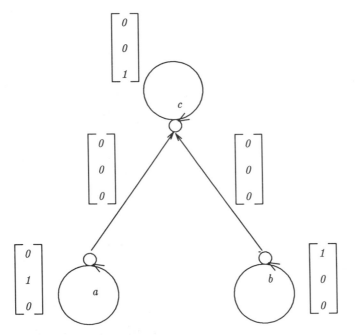

FIGURE 6.2 Reduced dependence graph (RDG) of the RIA for matrix multiplication. This is a very simple example of an RDG since it has no directed loops except for self-loops.

For example, the arc from a to c with weight $[0 \ \ 0 \ \ 0]^T$ indicates that c(I) is directly dependent on a(I − $[0 \ \ 0 \ \ 0]^T$) for all I, as is evident from the algorithm.

Given a single assignment algorithm, the traditional way to capture information regarding the parallelism in the algorithm is by means of a dependence graph. This graph has one node for each of the variables in the algorithm and a directed arc from node x to node y if and only if variable y is directly dependent on variable x in the algorithm. In an RIA, the variables can be represented in the form x(I), and the dependence graph has a node for each of the variable names {x} for each of the variable indices {I}. In some sense, the dependence graph of the RIA must be regular, since the direct dependences among the variables are regular. This regularity can be explicitly captured by embedding the dependence graph of the RIA in the index space as shown in Fig. 6.3 for the matrix multiplication example. Once such an embedding is done, one can now identify the computations that are associated with a particular index point of the RIA. Notice that the dependence information in the dependence graph

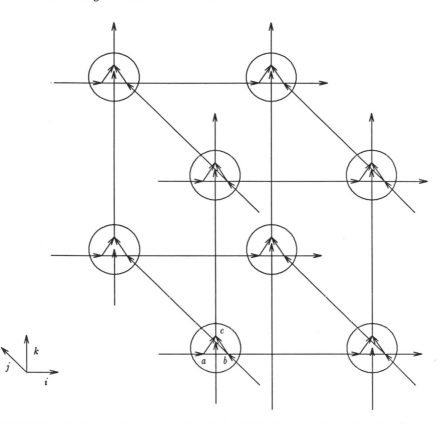

FIGURE 6.3 Dependence graph of the RIA for matrix multiplication.
The dependence graph, shown here for a (2 × 2 × 2) case, is embedded
in the index space (shown in dashed lines) to reveal the regular,
iterative nature of the graph. The arcs that are shown floating (with
no nodes at one extremity) are dummy arcs that are used to point to
the input nodes (nodes with no incoming arcs) and output nodes (nodes
with no outgoing arcs) in the algorithm. Note that in an RIA, the input
and output nodes are implicitly defined.

of an RIA is completely captured by the RDG itself. Hence one can
derive various properties of the dependence graph by examining the
RDG, and thus by spending effort that is independent of the "size"
of the algorithm.

The dependence graph of a single assignment algorithm specifies
a partial ordering among the computations in the algorithm; that is,
if there is a directed path in the dependence graph from node x to
node y, the computation represented by node y must be executed after

the computation represented by node x is completed, no matter how
many processors are brought to bear on the problem. From this observation, one can infer that the length of the longest path in the dependence graph is a lower bound for the total time required for executing
the algorithm, independent of the number of processors used in this
execution.

The main objective of the studies reported in this chapter is to
derive efficient techniques for determining processor array implementations for an RIA. One simple and brute-force method for achieving
this objective is to use a distinct processor for executing the computations at a particular index point of the RIA. This, in general, leads
to a very inefficient use of the computational resources, since each
processor is active only for a constant period of time, which could be
a minute fraction of the time required for completing the algorithm. To
achieve a better utilization of these resources, it is necessary to reuse
the same processor for handling a large number of computations. In
general, the set of computations that are assigned to processors can
be arbitrary disjoint subsets of the set of all computations. However,
the problem of optimally "scheduling" the computations that are assigned
to a given processor becomes extremely difficult if the partitioning is
arbitrary. To render this problem tractable, one can restrict attention
to linear (planar, hyperplanar) partitions. Here, a set of parallel lines
(planes, hyperplanes) are drawn through the index space of the RIA,
so that all computations that correspond to index points that lie on
the same line (plane, hyperplane) are handled by the same processor.
In this manner, the processor array itself can be obtained by projecting the embedded dependence graph of the RIA along these lines
(planes, hyperplanes) onto a lower-dimensional lattice of points, known
as the *processor space* (see, e.g., Fig. 6.4).

Once the processor space is decided on, one must attempt to
schedule the computations that are mapped onto a given processor.
This schedule must really be globally specified, since the result of
some computation performed by a particular processor may be used
by another for computing some other value. Furthermore, such a
specification is synonymous with assigning scheduling functions, which
are functions that map every computation in the algorithm into a
particular time slot. Obviously, the choice of the scheduling functions
is constrained both by the dependences in the algorithm and by the
choice of the processor space; if a variable, say $x(I)$, is computed
using the value of, say, $y(I - D)$, the scheduling function assigned
to $x(I)$ should be greater than that assigned to $y(I - D)$; furthermore,
if a given processor is expected to compute both $x(I)$ and $x(I - V)$,
these variables cannot be assigned to the same time slot.

Once again, determining the scheduling functions appears to be
difficult, unless one imposes further restrictions on the nature of
these functions. Specifically, in this chapter these scheduling functions

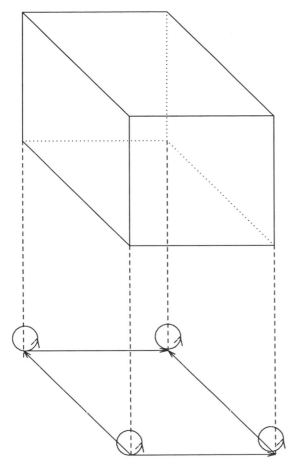

FIGURE 6.4 Processor array for implementing the matrix multiplication example. This processor array is obtained by drawing a set of parallel lines through the index space and then by assigning all computations that lie along the same line to the same processor.

will be assumed to be linear with respect to the index vectors. That is, a set of parallel lines (planes, hyperplanes) can be drawn through the index space so that all computations that correspond to index points that lie on the same line (plane, hyperplane) are executed at the same time (necessarily, by different processor) (Fig. 6.5). Thus linear scheduling functions correspond to isotemproal lines (planes, hyperplanes) that are drawn through the index space. However, one must emphasize here that this geometric interpretation of linear scheduling functions is of limited scope since such isotemporal lines may not exist for every RIA.

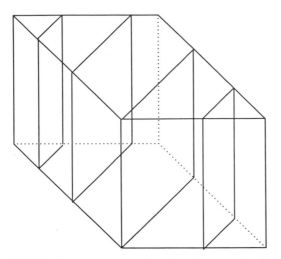

FIGURE 6.5 Illustration for linear scheduling fucntions. A set of
parallel planes are drawn through the index space so that all computations that lie on the same plane are executed at the same time.

With this brief overview, it is easy to visualize the sort of issues
that need to be addressed in attempting to provide a complete theoretical framework for the problem. These issues can be classified in
three categories as follows:

1. *Analysis issues*
   a. Is the given RIA computable; that is, is there some variable
      in the RIA that is circularly defined? If there is such a variable,
      the algorithm cannot be executed in the manner proposed.
   b. Given a computable RIA, what is the minimum achievable
      execution time for completing the algorithm? This information
      is crucial for the designer to decide how many processors
      should be used for executing the algorithm. For example, if
      the algorithm requires N^3 computations and the minimum
      achievable execution time is N, one would wish to use N^2
      processors, each working on disjoint sets of N computations,
      to complete the algorithm in N time.
   c. How much storage should be provided for executing the
      algorithm? This would provide an indication of how much local
      memory should be associated with each processor in the array.
2. *Implementational issues*
   a. Given a computable RIA, how should one choose the processor
      space? Is this choice restricted by some properties of the RIA?

b. Once the processor space is chosen, how should one determine the scheduling functions? Is it possible to choose these scheduling functions so that every processor is active most of the time? Is it always possible to use linear scheduling functions and still achieve the minimum execution time predicted by the analysis?

3. *Synthesis issues*
a. Given a problem, is it possible to find an RIA for solving it? Is there a theoretical basis for determining such RIAs?
b. Presuming that there are a lot of RIAs for solving a given problem, is there some systematic method for choosing one?

One would also like to answer all the questions concerning the analysis and implementations of RIAs by examining the reduced dependence graph of the RIA and not the dependence graph. This would ensure that the procedures designed for addressing these issues are of complexity independent of the size of the RIA (a rough measure of the size of the RIA is given by the number of nodes in the dependence graph of the RIA).

The analysis issues raised above are addressed in detail in Sec. 2. The implementation issues are addressed in Sec. 3. Here the implementations are assumed to be synchronous circuits, although the results are equally relevant to asynchronous interrupt-driven implementations as well. For brevity's sake, the synthesis issues are not addressed in this chapter; the interested reader is referred to Refs 2 and 3.

1.2 Literature Survey

The literature survey described in this section will pertain to those articles of immediate relevance to the analysis and implementation issues addressed in this chapter. As can be shown, Euclid's algorithm for computing the GCD of two integers and Gauss's algorithm for triangulating matrices can easily be reformulated as RIAs; thus the synthesis of RIAs dates back to ancient times and an exhaustive survey of past work on the synthesis of RIAs is beyond the scope of this section.

Some of the earliest studies on the properties of iterative arrays occurred in the field of switching circuits and automata theory, where the contributions of Atrubin [4], McCluskey [5], Hennie [6–10], Unger [11] and Waite [12] deserve special mention. In many ways, the iterative arrays (of combinatorial logic cells) studied by these authors correspond to the dependence graph of an RIA. Hennie's thesis [9] contains an extensive theoretical discussion on the "decidability" of various properties, such as equivalence, stability and so on, of these iterative arrays of logical circuits. Hennie proves that there exist no "finite" algorithms for solving any of these problems using reductions from the Post's correspondence problem. While Hennie's results may be drawn on to prove that there exists no finite algorithm for verifying whether two different RIAs are solving the same problem, their relevance to the analysis issues raised here is tenuous. However, Hennie's thesis [9] and later book [10] contain some remarks on the space-time representation of sequential circuits and on mapping an iterative array of combinational cells into a lower-dimensional array of sequential cells; this is precisely the problem of determining the processor space given the index space of an RIA. However, the heuristic techniques advocated by Hennie for clocking the resulting sequential circuit cannot be used for scheduling the operation of a processor array, in general. Furthermore, one need not restrict oneself to projections along the coordinate axis, as indicated in Refs. 9 and 10.

Waite [12] studied the problem of determining the existence of a path between two nodes in an iterative network. Once the iterative network of Waite is identified with the dependence graph of an RIA, it is easy to see that this problem is identical to that of checking whether a certain variable is dependent on another. In particular, one can use Waite's techniques to determine whether a certain variable is dependent on itself (i.e., if the algorithm is computable or not).

It is interesting to note that the reduced dependence graph (RDG), introduced in the preceding section, was used extensively by Waite for path detection in an iterative network. In Ref. 12 it is proved that there is an isomorphism between the paths in the RDG and paths in the dependence graph (i.e., iterative network) and consequently, the existence of a path in the dependence graph can be inferred by examining the RDG alone. Thereupon, Waite notes that the RDG is similar to the state diagram of a finite-state machine and therefore can be described by a regular expression in the manner described by McNaughton and Yamada [13]. Using these regular expressions, Waite then formulates a series of linear programming problems that must be solved in order to determine the existence of the path. However, this technique may, in general, lead to the necessity for solving an exponential (in the number of nodes in the RDG) number of polynomially sized linear programming problems.

The next and probably the most important contributions as far as the analysis issues are concerned appear in the paper of Karp et al. in 1967 [14]. Since the issues addressed in this paper are related to those studied in Sec. 2, a detailed description of the results in Ref. 14 is in order here. Translated using the terminology introduced in the preceding section, the results in Ref. 14 can be categorized as follows:

1. Given an RIA, defined over a semi-infinite index space (the set of all nonnegative integer vectors in S-dimensional Euclidean space), determine if the RIA is causal. (For the distinction between causality and computability, see below.)
2. For a causal RIA defined over a semi-infinite index space, show the existence of an upper bound for the scheduling functions.

Numerous other results given in Ref. 14, and some that are not explicitly stated in this paper, can be inferred from these two main contributions. For instance, one can easily deduce a procedure for determining the computability of an RIA defined over a finite index sapce, by appropriately modifying the procedure in Ref. 14 for determining the causality of an RIA defined over a semi-infinite index space.

To understand the implications of the results presented in Ref. 14, it is necessary to define the term "causality" and to point out the difference between causality and computability.

Causality An RIA defined over the set of all nonnegative integer points in S-dimensional Euclidean space is said to be *causal* if the following is true: The variable $x_k(I)$ is dependent on $x_k(I - D)$ for some $\{k,I\}$ implies that D has at least one strictly positive element.†

Notice that "computability" requires that there is no such dependence with D = 0 (i.e., that there is no directed loop in the dependence graph). A noncausal RIA could be computable whereas a noncomputable RIA is inherently noncausal.

If an RIA is noncausal, there is some integer vector D with nonpositive elements such that $x_k(I)$ is dependent on $x_k(I - D)$ for some k. Then, for any positive integer α, $x_k(I)$ is dependent on $x_k(I - \alpha D)$, due to the regular iterative nature of the dependences. Since D has nonpositive elements, one can choose α to be arbitrarily large and $(I - \alpha D)$ will still be within the confines of the index space, since the index space is assumed to be semi-infinite. Consequently, for any finite index point I, one will never be able to evaluate $x_k(I)$ in finite time. Furthermore, note that the existence of such a dependence implies that there is a directed loop in the RDG such that the sum of the vector weights on the edges that participate in this loop is a nonpositive vector (i.e., has nonpositive elements).

It is interesting to note that the RDG was also used by Karp and co-workers for determining various properties of the RIA. Their test for determining causality rests on examining a tree of subgraphs of

---

†This definition of causality is a natural extension of the definition used in the digital signal processing literature, wherein it is assumed that the inex space is one-dimensional.

the RDG. To construct this tree of subgraphs, they first find the set of all simple directed loops in the RDG (directed loops that pass through any node at most once). If G denotes the RDG, then let $C(G)$ be the set of all simple directed loops in the RDG and let $L(G)$ be the set of vector weights of these loops. For each vector weight in $L(G)$, say L_i, check to see if there exists a positive (not necessarily integral) combination of the other weights such that they add up to $(D - L_i)$, where D is some nonpositive vector. If so, mark the edges in G that participate in the loop C_i.

Once each of the simple directed loops are checked, consider the subgraph G' of the RDG obtained by deleting all unmarked edges. If there are no marked edges, the RIA is causal. Else, if this subgraph is connected, the RIA is noncausal. Otherwise, one must apply the same procedure to each of the connected components in G'.

With the procedure above, one can construct a tree of subgraphs of the RDG. For each of the subgraphs in the tree, a sequence of linear programming problems have to be solved. It can then be shown that the number of such problems that are formulated in the procedure is polynomial in the number of nodes in the RDG. However, some of these linear programming problems may involve exponentially many variables, since there can be 2^V simple directed loops in a graph with V nodes.

Karp and co-workers also show that for a certain subclass of RIAs if the tree of subgraphs generated by the foregoing procedure is of depth r (the length of the longest root to leaf path in the tree is the depth of the tree) and the RIA is causal, there exists a constant β and a vector X such that there is a schedule in which all variables at the index point I are computed at or before step $\beta(X^T I)^r$. This gives a uniform upper bound for the existence of a schedule. Furthermore, the authors show that the vector X has to be a nonnegative solution to a "dual" linear programming problem and that such a solution will always exist for the dual problem.

In this chapter we are concerned mainly with RIAs defined over a finite index space. Hence computability and not causality is of interest here. Nevertheless, the procedure above can be modified very easily to test for computability; just restrict attention to $D = 0$ instead of a nonpositive D and the resulting modified procedure is a test for the computability of the algorithm.

If the tree obtained in the modified procedure is of depth d, it is quite natural to expect that it must be possible to find a constant vector X and a scalar constant β such that there exists a schedule in which all variables at the index point I are computed at or before step $\beta(X^T I)^d$, since this result would correspond to an extension of the result in Ref. 14. However, it can be shown that it is impossible to find such constants (except under very special circumstances) and that one has to work harder to find a comparable result in the case of a computable RIA defined over a finite index space. This is done in Sec. 2 (see also Ref. 14).

A survey of the literature with regard to the implementation issues reveals that many authors have proposed the idea of mapping an iterative algorithm defined over a multidimensional index space into a lower-dimensional array of processors, by using linear transformations [9,10,16–33]. All these authors, with the exception of Hennie [9], restrict attention to one-dimensional projections, so that if the index-space is S-dimensional, the processor array is (S − 1)-dimensional. In addition, Hennie [9] presumes that multidimensional projections can be applied only along the coordinate axis, which is really not the case. Furthermore, in all of the papers above with the exception of Ref. 23, the authors assume that the scheduling of the computations on the processor arrays is accomplished using some standard scheduling techniques such as that of Leiserson et al. [34]. The drawback to this approach is that these scheduling techniques do not make use of the regularity of the processor array, and the existing scheduling techniques can be inefficient as well. This fact may not be of great concern if the processor array is intended to be implemented as a dedicated VLSI architecture. However, if this approach is to be used for programming a general-purpose array architecture, this is obviously not desirable.

Quinton [23] has derived a more formal approach that is quite similar to the techniques outlined in the previous overview. However, Quinton's techniques are valid only for a very special subclass of RIAs, referred to as "uniform recurrence equations." Uniform recurrence equations, as defined in Ref. 23, are RIAs in which all variables, except for one, are propagating variables. A propagating variable obeys a defining equation of the form

$$x(I) = x(I + D)$$

Hence the RDG of a uniform recurrence equation is a directed star network, in which every arc is either a self-loop or is directed from a boundary node to the central node. Many of the techniques presented in Sec. 3 of this chapter specialize to those given in Ref. 23 in the event the RIA is indeed of this form.

The techniques described in Moldovan [33] do not make use of the dependence information in the reduced dependence graph of the RIA. In equivalent terms, this would correspond to a situation in which the RDG of the RIA consists of a single node and self-loops—a drawback that limits the generality of these techniques.

2 ANALYSIS OF REGULAR ITERATIVE ALGORITHMS

2.1 Background

Any algorithm consists of several elementary instrucitons and an implicit or explicit partial ordering among them. For the class of algorithms under consideration in this chapter, an elementary instruction is assumed to correspond to the evaluation of an expression and its assignment to a set of variables. The values required for evaluating this expression, which will be referred to as inputs to the instruction, could be either external inputs or the results of previously evaluated expressions or both. The result of evaluating the instruction will be referred to as the output of the instruction. The output of an instruction may be assigned to many variables, although it will be assumed that every variable is assigned unique value in a particular instance of the algorithm. Thus assignment statements of the form (a := a + b) are not allowed since a appears on both sides of the statement. Such a restriction is imposed in many parallel languages, such as the single-assignment language (SAL) developed at Stanford University.

Given a single assignment algorithm (i.e., the dependence graph for it) one can infer several results:

1. One can check whether the algorithm has any values that are circularly defined by determining if there are any directed cycles in the dependence graph. If not, the algorithm is computable. Otherwise, some of the variables in the algorithm are implicitly defined, and thus the computation cannot proceed as dictated by the algorithm.

2. A lower bound on the latency of the algorithm, irrespective of the number of processors used in parallel, is given by the length of the longest path in the dependence graph. This is because every instruction represented by the nodes in this path has to be executed sequentially.

3. The amount of external communication required by the algorithm can be estimated by counting the number of external input and external output nodes in the dependence graph. Such nodes are identified by the absence of incoming edges (input nodes) or outgoing edges (output nodes).

4. The scheduling of all the instructions represented by the nodes in the dependence graph can also be performed, provided that the processors in which the instrucitons are to be executed have been selected. For the sake of simplicity, it can be assumed that all these processors are controlled by a global clock. Then a schedule for the algorithm corresponds to the assignment of the relative time (in terms of the global clock) at which each of these instructions has to be executed and the identity of the processor in which it is executed. This scheduling problem, under certain contraints of optimality, is in

general, extremely difficult; many variations have been shown to be NP-complete [35]. However, one can obtain reasonably simple formulations if "goodness" of the schedule, rather than optimality, is of interest. Furthermore, if the number of processors available for executing the algorithm is "unlimited" (i.e., one for each instruciton), one can apply PERT/CPM scheduling techniques for determining a schedule that is optimal with respect to the time required for completing the algorithm.

    5. One can also obtain a lower bound on the amount of memory required to execute an algorithm. Consider all the paths between a pair of nodes: A value that arrives along the shortest path has to "wait" until the value from the longest path arrives before the instruction represented by the node can be executed. Such a wait would correspond to an actual memory element if the delays are measured in terms of clock cycles. Thus a lower bound for the memory-time product can be obtained by adding all distinct waits between pairs of nodes in the dependence graph.

Example 1: Computable algorithm

    Consider the following linear state-space algorithm for a single-input, single-output, single-state digital filter:

For k = 1 to N do

$x(k + 1) = ax(k) + bu(k)$

$y(k) = cx(k) + du(k)$

The dependence graph of the algorithm consists of nodes (x,k), (y,k), and (u,k) for k = 1 to N. The directed arcs in the dependence graph are from node (x,k) to node $(x,k + 1)$, from node (u,k) to node $(x,k + 1)$, from node (x,k) to node (y,k), and from node (u,k) to node (y,k) for all k. The resulting graph is shown in Fig. 6.6.

    One can now check that the dependence graph of this algorithm is loop-free. Hence it is computable. It has $(N + 2)$ input nodes: (u,k) for all k and $(x,0)$ are nodes with zero in-degree (a node with no incoming arcs). It also has $(N + 2)$ output nodes: (y,k) for all k and $(x, N + 1)$ are nodes in the graph that have zero out-degree (a node with no outgoing arcs). The longest path in the dependence graph is of length $(N + 1)$: from $(x,0)$ to $(x, N + 1)$. The corresponding schedule for the computation of the algorithm can be obtained as

    compute x(k) at step k

    compute y(k) at step k + 1

    request u(k) at step k

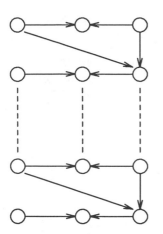

FIGURE 6.6 Dependence graph for the algorithm in Example 1. The dependence graph has $3N^2$ nodes denoted as (x,k), (y,k), and (u,k). Notice the extremely regular and periodic structure of the graph.

The amount of memory required is 1: at step k, store x(k) in memory. Use this to compute x(k + 1), y(k). Take the result x(k + 1) and store it in the same memory at step (k + 1). Since the value of x(k) is no longer needed for the computation of x(k + 2), and so on, one can use the same memory and overwrite x(k). In other words, the given algorithm is a single-state (memory) system.

In a noncomputable algorithm some variables are implicitly defined. One can, of course, determine the values of these implicitly defined variables by applying some iterative refinement methods. However, the steps in the iterative refinement procedures must also be included as part of the algorithm. In this way one could unravel the loops in the original dependence graph so that the new dependence graph is loop-free.

In Example [1], the properties of the dependence graph were deduced by inspection. Due to the regular (iterative) nature of the dependence graph, one was able to simplify considerably the graph-theoretic techniques referred to above. However, in general, the dependence graph of an algorithm may possess no structure whatsoever, in which case such simplifications may not be possible.

Even though the graph techniques are valid for any algorithm, they may not be useful in many circumstances since some of these procedures could be comparable in complexity to the algorithm itself. Furthermore, the PERT/CPM techniques used for scheduling do not take into account the cost for communicating between processors. These techniques are also based on the assumption that a sufficiently large number of processors are available. It turns out that if constraints on the number of processors and on the communciation between processors are incorporated into the scheduling problem, it becomes extremely difficult. It has been known for some time that optimal scheduling of an algorithm on an arbitrary number of processors is NP-hard (if the number of processors is also given as part of the input to the algorithm), even if the communication constraints between processors are not taken into account [35]. If communication constraints also need to be considered, the scheduling problem becomes even more intractable. Hence it is quite difficult to present a universal recipe for optimally mapping algorithms into multiprocessor architectures.

2.2 Regular Iterative Algorithms

A regular iterative algorithm consists of variables that are specified by a set of identifiers. This set of identifiers comprises a variable name, say x, and an S-dimensional integer vector, say I. The variable with this particular identifier is represented by the symbol $x(I)$ and the node in the dependence graph corresponding to this variable is represented as (x, I). Such variables are traditionally referred to as *indexed variables* and the integer vector I is referred to as the *index vector* of the variable.

Definition A *regular iterative algorithm* is defined by a triple $\{I, X, D\}$, where

1. $\underline{I}$ is the *index space,* which is the set of all lattice points enclosed within a specified region in S-dimensional Euclidean space.
2. $\underline{X}$ is the set of V variables that are defined at every point in the index space, where the variable x_j defined at the index point I will be denoted as $x_j(I)$ and takes on an unique value in any particular instance of the algorithm.
3. $\underline{D}$ is the set of direct dependences among the variables, restricted to be such that if $x_j(I)$ is directly dependent on $x_k(I - D)$, then
    a. D is a constant vector independent of I and the extent of the index space.
    b. For every J contained in the index space, $x_j(J)$ is directly dependent on $x_k(J - D)$ [if $(J - D)$ falls outside the index space, $x_k(J - D)$ is an external input to the algorithm].

An RIA is completely specified by the set of constraints that specifies the index space, together with the V defining instructions for the variables. This set of defining instructions will be referred to as the *iteration unit.* If the output variables of the defining instructions all have the same index vector, say I, the iteration unit is said to be in *output-standard form*; if all input variables in the iteration unit have the same index vector, the iteration unit is in *input-standard form*. It is easy to show that an iteration unit in one form can be converted to the other, possibly with the definition of additional variables. In case the defining instructions for a variable involve several conditional branches, the dependence will be assumed to include all the variables in every branch of the conditionals. For example, the set of expressions

For i,k = 1 to N do

$$x(i,k) = x(i - 1,k) + k^2 x(i,k - 1)^{y(i - 1,k)}$$

$$y(i,k) = \begin{cases} i^k y(i - 2, k + 1) + x(i - 1, k + 1) & \text{for even } i \\ -i^k y(i - 1, k + 1) + x(i - 1, k + 1) & \text{for odd } i \end{cases}$$

$$(6.1)$$

constitutes an RIA written in output-standard form. This can also be written in input-standard form as

$$x(i + 1, k) = x(i,k) + k^2 w(i,k)^{y(i,k)}$$

$$w(i - 1, k + 1) = x(i,k)$$

$$y(i + 1, k - 1) = \begin{cases} (i + 1)^{k-1} u(i,k) + x(i,k) & \text{for odd } i \\ - (i + 1)^{k-1} y(i,k) + x(i,k) & \text{for even } i \end{cases}$$

$$u(i + 1, k) = y(i,k) \qquad\qquad (6.2)$$

In the input-standard form, the number of inputs to the iteration unit is matched to the number of outputs (unless of course, some input is derived from the external environment for all I). This is a useful property, which will be featured in the discussion on systolic arrays in the next section. Of course, in the RIA defined by Eq. (6.1), it will be assumed that $y(i,k)$ is dependent on $y(i - 2, k + 1)$, $y(i - 1, k + 1)$ and $x(i - 1, k + 1)$ for all i,k in order to be able to apply techniques described below. Similarly, for the RIA defined in Eq. (6.2), $y(i + 1, k - 1)$ will be assumed to depend on $x(i,k)$, $u(i,k)$, and $y(i,k)$ for all i,k).

The dependence graph of an RIA has a special periodic structure that is useful both for efficiently deriving properties of interest, and for obtaining good array implementations. This structure in the dependence graph is best illustrated by embedding the graph within the index space so that node (x,I) that represents the variable $x(I)$ is situated at point I in the index space. Now, if the instruction for evaluating $x(I)$ to node (x,I) in the dependence graph. Correspondingly,

there is an outgoing arc from node (y, I) to node $(x, I + D)$. These
nodes are physically situated at the respective index points and the
arcs exist for evey index point, thereby creating a family of parallel
arcs.

The topology of the dependence graph is fully captured by
specifying the index space together with the information regarding
either the incoming our outgoing arcs at a particular point in the
index space. The information regarding an incoming arc consists of
the variable name and the relative displacement of the index point
from which it originates and the variable name at which it terminates
[e.g., $(y \rightarrow x, D)$ for the example above]. This can be depicted
succinctly with a reduced dependence graph (RDG). The RDG consists
of V nodes, one for each variable name. If there is an arc $(y, (I -$
$D)) \rightarrow (x, I)$, then a directed arc is drawn from node y to node x with
weight D in the RDG. If x and y are the same, a self-loop is created
with this weight. The weight vector D represents the relative displacement of the index points at the extremities of the corresponding
arc in the dependence graph. Therefore, it is referred to as the
index displacement vector for that arc. For example, the reduced
dependence graph for the RIA in Eq. (6.1) is shown in Fig. 6.7a.

Conversion of RIAs into standard forms

Given an RIA, it is easy to write it in the output standard form;
one merely displaces the index vector of every variable that partici-
pates in an instruction in the iteration unit until the output of the
instruction has the index I. However, there are other useful conver-
sions of the RIA that require a little more effort.

1. *Conversion to input standard form.* Any RIA can be converted
to input standard form by introducing additional variables, where
necessary. Now, in the input standard form, all inputs to every
instruction in the iteration unit have the same index vector. This
implies that the weights assigned to every incoming arc at any node
in the RDG must be the same. Conversely, and RDG that satisfies
this property can be made to correspond to an RIA in input standard
form. Given this information, one can work with the RDG to accomplish
the desired conversion. Let x be some node in the RDG and D_x be the
desired weight on all incoming arcs at node x. (Note that D_x is assumed
to be arbitrary, although some intelligent choice can be made so that
the number of additional variables introduced is minimized.) Let e be
an arc in the RDG from node y to node x with weight D_{yx}. If D_{yx} is
equal to D_x, there is no need to introduce additional variables.
Otherwise, create a new node, say u, and replace arc e with two arcs,
one from node y to node u with weight $(D_{yx} - D_x)$ and the other
from node u to node x weight D_x. In terms of the defining instructions,

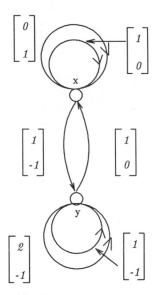

FIGURE 6.7a Reduced dependence graph for the example in Eq. (6.1).
The reduced dependence graph together with the specification of the
index space constitutes a complete description of the dependence graph
of the algorithm. For instance, the self-loop around node x with weight
$[1 \ 0]^T$ represents a family of parallel arcs from node (x,i,k) to node
$(x, i + 1, k)$ for all i,k in the full dependence graph.

$$x(I + D_x) = f(. . . , y(I - D_{yx} + D_x), . . .)$$

has been replaced by two instructions of the form

$$x(I + D_x) = f(. . . , u(I), . . .)$$

$$u(I + D_{yx} - D_x) = y(I)$$

and the equivalence of the two sets can be shown by eliminating $u(I)$
in the second set of instructions.

   As an example, consider the RIA in Eq. (6.1) for which the RDG
is shown in Fig. 6.7a. Suppose that for node x in the RDG, one wishes
to assign the weight $[1 \ 0]^T$ to all incoming arcs. The only incoming
arc at node x that does not have this weight is the self-loop with
weight $[0 \ 1]^T$. Hence create a new node, say w, and replace this
self-loop with an arc from node x to node w with weight $([0 \ 1]^T -$
$[1 \ 0]^T) = [- 1 \ 1]^T$ and an arc from node w to node x with weight

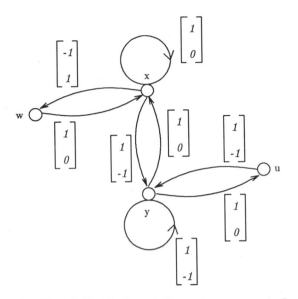

FIGURE 6.7b Reduced dependence graph for the example in Eq. (6.2) obtained by converting the RIA in Eq. (6.1) into input standard form.

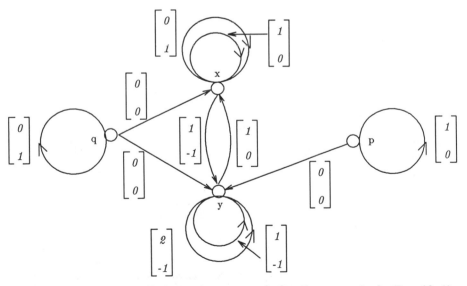

FIGURE 6.7c Reduced dependence graph for the example in Eq. (6.1) after converting the RIA into index-independent form.

$[1 \ 0]^T$. Similarly, all incoming arcs to node y can be assigned the weight $[1 \ -1]^T$, by introducing an additional variable, say u. The resulting RDG is shown in Fig. 6.7b and the RIA in input standard form is given by the set of instructions in Eq. (6.2).

2. *Eliminating variables that are not self-dependent.* This conversion process is precisely the opposite of the above and leads to an RDG with the minimum number of variable names. Here, suppose that x is a variable that is not self-dependent (i.e., the node x in the RDG does not have a self-loop). Let node x have incoming arcs from nodes $\{y_{in,1}, y_{in,2} \ldots\}$ with weights $\{D_{in,1}, D_{in,2}, \ldots\}$ and outgoing arcs to nodes $\{y_{out,1}, y_{out,2} \ldots\}$ with weights $\{D_{out,1}, D_{out,2} \ldots\}$. Now, delete node x in the RDG and for each pair $(y_{in,k}, y_{out,1})$, draw an arc from node $y_{in,k}$ to node $y_{out,1}$ with weight $(D_{in,k} + D_{out,1})$. The reasoning is that since

$$x(I) = f(y_{in,1}(I - D_{in,1}) \ y_{in,2}(I - D_{in,2}, \ \ldots ,$$

$$y_{in,k}(I - D_{in,k}), \ \ldots)$$

and $y_{out,1}$ has a dependence on x of the form

$$y_{out,1}(I) = g(\ldots , x(I - D_{out,1}), \ldots)$$

one can write

$$y_{out,1}(I) = g(\ldots , f(y_{in,1}(I - D_{in,1} - D_{out,1}), \ldots), \ldots)$$

and in the resulting equation, x does not appear at all since it was assumed that $\{y_{in,k}\}$ does not contain x. As an example, if one were to apply this conversion to the RDG given in Fig. 6.7b, the result would be the RDG in Fig. 6.7a.

3. *Conversion to index-independent RIA.* Frequently, it may be desirable to ensure that the iteration unit of the RIA is independent of the index point. Since the final objective is to implement the RIA on a processor array, this feature would save the designer from having to communicate to each processor what its actual position is in the array. Since this information differs from processor to processor in the array, one would otherwise have to code each processor separately. For instance, note that in the RIA given in Eq. (6.1), the variable x(i,k) is a function of k^2, and hence the iteration unit is indeed dependent on the index point. However, it turns out that every RIA can be converted into a form in which the iteration unit is independent of the index point, by using some so-called "propagating variables." This can easily be accomplished by introducing S propa-

gating variables, one for each of the coordinate axes of the index space. The kth propagating variable, say p_k, is such that

$$p_k(I) = i_k$$

and therefore can be defined in a regular iterative format as

$$p_k(I + [0 \ldots 0 \ 1 \ 0 \ldots 0]^T) = p_k(I) + 1$$

Here i_k is the kth component of the index vector I and the 1 in the vector above is in the kth position. The inputs to p_k must be appropriately initialized to correspond to the values on the respective surfaces of the index space.

Once these propagating variables are defined, one can replace the dependences on the index point with dependences on $\{p_k\}$ and thereby convert the RIA into index-independent form. Note that there may be better ways of achieving this purpose, although the procedure above can always be applied successfully. As an example, the RIA in Eq. (6.1) can be written in the form

for i, k = 1 to N do

$$x(i,k) = x(i - 1, k) + q(i,k)^2 x(i, k - 1)^{y(i-1,k)}$$

$$y(i,k) = \begin{cases} p(i,k)^{q(i,k)} y(i - 2, k + 1) + x(i - 1, k + 1) & \text{for even } p(i,k) \\ - p(i,k)^{q(i,k)} y(i - 1, k + 1) + x(i - 1, k + 1) & \text{for odd } p(i,k) \end{cases}$$

$$p(i + 1, k) = p(i,k) + 1$$

$$q(i, k + 1) = q(i,k) + 1$$

wherein the variables p and q correspond to the $\{p_i\}$ described above. The RDG for this modified RIA is shown in Fig. 6.13c. Notice that the two new nodes p and q introduced in this RDG are such that the only incoming arc at each node is a self-loop. All other arcs incident on these nodes are outgoing.

Two examples

The analysis procedures presented in this chapter to determine the computability of the RIA, the bounds on its storage, latency and I/O requirements, and a possible schedule for its implementation all rest on a careful examination of the RDG. To illustrate these procedures

and to highlight their important features, the following two examples
will be used.

Example [2]: Matrix Multiplication   Given two $(N \times N)$ matrices $\underline{A}$ and
$\underline{B}$, to determine the product matrix $\underline{C} = \underline{AB}$. An RIA for solving this
problem can be easily verified to be

> For $i,j,k = 1$ to N do
>
> $a(i, j + 1, k) = a(i,j,k)$
>
> $b(i + 1, j,k) = b(i,j,k)$
>
> $c(i,j, k + 1) = c(i,j,k) + a(i,j,k)b(i,j,k)$ (6.3)

Initialize this set of equations with

> $a(i,0,k) = \underline{a}_{ik}$
>
> $b(0,j,k) = \underline{b}_{kj}$
>
> $c(i,j,1) = 0$

Then the desired output is obtained as

> $c_{ij} = c(i,j, N + 1)$

The reduced dependence graph for this RIA is shown in Fig. 6.8.

Example [3]: Two-Dimensional Filtering Problem   Given an image
represented by the $(N \times N)$ matrix $\underline{U}$, find the filtered image $\underline{Y}$ such
that

$$y_{ij} = \sum_{k=1}^{n} a_k \underline{y}_{i-k,j-k} + \sum_{k=0}^{n} b_k \underline{u}_{i-k,j-k} \qquad (6.4)$$

There are many possible RIAs for solving this problem. The example
chosen below is a modification of an algorithm due to Deprettere and
Dewilde [34].

> For $i = 1$ to n do
>
> For $j,k = 1$ to N do
>
> $x(i, j + 1, k + 1) = f_{x,i}(x(i,j,k),y(i,j,k),w(i,j,k))$
>
> $y(i + 1, j,k) = f_{y,i}(x(i,j,k),y(i,j,k),w(i,j,k))$
>
> $w(i - 1, j,k) = f_{w,i}(x(i,j,k),w(i,j,k))$ (6.5)

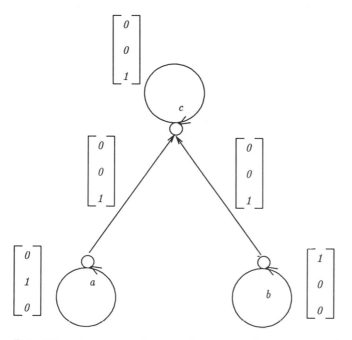

FIGURE 6.8 Reduced dependence graph of the regular iterative algorithm for matrix multiplication. This is a very simple example of an RDG since it has no directed loops except for self-loops.

where $f_{x,i}, f_{y,i}, f_{w,i}$ are some linear functions that are determined by a synthesis procedure.

This set of equations is initialized with $x(i,0,0)$ and $w(N,j,k)$ equal to zero for all values of i,j,k. The actual inputs are made available as $y(1,j,k) = u_{jk}$ and the output of the filter is obtained as $y(N + 1, j,k) = y_{jk}$. The reduced dependence graph for this RIA is shown in Fig. 6.9.

The RIA given in Example 2 for matrix multiplication falls into the class of uniform recurrence equations as defined by Quinton [23], whereas the one in Example 3 does not.

Some Assumptions

For ease of analysis, it will be assumed that the index space in question is the S-dimensional hypercube of length N and that the size of the index space is measured by the length N of the hypercube. However, the results can be easily extended to handle cases where the index space is not necessarily of this nature.

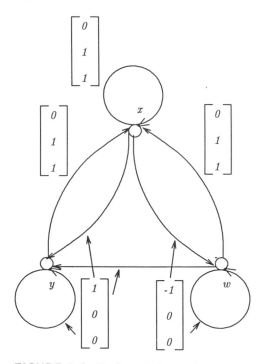

FIGURE 6.9 Reduced dependence graph of the regular iterative
algorithm for two-dimensional filtering.

Second, no attempt will be made to modify the algorithm itself.
Sometimes, it may be possible to transform the variables in the given
algorithm so as to obtain a better hardware implementation. (A classic
example of such a modification is the Schur versus Levinson algorithms
for inverting Toeplitz matrices.) However, this topic is not relevant
to the present discussion.

Third, it will be assumed that the computation of the value of a
variable at any given index point is performed exactly once. After
this computation is done, this value is stored by the relevant processor
at some appropriate local memory location until it is no longer needed
for the computation of other variable. At this point, this memory
location can be reused by the processor. Sometimes it may be possible
to recompute some values, and thus avoid having to store them,
thereby trading computation for memory. In such cases, the multiple
instances of the execution of a particular instruction must be represented explicitly by different variables, and it is questionable whether
the resulting algorithm remains an RIA or not.

Finally, it will be assumed that the number of variables V in the algorithm is independent of the size of the index space N. This assumption is necessary for determining asymptotic bounds on various properties of the algorithm. However, for the scheduling techniques developed in this section, this assumption need not be made.

2.3 Computability of Regular Iterative Algorithms

Any directed path in the dependence graph of an RIA corresponds to an equivalent path in its RDG. The displacement in the index space encountered by the path in the dependence graph is the sum of the arc weights in the corresponding path in the RDG. Therefore, if the dependence graph has a loop in it, there is a connected loop in the RDG such that the sum of the arc weights in the loop is zero.

An algorithm is *computable* if and only if its dependence graph has no loops, that is, if and only if its RDG has no zero-weight connected loops. To check whether an RDG has any zero-weight loops, define a matrix

$$\underline{G} = \left[\frac{\underline{C}}{\underline{D}} \right]$$

as follows. Let $\underline{C}$ be the *connection matrix* of the RDG with one row for each node and one column for each arc in the RDG. An element c_{ij} of $\underline{C}$ has value -1 if arc j is directed out of node x_i, value $+1$ if arc j is directed in to node x_i, and value zero otherwise. (If arc j is a self-loop, c_{ij} is 0 for all i.) Since each arc leaves exactly one node and arrives at exactly one node, each column of $\underline{C}$ (that corresponds to an arc that is not a self-loop) has exactly one $+1$ element in it and exactly one -1 element, with all the rest being zeros. $\underline{D}$ is the *displacement matrix* of the given RDG with one column for each arc and a total of S rows, where S is the dimension of the index space. Each column of $\underline{D}$ is the index displacement vector of the associated arc. Obviously, the ordering of the arcs must be the same in $\underline{C}$ and $\underline{D}$, so that each column of $\underline{G}$ has complete information about one arc of the RDG.

For example, the RIA for matrix multiplication given in Example 2 has three nodes (V = 3) in the reduced dependence graph, five arcs, and the weight on each arc is three-dimensional (S = 3). Consequently, the connection matrix is of dimension (3 × 5) and is given by

$$\underline{C} = \begin{bmatrix} 0 & 0 & 0 & -1 & 0 \\ 0 & 0 & 0 & 0 & -1 \\ 0 & 0 & 0 & 1 & 1 \end{bmatrix}$$

where the columns of the connection matrix correspond to arcs labeled 1, 2, 3, 4, and 5, respectively, in Fig. 6.8, and the rows to nodes a, b, and c. Similarly, the displacement matrix, $\underline{D}$ for this example, is

$$\underline{D} = \begin{bmatrix} 1 & 0 & 0 & 0 & 0 \\ 0 & 1 & 0 & 0 & 0 \\ 0 & 0 & 1 & 1 & 1 \end{bmatrix}$$

Similarly, for the RIA in Example 3, the connection matrix $\underline{C}$ can be determined to be

$$\underline{C} = \begin{bmatrix} 0 & 0 & 0 & 1 & -1 & -1 & 1 & 0 \\ 0 & 0 & 0 & -1 & 1 & 0 & 0 & 1 \\ 0 & 0 & 0 & 0 & 0 & -1 & -1 & -1 \end{bmatrix}$$

where the labels for the arcs in Fig. 6.9 are used to order the columns in the matrix. The rows correspond to ndoes x, y, and w, respectively. Similarly, the displacement matrix $\underline{D}$ is given by

$$\underline{D} = \begin{bmatrix} 0 & 1 & -1 & 0 & 1 & -1 & 0 & 1 \\ 1 & 0 & 0 & 1 & 0 & 0 & 1 & 0 \\ 1 & 0 & 0 & 1 & 0 & 0 & 1 & 0 \end{bmatrix}$$

The connection matrix of a directed graph has certain important properties. If Q is a right null vector of the connection matrix $\underline{C}$ (i.e., if $\underline{C}Q = 0$), with nonnegative integer entries, there exists a directed loop in the graph which is constructed by taking arc i, q_i times for all i. The loop so constructed may be a disjoint collection of several connected loops.

An arc may participate in a directed loop only if it exists within a tightly connected component of the graph, as defined next.

Definition: Tightly Connected Components The tightly connected components of an RDG are vertex disjoint subgraphs of the RDG that satisfy the following properties:

1. There exists a directed path from any node to any other node within the same component.
2. If such a component is an isolated node, it has a self-loop associated with it.
3. If there exists a directed path from a node in one component to a node in a different component, there cannot exist a return path.

These tightly connected components are related to what are called *maximally strongly connected components* in the graph theory literature [37]. In the set of maximally strongly connected components of a directed graph, every node in the graph belongs to exactly one such component. However, a node may not belong to any tightly connected component. This can happen, for example, if a node has no self-loop and is not strongly connected to any other component in the graph.

In order to detect the existence of a loop in the dependence graph, it is advantageous first to determine the tightly connected components of the RDG. Next, for each tightly connected component, form the connection matrix and the displacement matrix as described above. Now, if Q is a nonnegative right null vector for the connection matrix C of a tightly connected component and if Q is also a right null vector for the displacement matrix $\underline{D}$ (thus Q should be such that $\underline{G}Q = 0$), the loop described by Q could potentially be a zero-wegith conencted loop in the RDG and thus be representative of an actual loop in the dependence graph. Having determined Q, one should check to see if the arcs that have strictly positive entries in Q together form a single tightly connected component. If so, the RIA is not computable. If not, one should apply the same procedure to each of the individual tightly connected components.

A systematic procedure for determining computability can be obtained (see below) by examining a tree of subgraphs of the RDG such that each child is a subgraph of its parent and shares no nodes with any of its siblings. Since one works down the tree starting from its root and is dealing with smaller and smaller subgraphs of the RDG, the procedure is guaranteed to terminte.

In essence, this procedure is similar to the procedure described in Ref. 14 for determining the causality of an RIA defined over a semi-infinite index space. However, these authors never explicitly made use of the connection matrix.

Procedure 1 To determine the computability of a regular iterative algorithm given its RDG.

    1. *Initialization*. Construct a tree of subgraphs of the RDG as follows. Begin with the root, which represents the RDG itself, and mark the root *used*. Next, determine the tightly connected components of the RDG and create one child for the root for each such tightly connected component. Mark these nodes in the tree *unused*. Now the nodes representing the tightly connected components of the RDG are the only leaves in the tree.

    2. For the subgraph of the RDG corresponding to each unused leaf in the tree, form the matrix $\underline{G}$. Then construct the vector P as follows. For each i, check if a solution vector (nonnegative, but no necessarily integral) can be found for the set of constraints

$$\underline{G}Q = 0$$
$$q_j \geq 0 \qquad \text{for } j \neq i$$
$$q_i = 1$$

If a feasible solution exists for this set of constraints, set $p_i = 1$; otherwise, set $p_i = 0$. The zero elements of P correspond to arcs that can never participate in any zero-weight loops for this subgraph. Delete these arcs, and examine the remainder of the subgraph corresponding to this leaf.

3. If this remainder has a single tightly connected component, a zero-weight connected loop has been found and the algorithm is not computable. Stop. If the remainder has more than one tightly connected component, mark the leaf used, create new children for this leaf node to represent the constituent tightly connected components, mark these children (which are now leaves) unused, and go to step 2.

Proof of the Procedure: If the procedure terminates after detecting the existence of a zero-weight conencted loop in the RDG, then, clearly, the dependence graph has a cycle in it. On the other hand, if the procedure terminates normally by detecting the absence of unused leaves in the tree, it will be shown in Sec. 2.7 that there exists a finite schedule for the computation of the regular iterative algorithm represented by the dependence graph, thereby proving indirectly that the dependence graph is loop-free. It is not hard to obtain a direct proof if the reader so desires.

Step 2 of Procedure 1 can be accomplished in polynomial time since it is formulated as a series of at most E linear programming problems, each with at most E variables and (S + V) constraints, where E is the number of arcs in the RDG. Furthermore, the tree of subgraphs generated by Procedure 1 can have at most (S + 1) nodes, as shown below. Consequently, the procedure is of polynomial complexity.

If the index space is semi-infinite, the feasibility test in step 2 has to be modified to

$$\underline{C}q = 0$$

$$\underline{D}Q \leqslant 0$$

$$q_j \geqslant 0 \qquad \text{for } j \neq i$$

$$q_i = 1$$

since causality and not computability is the property of interest. Further , if the index space is of finite extent along some coordinate axes and semi-infinite along the rest, the feasibility test becomes

$$\underline{C}Q = 0$$

$$\underline{D}_1 Q = 0$$

$$\underline{D}_2 Q \leqslant 0$$

$$q_j \geqslant 0 \qquad \text{for } j \neq i$$

$$q_i = 1$$

where D_1 contains those rows of the displacement matrix $\underline{D}$ that correspond to the finite part of the index space, and $\underline{D}_2$, the semi-infinite part. However, unless otherwise specified, it will be assumed that the index space is finite and is enclosed by a S-dimensional hypercube of length N.

Given an RIA, one can always rewrite it in index-independent form using the procedure described earlier. This entails the introduction of additional propagating variables. Now the node corresponding to a propagating variable can only have outgoing arcs in the RDG, except for a single self-loop around it; hence every such node is in a tightly connected component all by itself. Consequently, if the original RIA is computable, the index-independent version is also computable, and vice versa. Furthermore, if the original RDG was not acyclic, the depth of the tree (the depth is the maximum number of nodes in a root to leaf path in the tree) of subgraphs generated by Procedure 1 remains the same for the index-independent RIA.

For the matrix multiplication example, there is no path from node c to either of a or b. Thus each of the three nodes in the RDG constitutes a tightly connected component by itself (along with the "self-loop" arc associated with it). The matrix $\underline{C}$ reduces to a scalar zero, and $\underline{D}$ to a vector that is the index displacement along the appropriate self-loop. Thus $\underline{G}$ is a nonzero vector for each of the three components, and no positive $\underline{Q}$ (now a scalar) can be found that can multiply it to give zero. Therefore, the algorithm is trivially seen to be computable. The corresponding tree of subgraphs constructed by the procedure consists of a root with three children as shown in Fig. 6.10. The depth of this tree is clearly 2.

For the filter example, there does exist a path from every node in the RDG to every other node; therefore, the entire RDG constitutes a single tightly connected component. Thus the leaf node in the initial tree corresponds to the RDG itself. To see if there exists a positive null vector for $\underline{G}$, one can simplify the problem a little by noting that every column of $\underline{D}$ has to be one of

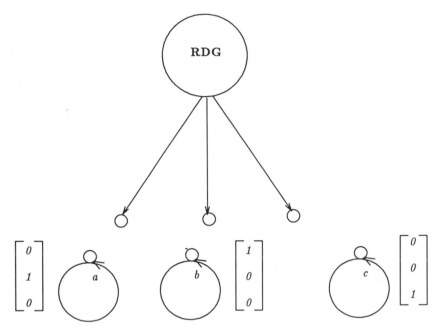

FIGURE 6.10 Tree of subgraphs of the RDG created by Procedure 1 for the matrix multiplication example.

$$d_1 = \begin{bmatrix} 1 \\ 0 \\ 0 \end{bmatrix} \quad d_2 = \begin{bmatrix} -1 \\ 0 \\ 0 \end{bmatrix} \quad d_3 = \begin{bmatrix} 0 \\ 1 \\ 1 \end{bmatrix}$$

A positive combination of d_1, d_2, and d_3 that leads to zero can only involve d_1 and d_2. Therefore, one can safely delete all arcs with weight d_3 from the RDG. The resulting "pruned" graph is shown in Fig. 6.11. The only zero-index loop in this pruned graph consists of two disjoint loops: the self-loop around y together with the self-loop around w. Hence the RDG splits into two tightly connected subgraphs, each of which is a single node with its associated self-loop arc. The corresponding tree, shown in Fig. 6.12, has depth 3.

Some Simplifications

The procedure for determining computability simplifies considerably in certain cases. For example, if the RIA is a uniform recurrent equation as defined by Quinton, the tree of subgraphs of the RDG is always of depth 2. Then the procedure simplifies to checking if there is a positive combination of weights of the self-loops around the central node (if there is more than one self-loop) that adds up to zero.

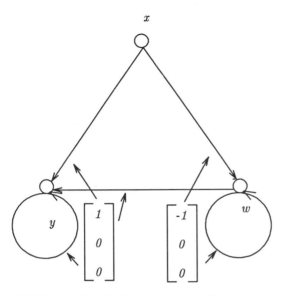

FIGURE 6.11 RDG for the two-dimensional filtering example, after deleting all arcs with weight $[0 \; 1 \; 1]^T$.

The *recursibility problem* considered by Mutluay and Fahmy [38] is the problem of determining whether an iterative equation of the form

$$y(I) = f_1(y(I - D_1), y(I - D_2), \ldots, y(I - D_k))$$

is computable or not. Here the RDG consists of a single node with multiple self-loops around it. The connection mtrix is zero, and therefore one needs to check if there is a nontrivial nonnegative vector Q such that DQ = 0. If such a vector exists, the corresponding sub-graph of the RDG is obviously tightly connected and the algorithm is not computable. If not, the algorithm is computable and the tree of subgraphs generated by Procedure 1 is necessarily of depth 2. Karp et al. have also studied this special case in detail in Sec. 4 of their paper [14]. Furthermore, the techniques outlined in Moldovan [33] can only be applicable to this case, since the concept of the reduced dependence graph is not utilized.

2.4 Minimum I/O Latency for a Regular Iterative Algorithm

It turns out that the procedure for determining computability indirectly provides an estimate of the minimum I/O latency. The minimum I/O

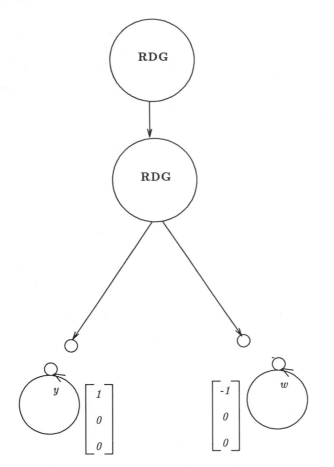

FIGURE 6.12 Tree of subgraphs of the RDG created by Procedure 1
for the two-dimensional filtering example.

latency is measured by the longest path in the dependence graph,
assuming that the computation of a variable, given the inputs to the
instruction, can be performed in time t where $t_{min} < t < t_{max}$, for
some positive constants t_{min}, t_{max}. It will now be shown that if the
depth of the tree of subgraphs of the RDG is d, the minimum I/O
latency of the algorithm is asymptotically equivalent to kN^{d-1}, where
k is some positive constant and N is the size of the index space. The
term "asymptotically equivalent to" refers to the fact that the minimum
I/O latency is equal to $[kN^{d-1} +$ (some lower-order terms)] for sufficiently large N. This equivalence is commonly indicated by the nota-
tion† $\Theta(N^{d-1})$.

Theorem 1 Let d be the depth of the tree of subgraphs of the RDG that is generated by Procedure 1. Then, as N, the size of the index space grows, the minimum I/O latency is asymptotically equivalent to kN^{d-1} for some positive constant k.

Proof: If the tree of subgraphs of the RDG generated by Procedure 1 has depth 1, the RDG is acyclic. The theorem is trivially proved.

If the depth of the tree is d, some value greater than 1, the RDG has cycles in it. It will be shown here that this implies the existence of a path, in the full dependence graph, of length $\Theta(N^{d-1})$ from some variable at some initial index point to the same variable at some different index point. Therefore, the minimum I/O latency of the regular iterative algorithm is $\Omega(N^{d-1})$. That the latency is also $O(N^{d-1})$ will be proved later by explicitly constructing a schedule for the computation of the regular iterative algorithm so that given sufficient number of processors, the algorithm can be completed in time $O(N^{d-1})$.

Delete the root of the tree to create a collection of trees (an outforest). For each node in this forest that is not the root of a tree, one can assign a displacement vector as follows. For the tightly connected subgraph associated with the parent of this node, let $\underline{C}$ be the connection matrix and $\underline{D}$ the matrix formed by arranging the displacement vectors of the edges in the subgraph. Form the matrix $\underline{G}$ and determine the vector P as in Procedure 1. Recall that p_i is 1 if the ith edge in the subgraph can be traversed a positive number of times in a zero-index loop in the subgraph and 0 otherwise. Now determine a non-negative integer vector Q such that the greatest common divisor of the elements of Q is 1, $\underline{G}Q = 0$, and $q_i > 0$ if and only if $p_i = 1$. Clearly, Q is a vector that is independent of N. Now the nonzero entries in P correspond to edges in some child of this node. For the edges corresponding to any particular child, form the sum of the displacement vectors of the edges in this child weighted by the corresponding entry in Q. This sum will be referred to as the displacement vector of the node represented by the child; the connected loop in the child given by the entries in Q will be referred to as the connected loop corresponding to the displacement vector of the node. Thus every node in the forest (except the roots) can be assigned a displacement vector and a connected loop associated with it.

For each of the roots, find any connected loop of constant length that uses all the edges in every one of its children at least once. It is possible to do this since each root is a tightly connected component. Use this loop as the connected loop for the root and the displacement vector of this loop as the displacement vector for the root.

---

†$f(N) = O(N^t)$ if there exists $k > 0$ such that $f(N) < kN^t$ for sufficiently large N. It is $\Omega(N^t)$ if for some $k > 0$, $f(N) > kN^t$ for sufficiently large N. Finally, $f(N) = \Theta(N^t)$ if it is both $O(N^t)$ and $\Omega(N^t)$.

From the construction above, an inductive proof can be derived
as follows. Let v be an interior node in the out-forest and let $\{v_1.$
$v_2, \ldots\}$ be its children. Let $\{D_V, D_{V,1}, D_{V,2} \ldots\}$ be the respective
displacement vectors and $\{\Theta(N^k), \Theta(N^{k1}), \Theta(N^{k2}), \ldots\}$ be the length
of some loops in the RDG (bounded by the extent of the index space)
involving only the edges corresponding to those in $\{v, v_1, v_2, \ldots\}$,
respectively, such that the displacement vectors for the loops are
$\{D_V, D_{V,1}, D_{V,2}, \ldots\}$. Now the claim is that given the loops for the
children of a node (i.e., the exponents $\{k_1, k_2, \ldots\}$), one can combine these loops using the unused arcs in v (the arcs that got deleted
and do not appear in any of its children) in such a manner that

$$k = \max_i (k_i + 1) \qquad\qquad (6.6)$$

Indeed, since $\Sigma D_{V,i} = 0$ and $D_{V,i} \neq 0$, one can achieve this as follows.
Traverse the connected loop corresponding to node v in some ordered
fashion, and for this traversal, determine the first node that is encountered in every child. When these nodes are reached for the first
time, branch off into the local loop [with length $\Theta(N^{ki})$] and go
around it cN times for some constant c that is chosen so that the
equivalent path in the full-dependence graph stays within the bounds
of the index space throughout the traversal. This loop now has the
length predicted above and has displacement vector D_V as desired.

By induction, a loop can be found using the edges in the subgraph of the RDG corresponding to a root r with displacement vector
D_r and of length $\Theta(N^{d_r-1})$, where d_r is the depth of the tree (in
the forest) of which r is the root. Since D_r is a constant vector, this
loop can be traversed $\Theta(N)$ times so that a path of length $\Theta(N^{d_r})$ can
be found in the dependence graph using only the edges corresponding
to the subgraph of the RDG represented by this root. Taking the
maximum over all such roots, it is easy to see that the minimum I/O
latency is at least $\Omega(N^{d-1})$, as stated in the theorem.

In the matrix multiplication example, the depth of the tree of
subgraphs of the RDG is 2, as mentioned earlier. Therefore, the
minimum I/O latency for the computation is $\Theta(N)$. For the filtering
problem, by the same token, the minimum I/O latency is $\Theta(N^2)$. However, note that for the same I/O properties of the filter, other regular
iterative algorithms can be derived for which the minimum I/O latency
is $\Theta(N)$. As far as the parallelism is concerned, these algorithms
would be clearly superior to the one under consideration.

2.5 On the I/O Requirements for a Regular Iterative Algorithm

A node in the full-dependence graph is an input node if its in-degree is zero (i.e., if it has no incoming arcs). Similarly, it is an output node if its out-degree is zero. Now, by definition of an RIA, each node in the dependence graph at a given index point has incoming arcs from and outgoing arcs to the same set of "directions" irrespective of the point's locaiton in index space. At or near the surface of the index space, the other extremities of some of these incoming arcs may be located at points that are outside the range of the index space. The nodes corresponding to these extremities are clearly input nodes, since the values of the variables represented by these nodes cannot be computed by the algorithm but have to be provided externally. Furthermore, every incoming arc at a given index point defines input variables at one or more surfaces of the index space. If the index space is an S-dimensional hypercube of length N, the number of inputs that have to be provided externally to a regular iterative algorithm is $\Omega(N^{S-1})$. Clearly, these considerations apply for output nodes as well.

Now, consider a node that is at some interior point in the hypercube that is sufficiently distant from the surface. If such an interior node is an input to (output for) the algorithm, then, by regularity, all interior nodes for the same indexed variable are input (output) nodes for the algorithm. In this case, the number of inputs to (outputs for) the algorithm is $\Theta(N^S)$; else it is $\Theta(N^{S-1})$. This is clearly reflected in the RDG of the algorithm since the node for the corresponding indexed variable has zero in-degree (out-degree).

In the matrix multiplication example, the algorithm requires $3N^2$ inputs and could generate $3N^2$ outputs. Since the input c values are all zero, and since the a and b values output are not of interest, there effectively are only $2N^2$ inputs and N^2 outputs.

In the filtering example, all the x and w inputs are zero; however, $y(0,j,k)$ has to be input for every value of j and k. At the output, the x, the w and the y values are potentially useful quantities depending on the application at hand, leading to a total of $3N^2$ outputs.

2.6 Lower Bound for the Memory-Time Product

Consider two nodes in the acyclic dependence graph, say (x_j, I_J) and (x_k, I_K) of a computable regular iterative algorithm such that the shortest path from one to the other is of constant length and the longest path (there has to be a longest path since the dependence graph is loop-free) is of length $\Theta(N^t)$. Then at least one variable that lies along the short path has to be stored for time that is $\Theta(N^t)$ because of the assumption that memory is never traded for computation. Due to the structure of the regular iterative algorithm, it follows that the index displacement between I_J and I_K is a constant independent of N, and therefore $\Theta(N^S)$ such pairs of nodes can be found. This implies that the total amount of storage required (memory × time) for the execution of the algorithm is $\Omega(N^{S+t})$.

The estiamte above has some important implications. Since there are $\Theta(N^S)$ values that are to be computed by the processors, clearly [(number of processors) × time] should be at least $\Omega(N^S)$. One would, of course, like to achieve this lower bound, in which case, in a distributed memory environment, each processor should have at least $O(N^t)$ memory units. To determine this storage bound, one needs to construct a tree of subgraphs of the RDG in exactly the same fashion as the procedure for determining computability, except that the initialization of the tree is different.

Procedure 2 To determine the (memory × time) lower bound of a regular iterative algorithm given its RDG.

1. *Initialization.* Determine the tightly connected components of the RDG. Construct a graph by collapsing each tightly connected components in the RDG into a single node, by retaining only those edges between different tightly connected components, and by deleting all parallel edges that may result in the process. In this graph, list out all the maximal directed paths. (Thus, given that the path ABC is included in the list, the paths AB and BC should not be. Further, if AC is a path by itself, that too should not be included. However, it should be noted that if one works with all the paths instead of just the maximal ones, the result would not be altered, even though the computation required for the procedure is increased unnecessarily.) Now construct the tree of subgraphs of the RDG as follows. Begin with the root, which represents the RDG itself, and mark the root used. Next, for each path found above, obtain the union of the tightly connected components of the RDG that participate in the path. Call this the subgraph for the path. Corresponding to each subgraph for the paths in the list, create a child for the root node. Mark all these children of the root unused.

2. Execute Procedure 1 using the tree above as the initial tree.

Theorem 2 Let d_m be the depth of the tree created by Procedure 2. Then, for any implementation of the RIA, the (memory × time) product is at least the larger of $\Omega(N^{S+d_m-2})$ and $\Omega(N^S)$.

Proof: The proof is similar to the proof of Theorem 1. Thus only the differences in the steps involved will be pointed out.

Since there are $\Theta(N^S)$ variables in the algorithm and the value of each variable should be latched for at least some constant time, it is clear the (memory × time) product is at least $\Omega(N^S)$. Now consider the case where the depth of the tree, d_m is strictly greater than 2. Just as in the proof of Theorem 1, delete the root of the tree and consider the collection of trees so obtained. For each node in this forest that is not the root of a tree, define its displacement vector and its connected loop as before.

For each root node, r, in the forest, determine a connected path of constant length (i.e., independent of N) that involves at least one node in every tightly connected component of its subgraph, together with the necessary arcs that interconnect them. By construction, such a connected path can be found. Furthermore, the displcement vector for this path, D_r, is a constant vector.

Let d_r be the depth of the tree of which r is the root. Then, using a similar argument as above, it can be shown that there exists a path in the dependence graph with total index displacement D_r and total length $\Theta(N^{d_r-1})$. Moreover, there also exists a path of constant length between the two terminal nodes (i.e., the connected path constructed above).

By taking the maximum of d_r over the trees in the forest, one can deduce that there exists a pair of nodes in the dependence graph that are located a constant distance away in the index space and that are connected by a path of constant length and by a path of length $\Theta(N^{d_m-2})$. Consequently, the (memory × time) product is at least $\Omega(N^{S+d_m-2})$.

The lower bound for the storage requirements of the algorithm is related to the minimum I/O latency since the procedures used for generating the relevant trees are almost identical. Indeed, d_m is related to d by the following theorem.

Theorem 3 The depth of the two trees generated by Procedure 1 and 2 are related by the inequality $d \leq d_m \leq (d + 1)$.

Proof: The fact that d_m is less than or equal to $(d + 1)$ is obvious. Otherwise, there exist two nodes in the dependence graph that are connected by a path of length greater than $\Theta(N^{d-1})$, which is impossible since the longest path in the dependence graph is $\Theta(N^{d-1})$.

Thus it just needs to be shown that $d_m \geq d$. Consider the initial tree constructed for Procedure 2. The children of the root in this tree are associated with subgraphs of the RDG in such a manner that for any tightly connected component of the RDG, there is at least one child that has this tightly connected component as a subgraph. Now, in the initial tree constructed for Procedure 1, the children of the root as associated with the tightly conencted components of the RDG.

Consider any two vertex disjoint subgraphs of the RDG with associated $\underline{G}$ matrices, $\underline{G}_1$, $\underline{G}_2$, respectively. If Q_1, Q_2 are two non-negative null vectors of $\underline{G}_1$ and $\underline{G}_2$, then

$$[\underline{G}_1 \quad \underline{G}_2]\begin{bmatrix} Q_1 \\ Q_2 \end{bmatrix} = 0 \tag{6.7}$$

Therefore, for the union of these subgraphs, the union of the zero-index loops for the constituent subgraphs is a valid zero-index loop. Using an induction on the subgraphs at each level of the trees, it is easy to infer that $d_m \geqslant d$.

An important implication of this theorem is that if the RIA is inherently sequential (i.e., with large d), it also requires a large amount of memory for its implementation.

For the two examples, the respective trees and the subgraphs assocaited with the nodes of the tree are shown in Figs. 6.13 and 6.14. Thus the (memory × time) lower bounds for these examples are $\Omega(N^3)$ and $\Omega(N^4)$ respectively.

The lower bound obtained using the procedure above may not be tight. That is, if the meory-time product is found to be $\Omega(N^{S+d-2})$ using the procedure, one may still be able to find a better lower bound, $\Omega(N^{S+d-1})$. An example illustrative of this fact is

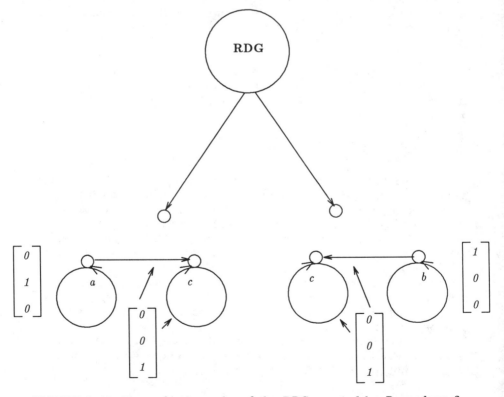

FIGURE 6.13 Tree of subgraphs of the RDG created by Procedure 2 for the matrix multiplication example.

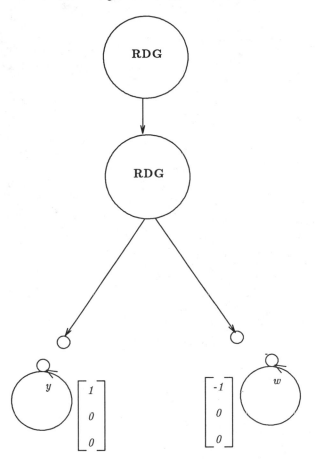

FIGURE 6.14 Tree of subgraphs of the RDG created by Procedure 2 for the two-dimensional filtering example.

For i = 1 to N do

$x(i) = x(i - 1)^2$

$y(i) = y(i + 1)^2$

$w(i) = x(i) + y(i)$

For this RIA, it is easy to check that the lower bound on I/O latency is $\Theta(N)$ and that for the memory-time produce is $\Omega(N)$, according to the procedures above. The dependence graph of this RIA is shown in Fig. 6.15. Using this graph, one can heuristically argue that the memory-time product should at least be $\Omega(N^2)$.

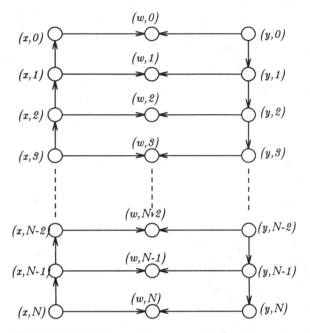

FIGURE 6.15 Dependence graph for the example that proves the inadequacy of the (memory × time) lower bound.

Let $x(0)$ be input at time 0. Then the time $L_x(i)$ at which $x(i)$ is computed must be greater than i, assuming that each squaring operation takes 1 time unit. That is,

$$L_x(i) \geqslant i$$

Similarly, the time $L_y(i)$ at which $y(i)$ is computed must satisfy

$$L_y(i) \leqslant \alpha(N) - i$$

where $\alpha(N)$ is some constant independent of i but possibly dependent on N. But for each i, both $x(i)$ and $y(i)$ are needed simultaneously in order to compute $w(i)$. Hence

$$\text{(memory} \times \text{time)} \geqslant \sum_{i=0}^{N} |L_x(i) - L_y(i)| \geqslant \left[\sum_{i=0}^{\alpha(N)/2} (2i - \alpha(N)) \right]$$

$$+ \left[\sum_{i=0(\alpha(N)/2+1)}^{N} (\alpha(N) - 2i) \right]$$

which is clearly $\Omega(N^2)$.

Once again, one must emphasize the fact that the above lower bound is invalid if each instruction in the algorithm can be executed many times. However, since in a single-assignment algorith, the output of an instruction must be assigned a distinct name, each instance of this reevaluation must correspond to a distinct node in the dependence graph. That is, by recomputing some values, one is really executing a "different" algorithm, and this algorithm need not be an RIA.

One might have expected the above lower bound since there were two self-loops in the RDG with equal but opposite weights. However, the result is difficult to prove in its generality, although the following conjecture arises naturally:

Conjecture If the reduced dependence graph of the RIA is not disjoint, a tight lower bound for the memory-time product can be obtained by applying Procedure 1 with an initial tree consisting of a single node that represents the RDG itself. If the depth of this tree is d_r, the memory-time lower bound is $\Theta(N^{S+d_r-1})$.

2.7 Linear Scheduling Functions for Hypercubic Index Spaces

For the present, assume that the processor resources are unlimited, so that a schedule S for the RIA is the assignment of nonnegative integer values to the variables in the algorithm so that the precedence constraints among the variables are satisfied. Furthermore, suppose that the processors are identical and require one time unit for computing any instruction in the algorithm. Therefore, if $y(I - D)$ is an input to the instruciton for evaluating $x(I)$, the integer $S_x(I)$ assigned to $x(I)$ should be such that

$$S_x(I) \geq S_y(I - D) + 1 \tag{6.8}$$

A simple interpretation of a schedule assignment is in terms of execution steps. The computation of the algorithm proceeds in the increasing order of the assigned values; at step t, all variables that are assigned the value t in the schedule are executed. By definition, the inputs required for the computation of a variable that is assigned to step t would have been computed at or before step $t - 1$. Thus this schedule provides a valid sequencing of the computations in the algorithm.

In a schedule, several variables may be assigned the same value. The computation of all these variables can be executed concurrently or sequentially or in any order whatsoever, as long as all these variables are computed before the next step begins. The steps themselves need

not be uniformly populated. Therefore, with a fixed number of
processors, steps may require different execution times for completion.
This distinction is important, since schedules are frequently con-
fused with the timing of an implementation.

Given the loose constaints that have to be met in the schedule
assignment above, there are usually a myriad of valid schedules for
a given algorithm. For RIAs, the preceding analysis provides some
rough measures for estimating the goodness of a schedule. Thus if the
minimum I/O latency of the algorithm is determined to be $\Omega(N^c)$, one
would hope to find a schedule for the RIA in which the computation of
the variables are assigned among a total of $\Theta(N^c)$ execution steps.

Karp and co-workers showed that if the tree of subgraphs of the
RDG obtained while determining causality is of depth $(r + 1)$, there
exists a positive constant β and a nonnegative row vector X^T such
that all variables at the index point I are computed at or before step
$\beta(X^T I)^r$. Naturally, one might wonder whether such a statement can
be made in the present situation also, for the case of a computable
RIA. However, computability does not imply causality; thus in a
computable RIA it is likely that there exists a noncausal dependence,
in which case at some index point I that is independent of N there
exists some variable $x_k(I)$ which cannot be computed at a step that
is not a function of N. Hence the statement of Karp et al. is no longer
valid. Next, even if a computable RIA is also causal, the tree of subgraphs obtained while determining causality can be of depth larger
than that for the tree of subgraphs obtained while determining computability. In this case, the statement above is valid only with respect
to the former tree and a tighter schedule other than the one implied
in their statement can be obtained, as shown in the remainder of this
section. The only situation in which this statement can still be used
is when the depth of the two trees coincide.

In this section the class of linear schedules is examined. A linear
schedule L for a regular iterative algorithm is a schedule in which the
nonnegative integer $L_x(I)$ assigned to the variable $x(I)$ satisfies the
functional relation

$$L_x(I) = \Lambda_x^T I + \gamma_x \tag{6.9}$$

where Λ^T_x is an integer row vector of length S and γ_x is some constant. In other words, for each variable a linear scheduling function
is defined as a linear combination of the indices (plus a constant).
Thus a linear schedule for the regular iterative algorithm can be
presented in a few concise expressions (one for every variable).
Moreover, these expressions can be obtained from a study of the
RDG, and hence with little computational effort. In addition, linear
schedules are useful for ensuring regular timing patterns in an implementation, and thus for systolic array designs (Sec. 3).

Every arc in the RDG imposes a constraint on the choice of the linear scheduling functions for the RIA. For example, if there is an arc from node y to node x in the RDG with weight D, the precedence constraint corresponding to this arc can be written as

$$\Lambda_x^T I + \gamma_x \geqslant \Lambda_y^T I - \Lambda_y^T D + \gamma_y + 1 \qquad (6.10)$$

Of course, this constraint needs to be satisfied only for all $I \in$ index space. If the index space itself is defined by a set of convex constraints, it is easy to see that the problem of determining linear scheduling functions for the RIA reduces to that of finding a feasible solution to a set of convex constraints. One can easily extend this to include some sort of optimization criterion, such as minimizing the range of the linear scheduling functions over the index space, and thereby formulate a constrained optimization problem.

The technique suggested above is applicable in a practical situation where the index space is defined in numerical terms. In the present analysis, the index space is assumed to be an S-dimensional hypercube that is parametrized by its size N, and the objective is to restrict the growth of the range of the linear scheduling functions with respect to the size of the index space. It is shown below that there exist feasible solutions to the set of convex constraints defined above, such that the growth of the range of the linear scheduling functions is restricted to the predicted asymptotic lower bound derived in Sec. 2.4. Moreover, these feasible solutions can be determined by solving a series of linear programming problems.

To reduce the set of nonlinear constraints obtained above to a set of linear constraints, assume that $\Lambda_x = \Lambda_y = \cdots = \Lambda$. Under this assumption, the term involving the index vector I in Eq. (6.10) cancels out, leaving

$$\Lambda^T D + \gamma_x - \gamma_y \geqslant 1 \qquad (6.11)$$

Writing such inequalities out for every arc in the RDG, and collecting them in an appropriate form, one obtains

$$Y^T \begin{bmatrix} \underline{C} \\ \underline{D} \end{bmatrix} = Y^T \underline{G} \geqslant [1 \ 1 \ 1 \cdots] \qquad (6.12)$$

where $\underline{C}$, $\underline{D}$ are the connection matrix and displacement matrix, respectively, and Y^T is the vector of unknowns,

$$Y^T = [\gamma_x \ \gamma_y \ \cdots \ \Lambda^T]$$

$$(6.13)$$

If a feasible vector exists for the set of constraints given in Eq. (6.13), it can be chosen to be independent of the size of the index space. This implies that the RIA can be executed in $\Theta(N)$ steps, which is possible only if the depth of the tree of subgraphs of the RDG generated by Procedure 1, d, is at most 2. Furthermore, the existence of such a feasible vector also implies that a variable needs to be stored only for a constant number of steps. This is because if $y(I - D)$ is an input to the instruction for evaluating $x(I)$, then $[L_x(I) - L_y(I - D)]$ will be equal to $[\Lambda^T D + \gamma_x - \gamma_y]$, which is a constant independent of N. Therefore, the depth of the tree of subgraphs generated by Procedure 2, d_m, should also be at most 2.

The subclass of RIAs for which Eq. (6.13) has a feasible solution is extremely important. It will be shown in the next section that this subclass is a complete characterization of the class of algorithms that can be mapped onto systolic architectures. To identify this subclass of RIAs, the following variation of the duality theorem in linear programming (this theorem is based on the Minkowski-Farkas lemma; see, e.g., [39]) is necessary.

Theorem 4 If the primal linear programming problem

$$\max \ H^T Q$$

subject to $\underline{G}Q = 0$ and $q_i \geqslant 0$ (6.14)

has an optimal feasible solution with a finite objective function, the dual set of constraints

$$Y^T \underline{G} \geqslant H^T$$

$$(6.15)$$

has a finite feasible solution. Otherwise, this set of inequalities is inconsistent. Furthermore, the vector of Lagrange multipliers obtained in the optimal solution to the primal problem is a feasible vector for the dual set of constraints.

To apply Theorem 4 to the set of constraints in Eq. (6.12), (6.13), one should use $H^T = [1 \ \ 1 \ \ 1 \ \cdot \cdot \ \cdot]$. Here the objective function for the primal problem is finite only when $Q = 0$ is the unique feasible vector for the primal problem. Otherwise, one could increase the objective function arbitrarily. More formally:

Theorem 5 Let the RDG of a regular iterative algorithm be such that the only feasible solution to the set of constraints

$$\underline{G}Q = 0 \qquad q_i \geqslant 0$$

is Q = 0. Then linear scheduling functions can be found for the indexed variables in the algorithm such that

1. They are the same for all variables up to an additive constant.
2. The multiplicative coefficients of the linear scheduling functions, $\{\lambda_i\}$, are all constants independent of N.

The class of algorithms for which this theorem applies includes the *uniform recurrent equations* in Quinton [23] as a special case. The linear scheduling functions for this special case reduce to what is referred to in Quinton's paper as *affine timing functions*. However, one should emphasize at this stage that linear scheduling functions merely determine at what "step" a particular variable should be computed. This does not imply that step t should be executed at time t, and therefore the scheduling function does not posses the usual connotations associated with the term "timing function."

Now suppose that there are nontrivial nonnegative solutions to the equation $\underline{G}Q = 0$. Clearly, one is now forced to relax the assumption that the scheduling functions for all variables have the same coefficients. It would then be advantageous to do so in such a manner that the basic duality theorem can still be applied. That is, suppose that for the equation $\underline{G}Q = 0$, there exists a nonnegative solution vector Q such that q_i is strictly greater than zero for some i. Now suppose that the linear scheduling functins are modified in such a manner that for every such value of i, in the vector H^T, the ith element h_i is less than or equal to zero and in all other respects, the set of scheduling constraints is the same. Then the optimal solution for the primal problem is still finite (in fact, it is zero) and the duality theorem can now be applied. To develop this more rigorously, the concept of linear scheduling functions for a subgraph of the RDG is introduced.

Definition: Linear Scheduling Functions for a Subgraph of the RDG
Let R be a subgraph of the RDG with n nodes. Then a set, $\{L_{Rj}(I)\}$, of n linear functions, one per node in the subgraph, constitues a set of linear scheduling functions for the subgraph if and only if these functions satisfy the precedence constraints imposed by the arcs in the subgraph.

The following theorems are based on this concept, and they complete the tools necessary for designing a systematic procedure for obtaining linear scheduling functions.

Theorem 6 Let $\{\bar{L}_{ij}(I)\}$ be a set of linear scheduling functions for the ith tightly connected component of the RDG. Let $\bar{L}_{ij}(I)$ assume nonnegative integer values that are at most $O(N^k)$ within the index space. Then a set of linear scheduling functions for the RDG can be found such that for the node v:

1. $L_v(I) = (\bar{L}_{ij}(I) + \alpha_i)$ if the node v is also the jth node in the ith tightly connected component of the RDG for some i, j; or
2. $L_v(I) = \beta_v$, otherwise.

Furthermore, the constants α_i, or β_v, as the case may be, can be chosen to be nonnegative, integral, and at most $O(N^k)$.

Proof: In the RDG, collapse every tightly connected component into a single node, retaining only the arcs that connect nodes in different tightly connected components. The resulting graph (known as a *condensation* [37]) is, be construction, acyclic. Therefore, a schedule can be found for this acyclic graph using standard PERT/CPM techniques. While determining this schedule, assume that each node that represents a tightly connected component takes a certain amount of time for execution that is precisely the range of the scheduling functions for this component. Since there are a constant number of nodes in this acyclic graph, and since the time required for completing the tasks in each tightly connected component is at most $O(N^k)$, the total execution time can be made to be $O(N^k)$.

The numbers obtained above are then used as the constants α_i or β_v respectively, as the case may be . This clearly results in a valid schedule for the RDG.

The importance of Theorem 6 is that it allows one to concentrate on the tightly connected components of the RDG independently (note the similarity with the initialization step in Procedure 1). Therefore, if linear scheduling functions are determined for each tightly connected component of the RDG assuming that all arcs that exist between nodes in different components are deleted, Theorem 6 can be applied to determine the set of linear scheduling functions for the RDG itself. To determine linear scheduling functions for the tightly connected components, the tree of subgraphs of the RDG generated Procedure 1 can be used.

Theorem 7 Consider the tree of subgraphs of the RDG generated by Procedure 1. Let a be an interior node in this tree and let $\{a_1, a_2, \ldots\}$ be its children. Let $\{\bar{L}_{ij}(I)\}$ be a set of linear shceduling functions for the subgraph of the RDG corresponding to a_i and for all i, j, let $\bar{L}_{ij}(I)$ be nonnegative, integral, and at most $O(N^k)$ within the index space. Let $\{L_v(I)\}$ be the set of linear scheduling functions for the subgraph of the RDG corresponding to a. Then for some linear function $L(I)$ over the index space,

$$L_v(I) = L(I) + \overline{L}_{ij}(I) + \gamma_v$$

if v is also the jth variable in the subgraph corresponding to a_i for some i, j, or

$$L_v(I) = L(I) + \gamma_v$$

otherwise. Furthermore, the linear function $L(I)$ can be chosen that it is nonnegative integral and at most $O(N^{k+1})$ within the index space, while the constants $\{\gamma_v\}$ can be chosen to be at most $O(N^k)$.

Proof: Given the scheduling functions $\overline{L}_{ij}(I)$ for all i, j, assume that there exists a linear function $L(I)$ and the constants γ_v as given in the statement of the theorem. The linear function $L(I)$ can be expressed as

$$L(I) = \Lambda^T I \tag{6.16}$$

Now, for any variable node v in the subgraph of a, choose $L_v(I)$ according to the statement of the theorem. Then the set of constraints that $L(I)$ and γ_v have to satisfy is given by

$$[\gamma_1, \gamma_2, \ldots, \Lambda^T]\underline{G} \geqslant H^T \tag{6.17}$$

where $\underline{G}$ is the $\underline{G}$ matrix for the subgraph of the RDG corresponding to a. Now let h_m be the mth entry of H^T. If the mth arc of the subgraph of the RDG corresponding to a is also found in the subgraph corresponding to the ith child of a (i.e., a_i), it can be shown that h_m is either 0 or negative. Indeed, let u and v be the variable nodes at the two extremities of the arc m. Then

$$L_u(I) = L(I) + \overline{L}_{iu}(I) + \gamma_u$$
$$L_v(I) = L(I) + \overline{L}_{iv}(I) + \gamma_v \tag{6.18}$$

and by the definition of $\overline{L}_{iu}(I)$, $\overline{L}_{iv}(I)$, for arc m, one should have

$$\overline{L}_{iv}(I + D_m) \geqslant \overline{L}_{iu}(I) + 1 \tag{6.19}$$

Moreover, the precedence constraint imposed by this arc in the subgraph of the RDG corresponding to a reduces to

$$L(I + D_m) + \bar{L}_{iv}(I + D_m) + \gamma_v \geqslant L(I) + \bar{L}_{iu}(I) + \gamma_u + 1 \qquad (6.20)$$

and using the preceding equation, it can be easily inferred that

$$h_m \leqslant 0 \qquad (6.21)$$

In the other situation, when arc m is not found in any of the children of a, the value of h_m is obtained by evaluating the worst-case situation in the inequality. Since $\bar{L}_{ij}(I)$ is at most $O(N^k)$, the value of h_m can at most be $O(N^k)$.

Now, consider the primal problem of Theorem 3 with the coefficients of the objective function given by H^T. In the set of constraints, $\underline{G}Q = 0$ with $q_i \geqslant 0$, suppose that there exists a feasible vector with $q_i > 0$ for some i. Then, by the argument above, h_i is less than or equal to zero, since this arc is found in some child of a. Therefore, the objective function for the primal problem has the optimal value of 0, thereby implying that there exists a finite feasible solution to the set of constraints in Eq. (6.17). Moreover, the coefficients of H and consequently λ_i, γ_v are at most $O(N^k)$, thereby proving the theorem.

Using the constructions given in the proofs of these theorems, a systematic procedure can be devised for obtaining linear scheduling functions.

Procedure 3 To obtain linear scheduling functions for a computable regular iterative algorithm given its RDG.

1. Derive the tree of subgraphs of the RDG via Procedure 1.
2. For each leaf node in the tree, obtain linear scheduling functions with constant coefficients for the subgraph of the RDG corresponding to this node. (Use Theorem 5.). For each interior node in the tree for which linear shceduling functions have not been assigned, check if scheduling functions have been assigned to all its children. If so, use the construction in the proof of Theorem 7 to derive linear scheduling functions for the subgraph corresponding to this node. Repeat until all interior nodes have been assigned linear scheduling functions.
3. Use the construction in the proof of Theorem 6 to derive linear scheduling functions for the subgraph of the RDG corresponding to the root (i.e., the RDG itself).

The determination of linear scheduling functions as above requires the solution of a series of integer programming problems. These integer programming problems have some special structure and consequently, some simplifications can be obtained [2].

If the depth of the tree of subgraphs of the RDG generated by Procedure 1 is d, then, by construction, the linear scheduling functions for all the variables in the algorithm can only assume integer values in the range 0 to $\Theta(N^{d-1})$. This is consistent with the asymptotic bound on the minimum I/O latency predicted by Theorem 1.

At each step in this procedure, one has to exercise a choice for a feasible solution to a set of linear constraints of the form

$$Y^T G \geq H^T$$

Clearly, there may exist several different possibilities for Y^T and it may be difficult to select one among them. Instead, one could also impose an objective criterion such as

$$\text{minimize } Y^T G \begin{bmatrix} 1 \\ 1 \\ \vdots \\ 1 \end{bmatrix}$$

subject to the constraints above. This serves to simplify the choice and in addition provides a solution that utilizes the minimum total "wait" between steps. There may be several other possibilities for the objective function, depending on the requirements of the designer.

The nonuniqueness of the linear scheduling functions has important consequences. For the matrix multiplication example, if one obtains these fucntions using Procedure 3 and tries to meet the constraints as tightly as possible at every stage, one gets

$$L_a(i,j,k) = j$$

$$L_b(i,j,k) = i$$

$$L_c(i,j,k) = k + N$$

However, for this example, Theorem 5 is directly applicable and the result is

$$L_a = L_b = L_c = (i + j + k)$$

and this is also as tight as possible for the assumptions made. However, the values of the scheduling function now range from 0 to 3N, as opposed to the range 0 to 2N obtained for the schedule given above, although the reduction in number of steps is accompanied by an increase in the required storage.

For the filter example, for the leaf nodes in the tree of subgraphs of the RDG, the linear scheduling functions can be obtained as

$$\bar{L}_y = i$$

$$\bar{L}_w = N - i$$

Moving up the tree to the interior node, the linear shceduling functions can be calculated to be

$$L_x = (2N)j + k$$

$$L_y = i + (2N)j + k + N$$

$$L_w = -i + (2N)j + k + N$$

Notice again that the range of these functions is from 0 to $(2N^2 + 3N)$, which meets the predicted lower bound for the algorithm. Again, a whole slew of valid scheduling functions can be generated for the same range. One can, as remarked earlier, impose some objective criterion at each step of the procedure in order to limit the choice of the scheduling functions.

The scheduling functions determined by applying Procedure 3 are optimal with respect to the lower bound on the I/O latency. However, the lower bound on the memory time may not be satisfied. This can happen even if the memory time lower bound is itself tight, as evidenced by the matrix multiplication example given above.

2.8 Linear Scheduling Functions for Nonhypercubic Index Spaces

The assumption made throughout this section that the index space is an S-dimensional hypercube of length N can be relaxed while determining linear scheduling functions. The necessary modification of Procedure 3 lies in the technique for determining the H vector at each stage of the procedure.

Consider the tree of subgraphs of the RDG generated by Procedure 1. Let a be an interior node in this tree and let $\{a_1, a_2, \ldots\}$ be its children. Let $\{\bar{L}_{ij}(I)\}$ be a set of linear scheduling functions for the subgraph of the RDG corresponding to a_i. Let $\{L_v(I)\}$ be the set of linear scheduling functions for the subgraph of the RDG corresponding to a. Then, for some linear function $L(I)$ over the index space,

$$L_v(I) = L(I) + \bar{L}_{ij}(I) + \gamma_v$$

if v is also the jth variable in the subgraph corresponding to a_i for some i, j, or

$$L_v(I) = L(I) + \gamma_v$$

otherwise. Now, let the ith edge in the subgraph corresponding to a be from node x to node y with weight D. Both these nodes may not be in any of the subgraphs corresponding to the children of a, in which case $h_i = 1$. Otherwise, one can write

$$L_x(I) = \bar{L}_{jk}(I) + L(I) + \gamma_x$$

$$L_y(I) = \bar{L}_{lm}(I) + L(I) + \gamma_y \qquad (6.22)$$

where either $\bar{L}_{jk}(I)$ or $\bar{L}_{lm}(I)$ could be zero. The precedence constraint corresponding to the ith edge can now be written as

$$\gamma_y - \gamma_x + \Lambda^T D \geqslant \bar{\Lambda}_j^T I - \bar{\Lambda}_l^T (I + D) + \bar{\gamma}_x - \bar{\gamma}_y + 1 \qquad (6.23)$$

where I is an integer vector that ranges over the index space. Therefore, h_i can be determined by solving the following constrained maximization problem:

$$\max h_i = \bar{\Lambda}_j^T I - \bar{\Lambda}_l^T (I + D) + \bar{\gamma}_x - \bar{\gamma}_y + 1$$

$$\text{subject to } I \; \varepsilon \; \text{index space} \qquad (6.24)$$

If the set of constraints defining the index space is linear, the above reduces to a linear (integer) programming problem.

Applying this at each stage of Procedure 3, one can obtain linear scheduling functions for nonhypercubic index spaces. In the filter example, the index space was given to be

$$1 \leqslant i \leqslant n$$

$$1 \leqslant j, k \leqslant N$$

which is not hypercubic. Using Procedure 3 and the modification suggested above, a set of linear scheduling functions can be determined to be

$$L_x = (2n)j + k$$

$$L_y = i + (2n)j + k + n$$

$$L_w = -i + (2n)j + k + n$$

which range from 0 to $(2nN + 2n + N)$. In a practical filtering situation, the order of the filter n is much less than the size of the image N, and therefore, this is an improvement on the set of scheduling functions deduced earlier.

3 IMPLEMENTATIONS ON SYSTOLIC ARRAYS

Since the dependence graph of an RIA can be embedded in a multi-dimensional lattice of points (i.e., the index space), it is only natural to seek implementations of regular iterative algorithms in a similar, possibly lower-dimensional lattice of "processors." This task of mapping an RIA onto a processor array of this sort can be accomplished systematically by using simple projection techniques. The index space of an RIA is enclosed by a region in the S-dimensional Euclidean space. A projection of this region on to a lower-dimensional subspace of the Euclidean space maps the index space onto a lower-dimensional lattice of points, each point representing a processor in the array. The variables whose indices are projected on to a given processor are evaluated by this processor in an appropriate order.

The term *processor* is used in this chapter to denote a black box that is capable of evaluating all the expressions in the iteration unit of the RIA. Within this black box, there may be several computational modules that are shared for the actual evaluations. For instance, if the iteration unit of the RIA is given by the set of equations

$$a(i, j + 1, k) = a(i,j,k)$$

$$b(i + 1, j,k) = b(i,j,k)$$

$$c(i,j, k + 1) = c(i,j,k) + (a(i,j,k)b(i,j,k))$$

then the processor may consist of an adder, a multiplier (i.e., two different computational modules). On the other hand, one may also wish to implement this iteration unit on a general-purose micro-processor, in which case the processor consists of a single computational module.

For the most part, in this chapter, the internal details of the processor turn out to be not as important. However, the terminal information regarding the computational delays through the processor is of interest. This terminal information, referred to by Jagadish et al.

[39,40] as the *intrinsic schedule* of the processor, provides the relative arrival times and departure times for the various inputs and outputs in the iteration unit. These arrival and departure times will then be assumed to be such that if an output $x_j(I + D)$ is computed using $x_k(I)$ in the iteration unit, it will be meant to imply that the departure time for x_j in the intrinsic schedule is at least one clock cycle later than the arrival time for x_k. It must be emphasized here that time is always assumed to be measured in terms of clock cycles and only integer times are allowed. Thus a computational delay through the processor represents "register delays" or "clocked delays" and not "propagation delays" that one usually associates with combinational chains.

The intrinsic schedule of the processor also specifies the intrinsic iteration interval of the processor (this concept is also discussed by Jagadish et al. [39,40]). The intrinsic iteration interval δ_{int} is representative of the amount of pipelining that can be achieved within a processor. That is, if the arrival times for the current iteration unit are given by the set $\{a_j\}$ and the departure times by $\{b_j\}$, the processor can potentially operate on the next set of data arriving at $\{a_j + \delta_{int}\}$ and provide the next set of outputs at $\{b_j + \delta_{int}\}$. The intrinsic iteration interval of the processor can be substantially less than the computational delays through the processor.

Given the RIA and an intrinsic schedule for the processor, the main objective in this section is to show how one would obtain a regular interconnection of such processors together with shift registers of appropriate lengths so as to execute the RIA correctly. The shift registers are necessary since it is assumed that the processor does not know how to determine the indices of the data values that appear at its input. (Hence the RIA must also be converted into index-independent form.) Thus one does not expect the processor to store values until they are needed. On the other hand, it will be assumed that the appearance of new data at any input automatically activates the processor into doing "whatever it can" with the data. In this sense, one can model a processor as being composed of V computational modules, one for each of the instructions in the iteration unit; then each computational module is activated independently by a clock signal. This assumption is necessary since, in general, not every instruction within an iteration unit can be executed simultaneously.

Among the dedicated VLSI architectures that can be designed for implementing RIAs, systolic arrays form an important special class. The algorithm executed by a systolic array can be shown to be an RIA with some characteristic properties. Conversely, every RIA with these properties can be shown to be executable on a systolic array.

Systolic arrays not only correspond to a simple subclass of RIAs, but they also correspond to a one-dimensional projection of the index

space in the manner described above. Hence the techniques necessary
for the design of systolic arrays turn out to be extremely simple as
shown in the rest of this section. The design of processor arrays
that implement RIAs in general using multidimensional projections of
the index space, is relatively more complicated. Nevertheless, the
resulting processor arrays can be very efficient and they use considerably fewer computational resources at the expense of execution
time. Furthermore, these arrays can also be described as a regular
interconnection of processors and shift registers, just as in the case
of systolic arrays. For the purposes of this chapter, we restrict
our attention to the design of systolic arrays; the more general cases
have been dealt with in detail in Ref. 2.

3.1 Model for Systolic Arrays

A notable inconsistency in the literature on systolic arrays is the
incompatibility between the qualitative definitions of systolic arrays
and the quantitative models that have been proposed to describe them.
Almost every qualitative definition of systolic arrays involves such
terms as regularity, temporal locality, efficiency, and spatial locality
[41—44]. On the other hand, the graph-theoretic quantitative models
described in Refs. 34, 45, 46 are much too general. Indeed, these
modesl are so all-encompassing that it has been shown that every
synchronous circuit can be made to conform to these models by using
a sequence of simple equivalence transformations [34,48]. Thus if
one were to consider these models to be accurate representations of
systolic arrays, every algorithm is a systolic algorithm or is equivalent
to one. However, the resulting circuits need not be locally interconnected, regular, or worse, perform as inefficiently as a sequential
implementation. Hence, these circuits would violate the qualifications
attributed to systolic arrays in [41,42,47].

   Chen and Mead [48] describe a semantic model for systolic arrays
that is based on the familiar finite-state representation of a synchronous
circuit:

$$X(k + 1) = f\{X(k), U(k)\}$$
$$Y(k) = g\{X(k), U(k)\} \tag{6.25}$$

Once again, in this description, the iterated nature of "systolic arrays"
is not explicitly brought forth.

   Although it appears to be a matter of choice whether to subscribe
to one view or the other, for the purpose of this section, it will
henceforth be assumed that every systolic array does indeed possess
all the regular topological and computational features usually attributed
to them in the qualitative definitions. There are several compelling
reasons for advocating this view:

1. A vast majority of the systolic array designs that have appeared in the literature possess these regular features.
2. There are definite advantages to an array architecture with regular features as opposed to one with haphazard topological and computational characteristics.
3. No matter what the terminology used to describe these arrays, it is of interest to assess the capabilities of such regular, locally interconnected, globally synchronous, fully pipelined processor arrays.

There is one major drawback, however, in basing studies on the existing equalitative definitions of systolic arrays. This arises due to fact that even though these definitions are intuitively appealing, their interpretations are highly subjective. For instance, what does "local" mean in the attributes "temporal locality" and "spatial locality?"

To resolve such issues, it will be presumed that systolic arrays are designed to solve a family of problems rather than a single instance of a problem. A family of problems consists of an infinite number of problem instances, each of which is uniquely represented by a set of integers known as the *size parameters*, which have some natural ordering (see, e.g., Ullman [46, pp. 42]). For instance, consider the problem of multiplying an (m × n) matrix by an (n × p) matrix. One can consider (m,n,p) to be the size paramters of a family of matrix multiplication problems; the systolic array introduced by S. Y. Kung [42] to solve this problem consists of a rectangular array of processing elements arranged in a grid of dimension (m × p), and that solves the problem in time (max(m + n, p + n)). Thus the size parameters of a problem can also be considered to be the size parameters of the systolic array itself. These parameters play an important role in the following characterization since the term "local" is interpreted relative to them.

As a first step toward developing the formal model, a more rigorous interpretation of the qualitative features of systolic arrays will be developed here. However, a few preliminary definitions are necessary:

Processor Space. The processor space is a multidimensional lattice of points whose extent is dependent on the size of the problem. For the sake of simplicity, we shall assume that the lattice points all have integer coordinates in their multidimensional vector representation.

Iteration Unit. The iteration unit is a finite set of instructions, with cardinality independent of the size of the problem that defines an I/O map between V input variables, labeled $\{x_{j,in}\}$, and V output variables, labeled $\{x_{j,out}\}$, where each instruction may have finitely many conditional branches. Here V is necessarily independent of the size of the problem.

With these definitions in place, a formal characterization of systolic arrays can be obtained as follows.

Characterization of a systolic array

A systolic array is a network of processors in which the processors can be placed at the grid points of the processor space so that

1. If there is a directed link from (to) the processor at location I to (from) the processor at location (I + D), then
   a. D is independent of the size of the problem.
   b. There is such a link for every I in the processor space [if (I + D) falls outside the processor space, this is an output (input) link to (from) the external environment].
2. Every processor repeatedly executes the iteration unit with period Δ clock cycles, where Δ is the fundamental time period of the systolic array that is independent of the size of the array. If a processor receives the value of $x_{j,in}$ at time t on an incoming link, it places the value of $x_{j,out}$ at time $(t + \Delta)$ on the corresponding outgoing link to be used as $x_{j,in}$ by the processor at the other extremity of this link. Further, the subsequent value of $x_{j,in}$ arrives (if necessary) on the same incoming link at time $(t + \Delta)$, thereby ensuring the periodicty of the execution.

The first property ensures "spatial locality" and partially ensures the "regularity" of the array. The second property completes the requirement of "regularity" and also ensures "temporal locality." However, it remains to be proven that the resulting array is pipelined and efficient.

As a justification for this qualitative characterization, one can easily verify that most of the systolic arrays found in the literature can be described in this format, mainly because each instruction in the iteration unit is allowed to have a finite number of conditional branches. There are some exceptions, notably the arrays for solving the transitive closure problem derived by Guibas et al. [49] and by S. Y. Kung and Lo [87]. However, it can be shown that these arrays are really simulating a systolic array of larger dimensions, and that this larger array does conform to the foregoing characterization [51, 52]. To illustrate the terms used in this characterization, consider, for example, the systolic array proposed by Kung for solving the matrix multiplication problem, shown in Fig. 6.16. This consists of a square array of (n × n) processors wherein each processor receives the $\underline{B}$ and $\underline{A}$ matrices from above and from the left, respectively. The elements of the product matrix $\underline{C} = \underline{AB}$ are accumulated within each processor. The processor space of this array is precisely the domain

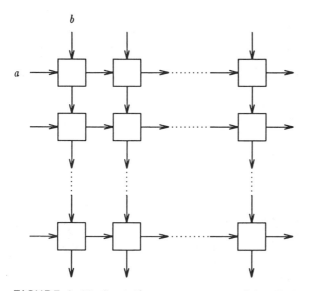

FIGURE 6.16 Systolic array proposed by S. Y. Kung for matrix multiplication.

in which the processors reside. It is represented by the lattice points that are enclosed within a square region of side n, as shown in Fig. 6.17.

The iteration unit of the array is the set of instructions that are repeatedly executed by each processor in the array. For the array in Fig. 6.16, it is given by

$$a_{out} = a_{in}$$

$$b_{out} = b_{in}$$

$$c_{out} = c_{in} + a_{in}b_{in} \qquad\qquad (6.26)$$

3.2 Systolic Algorithms

From the characterization above, it is easy to deduce that a systolic array operates in a globally synchronized fashion. Suppose that P_1 and P_2 are adjacent processors in the array and that the data $x_{j,out}$ from processor P_1 is received as $x_{j,in}$ by processor P_2. Then the time at which $x_{j,in}$ is received by processor P_2 is the same as the time at which $x_{j,out}$ is output by processor P_1, which is also the same as the time at which a new value of $x_{j,in}$ is being received by processor P_1.

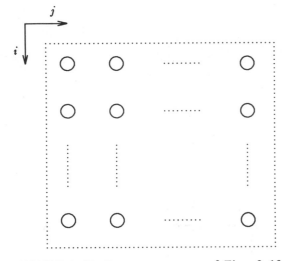

FIGURE 6.17 Processor space of Fig. 6.16.

Tracing this pattern over the array, it can be seen that at this particular instant of time, all processors are receiving values corresponding to $x_{j,in}$, assuming that the array is not disjoint.

The globally synchronous operation of a systolic array considerably simplifies the task of identifying the algorithm executed by the array. This can be done by assigning a unique index vector to any particular iteration unit executed by any specific processor in the array. This index vector contains two components: one component corresponding to the location of the processor in the processor space (the *processor index*), and another corresponding to the time at which the iteration unit is executed (the *time index*). Suppose that at time 0, the first input, say $x_{1,in}$, is input to the array at some processor. At this time, many other processors may also be receiving their first $x_{1,in}$ input. The iteration units to which all these $x_{1,in}$s belong is assigned the time index 0. Similarly, the iteration units to which the $x_{1,in}$s are input at time $k\Delta$ are assigned the time index k.

Now, for every $j = 1 \ldots V$, let $x_{j,in}$ input to processor I for the iteration unit with time index k be assigned the index vector (I,k). Then the $x_{j,out}$ output by this processor corresponding to this iteration unit is really the $x_{j,in}$ input to the processor at location $(I + D_j)$ with time index (k + 1), where D_j is some constant vector independent of I (and independent of the size of the array). Hence one can write

$$x_j(I + D_j, k + 1) = f_j\{x_1(I,k), \ldots, x_j(I,k), \ldots\}$$
$$j = 1, \ldots, V$$

$$(6.27)$$

This set of equations, together with the specification of the extent of the processor space and the time index, defines the algorithm executed by the systolic array (i.e., a systolic algorithm). More formally:

Theorem 8 Every systolic algorithm is a regular iterative algorithm with the following property: The connection matrix $\underline{C}$ and the displacement matrix $\underline{D}$ of the reduced dependence graph of the RIA are such that there exists a finite feasible solution to the set of constraints

$$\underline{Y}^T \begin{bmatrix} \underline{C} \\ \underline{D} \end{bmatrix} \geqslant [1 \quad 1 \quad 1 \cdots]$$

Equivalently, there exists no disjoint collection of loops in the reduced dependence graph of a systolic algorithm, the sum of whose displacement vectors is zero.

Proof: That a systolic algorithm is an RIA is evident from the argument above. The rest of the theorem is proved by noting that the index-displacement vectors of the RIA are always of the form $\begin{bmatrix} \underline{D} \\ 1 \end{bmatrix}$. Hence $\underline{Y}^T = [0 \quad 0 \quad \cdots \quad 0 \quad 1]$ is a finite feasible solution to the set of constraints given in the equation. Since this set of constraints has a feasible solution, by the duality theorem (Theorem 4), the linear programming problem

$$\text{maximize } \{[1 \quad 1 \cdots 1] \underline{Q}\}$$

$$\text{subject to } \begin{bmatrix} \underline{C} \\ \underline{D} \end{bmatrix} \underline{Q} = 0$$

$$q_i \geqslant 0$$

has a feasible solution with a finite objective function. However, if there exists any nontrivial feasible solution to $\underline{Q}$ subject to these constraints, one can multiply this solution by an arbitrarily large number. Consequently, the objective function is unbounded. Hence there can be no feasible solution for $\underline{Q}$, except for the trivial one $\underline{Q} = 0$. Using the interpretation in Sec. 2.3, this is equivalent to stating that there is no collection of loops in the RDG such that the sum of their displacement vectors is zero.

As an illustration of this procedure for identifying the algorithm executed by a systolic array, consider the systolic array shown in Fig. 6.16. Let the processor space be indexed by the integers i, j as shown in Fig. 6.17. Let k be the time index. Then since the a values are transferred from left to right and the b values from top to bottom, one must have

$$a(i, j + 1, k + 1) = a(i,j,k)$$

$$b(i + 1, j, k + 1) = b(i,j,k)$$

$$c(i, j, k + 1) = c(i,j,k) + a(i,j,k)b(i,j,k) \qquad (6.28)$$

Next, to complete the algorithm, one must specify the range of the indices (i,j,k) and the initialization for the algorithm. By inspection,

$$i, j = 0 \text{ to } (n - 1)$$

$$k = 0 \text{ to } (3n - 1) \qquad (6.29)$$

and

$$a(i,0,k) = a_{i,k-i}$$

$$b(0,j,k) = b_{k-j,i}$$

$$c(i,j,0) = 0 \qquad (6.30)$$

which completes the algorithm. Note that the algorithm defined in Eqs. (6.28–6.30) is almost identical to the RIA in Example [1], but for some discrepancies in the index-displacement vectors.

Thus a systolic algorithm is an RIA for which there exists a solution to the set of constraints

$$Y^T \begin{bmatrix} C \\ \underline{D} \end{bmatrix} \geqslant \begin{bmatrix} 1 & 1 & 1 & \cdots \end{bmatrix} \qquad (6.31)$$

In the remainder of this section, the converse statement will be proved: Given such an RIA, it is possible to identify the processor space and time index of a systolic-array implementation of the algorithm. In this way, the characterization of systolic arrays in terms of RIAs would be complete.

Identifying the processor space of a systolic array

In Refs. 23 and 33, for example, the processor space of a systolic array is identified by applying a multiplicative transformation to the index space of the RIA so that

$$\underline{T}I = \overline{I} = \begin{bmatrix} \overline{I}_P \\ \underline{\tau}_I \end{bmatrix} \qquad (6.32)$$

Next, the range of $\bar{I}_P$ is assumed to represent the processor space and η is a scalar that represents the iteration index. Clearly, one can choose any constant nonsingular matrix $\underline{T}$ so that its first $(S - 1)$ rows determine the processor space. However, the drawback in this approach is that the number of degrees of freedom in determining the choice of the processor space is not $(S^2 - S)$, as it may appear at first glance. Given the processor space, one can choose any basis one pleases in order to identify the processors. The choice of the basis does not change the topology of the array, but only modifies the labels assigned to the processors. Thus any transformation of the form

$$\underline{T}_1 = \begin{bmatrix} \underline{R} & 0 \\ 0 & 1 \end{bmatrix} \underline{T} \tag{6.33}$$

is equivalent to $\underline{T}$, where $\underline{R}$ is any nonsingular matrix. Conversely, any two transformations $\underline{T}$ and $\underline{T}_1$ identify the same processor array if they are related by Eq. (6.33). Therefore, the number of degrees of freedom in the choice of the processor space is $(S^2 - S) - (S - 1)^2$, which is equal to $(S - 1)$.

To avoid such redundant choices, the concept of an iteration vector is useful. Suppose that $\underline{P}^T$ is the $((S - 1) \times S)$ matrix that represents the processor space. That is

$$\bar{I}_P = \underline{P}^T I \tag{6.34}$$

and $\underline{P}^T$ is obviously the submatrix of $\underline{T}$ obtained by deleting its last row. Then the iteration vector, U, is precisely the right null vector for $\underline{P}^T$. That is,

$$\underline{P}^T U = 0 \tag{6.35}$$

Of course, there are several solutions to this equation, but all these solutions are along the same "direction." However, one can ensure the uniqueness of the iteration vector by requiring the following:

Reducedness. The greatest common divisior of the elements of U is 1.
    In other words, the elements of U must be mutually coprime.
Positive leading element. The first nonzero element of U must be
    greater than zero.

The iteration vector has many useful properties. Three of these are listed below.

1. Two index points I and J are mapped on to the same processor
   (i.e., have the same image in the processor-space) if and only if

   $$(I - J) = \alpha U \tag{6.36}$$

   where α is some scalar.

2. The iteration vector represents the direction in the S-dimensional
   Euclidean space along which the index space is projected to obtain
   the processor space.

3. If two different processor-space mappings $\underline{P}_1^T$ and $\underline{P}_2^T$ are related
   by

   $$\underline{P}_2 = R\underline{P}_1^T \tag{6.37}$$

   where $\underline{R}$ is some nonsingular matrix, both these mappings correspond to the same iteration vector U. If $\underline{P}_1^T$ and $\underline{P}_2^T$ are
   related as above, they represent two different labeling conventions
   for the same processor array. Hence it is not surprising that the
   iteration vector remains invariant.

   The first and third properties given above follow from the definition of U. The second property is an interpretation of the first.
   The processor array is obtained by projecting the index space
along the direction of the iteration vector. (Thus the concept of an
iteration vector is more in line with the geometric methods advocated
in Refs. 24–26.) The interprocessor communication links in the
resulting array are obtained by projecting the dependence graph of
the RIA in the same fashion. Of course, the dependence graph of the
RIA has to be embedded in the index space as usual. Thus if in iteration unit located at the index point I, the variable $x_j(I)$ is computed
using the value of $x_k(I - D)$, there should occur data transfer from
the processor corresponding to the index point $(I - D)$ to the
processor corresponding to the index point I. Since D is a constant
vector, the data transfer required will be local. If the vector D is
parallel to the iteration vector, the value of $x_k(I - D)$ should be
stored locally as a state within each processor.
   As an illustration of this procedure, consider the matrix multiplication example introduced in the preceding section. The RIA for
this example clearly has a feasible solution to the set of constraints
given in Eq. (6.31). Now, assume that the iteration vector is chosen
tobe $(1 \ 1 \ 1)^T$. Then $\underline{P}^T$ can be obtained as

$$\underline{P}^T = \begin{bmatrix} 1 & -1 & 0 \\ 0 & 1 & -1 \end{bmatrix}$$

The processor array is obtained by assigning one processor to every distinct point in the range of $\underline{P}^T I$. One can also obtain the processor array geometrically by projecting the index space (a cube of side N) along the direction of the iteration vector.

Now, for the interconnecting links between the processors. Since there are five arcs in the reduced dependence graph of the RIA, one may be inclined to presume that there should be five distinct incoming links and five outgoing links at processor. However, the given algorithm is in input-standard form. Consequently, there are only three distinct inputs and three distinct outputs for each iteration unit, one for each of the (V = 3) variable names. Hence there must be three incoming links and three outgoing links at each processor. The resulting processor array is shown in Fig. 6.18. Topologically, it is identical to the systolic array of Kung and Leiserson shown in Fig. 6.19, except for the fact that the input and output data lines are drawn across the array to indicate that some inputs are fed into the processors at the interior of the array and some outputs are tapped from the interior as well. Even though the input/output nodes of the dependence graph of the RIA are on the boundary surfaces of the index space, a majority of these points map inot interior processors under the projection rule, since the iteration vector is not parallel to

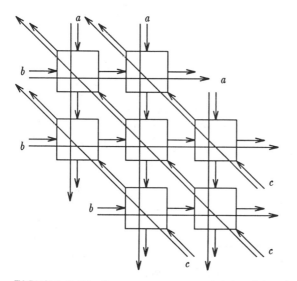

FIGURE 6.18 Processor array obtained by projecting the dependence graph of the matrix multiplication example along the direction $[1 \ 1 \ 1]^T$. The external I/O lines are drawn across the array to indicate the fact that some of the input/output nodes in the dependence graph map onto processors at interior points in the array.

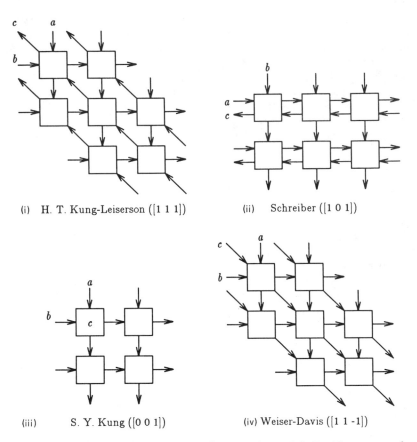

(i)   H. T. Kung-Leiserson ([1 1 1])   (ii)   Schreiber ([1 0 1])

(iii)   S. Y. Kung ([0 0 1])   (iv) Weiser-Davis ([1 1 -1])

FIGURE 6.19a Systolic arrays for matrix multiplication: some known arrays derived by applying Procedure 4 with (i) iteration vector $[1 \ 1 \ 1]^T$ (H. T. Kung-Leiserson), (ii) iteration vector $[1 \ 1 \ 0]^T$ (Schreiber), (iii) iteration vector $[1 \ 0 \ 0]^T$ (S. Y. Kung), and (iv) iteration vector $[1 \ 1 \ -1]^T$ (Weiser-Davis).

these surfaces. To ensure that the inputs to the processor array and the outputs from the array appear at the boundary processors, one must expand the index space appropriately such that the corresponding boundaries of the index space are parallel to the direction chosen for the iteration space. Note that the expanded region of the index space represents iteration units that are used to propagate the relevant inputs to the appropriate locations in the true index space. If the index space is expanded, it is very likely that the range of the linear scheduling functions (Sec. 2.6) also increases. This could result in a less efficient implementation, as will be shown presently. However, if

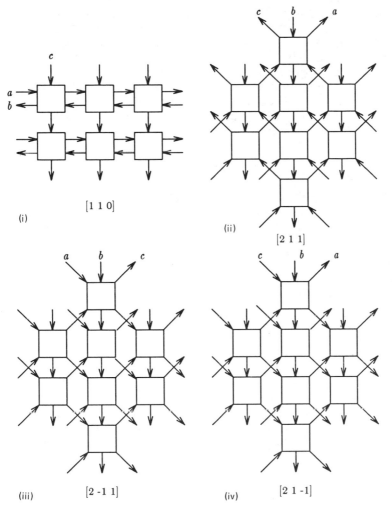

FIGURE 6.19b   Systolic arrays for matrix multiplication: some new arrays derived by applying Procedure 4 with (i) iteration vector $[2\ 1\ 0]^T$, (ii) iteration vector $[2\ -1\ 0]^T$, (iii) iteration vector $[2\ -1\ 1]^T$, and (iv) iteration vector $[2\ 1\ -1]^T$.

inputs can be fed directly to the interior processors where necessary, one can show that every implementation of the algorithm can be scheduled to finish its execution at the same time. Thus, for example, the variation on the Kung-Leiserson matrix multiplication array shown in Fig. 6.18 can be as efficient as the systolic array due to S. Y. Kung [42] or the one due to Weiser and Davis [53]. Of course, in this situation, one would wish to use the array of Fig. 6.18 only in the case of band-matrix multiplications. For the multiplication of square matrices, S. Y. Kung's array requires considerably fewer processors and is hence advantageous.

In general, if the given algorithm is in input-standard form with V variable names, one would need only V incoming links and V out-going links at each processor. The interprocessor communication links are determined by the columns of the matrix $\underline{P}^T\Delta$, where each column of Δ is the index-displacement vector that is a common weight on all incoming arcs at a distinct node in the RDG.

Identifying the time index

Suppose that for the given RIA, a constant vector Y^T has been found using standard linear programming techniques that satisfies the constraints in Eq. (6.31). Partition this vector as

$$Y^T = [\Gamma^T \quad \Lambda^T] \tag{6.38}$$

where Γ^T is a row vector of dimension V and Λ^T is a row vector of length S. Suppose for the moment that Γ^T is zero. Then

$$\Lambda^T\underline{D} \geqslant [1 \quad 1 \quad \cdots \quad 1] \tag{6.39}$$

That is, $\Lambda^T I = t(I)$, the required time index, provided that this choice of the time index does not conflict with the previous choice of the iteration vector.

In terms of the global clock, the time index $\Lambda^T I$ represents the clock cycle in which all variables with index I will be computed by the processor at location $\underline{P}^T I$ in the array. Thus the variable x_k at two index points I and J are scheduled to be executed at the same time if and only if

$$\Lambda^T(I - J) = 0 \tag{6.40}$$

Furthermore, the same processor will be responsible for these compu-tations if $(I - J) = \alpha U$ for some scalar α. Hence an unacceptable situation arises if

$$\Lambda^T U = 0 \tag{6.41}$$

The implication is that the designer does not have the freedom to choose the iteration vector and the linear scheduling function inde-pendently; one is constrained by the other. Given this fact, it would appear to be more desirable to let the choice of the iteration vector be free, so that the topology of the array is given first preference. The choice of the scheduling function can then be constrained by the iteration vector, as shown below.

The assumption that the vector $\underline{\Gamma}^T$ is zero can be easily relazed. Suppose that $\underline{\Gamma}^T$ is nonzero. Let γ_j be the jth element of the vector $\underline{\Gamma}^T$ so that it corresponds to the variable name x_j. Now, consider the following additive transformation:

$$x_j(I) \rightarrow x_j(I + \gamma_j W) \tag{6.42}$$

where W is any vector of appropriate dimensions such that

$$\Lambda^T W = 1 \tag{6.43}$$

Then, by some simple algebraic manipulations, it can be shown that the effect of such a "renaming" of the variables in the RIA is to modify the displacement matrix so that in the new domain

$$\underline{D}' = \underline{D} + W \Gamma^T C \tag{6.44}$$

Hence, in the new domain, a solution to Eq. (6.31) can be found with the required condition, since

$$\Lambda_T \underline{D}' = \Lambda^T \underline{D} + (\Lambda^T W) \Gamma^T \underline{C} = \Lambda^T \underline{D} + \Gamma^T C \tag{6.45}$$

The steps in involved in the foregoing process of identifying a systolic array, given an RIA, can be concisely expressed in the form of a procedure as follows.

Procedure 4: To transform an RIA to a systolic algorithm Our assumption is that the given regular iterative algorithm is such that there exists a finite feasible solution to the set of constraints

$$Y^T \begin{bmatrix} C \\ \underline{D} \end{bmatrix} = Y^T \underline{G} \geqslant [1 \quad 1 \quad \cdots] \tag{6.46}$$

where C, D have the usual significance. Choose one such solution (possibly by minimizing some objective function) and partition it as

$$Y^T = [\Gamma^T | \Lambda^T] \tag{6.47}$$

where Γ^T is a row vector of length V and Λ^T is of length S.

1. Choose a constant S-dimensional column vector U such that it is not orthogonal to Λ^T. Call this the iteration vector of the processor array.

2. The topology of the systolic array is determined by projecting the dependence graph of the RIA along the direction of the iteration vector onto a plane that is orthogonal to it. If need be, check to see if the input nodes in the dependence graph do indeed map onto processors that are on the boundary of the array. If not, expand the index space appropriately to ensure this property (see Examples 4 and 5).

3.   For j = 1 to V, apply additive transformations of the form

$$x_j(\bar{I}) \rightarrow x_j(\bar{I} + \gamma_j W) \tag{6.48}$$

where W is any column vector such that $\Lambda^T W = 1$.

4. Apply a multiplicative transformation to the index space using the transformation matrix

$$\underline{T} = \left[\frac{P}{\Lambda^T} \right] \tag{6.49}$$

where $\underline{P}$ consists of a set of (V − 1) linearly independent row vectors that are all orthogonal to U. The transformed index space is now the span of $\bar{I}$, where

$$\bar{I} = \underline{T} I \tag{6.50}$$

5. The given RIA has now been transformed into a systolic algorithm. Determine the displacement matrix of the systolic algorithm using the relationship

$$\bar{D} = \left[\frac{PD}{Y^T \underline{G}} \right] \tag{6.51}$$

6. Determine the range of the transformed index space using standard techniques in analytical geometry (see Examples 4 and 5).

Consider, for example, the matrix multiplication example presented in the preceding section. The RIA corresponding to this example clearly satisfies the necessary condition, and hence it can be converted into a systolic algorithm using Procedure 4. A particular choice of the linear scheduling function for this example was deduced to be given by $\Gamma^T = 0$ and $\Lambda^T = [1 \ 1 \ 1]$. Now, if the iteration vector is chosen to be $[0 \ 0 \ 1]^T$, the processor space is span $\{[1 \ 0 \ 0]^T, [0 \ 1 \ 0]^T\}$.

Thus a and b, respectively, are transmitted horizontally and vertically in the processor array, while c corresponds to a storage node within each processor. The resulting matrix multiplier is the well-known array first presented by S. Y. Kung in Ref. 42. Similarly, through picking different choices of the iteration vector, one can obtain the arrays of H. T. Kung and Leiserson [54], Weiser and Davis [53], and so on. In Fig. 6.19, these arrays are presented together with a few others that have hitherto not been derived elsewhere in the literature.

It is clear that there are a multitude of different choices for the iteration vector. Each of these will correspond to a different systolic array. Thus one can systematically generate an "infinity" of different systolic arrays for matrix multiplication! In practical terms, however, one may wish to limit the arrays generated to have not just local but nearest-neighbor communication.

It should be obvious to the reader that the existence of a variety of choices for the iteration vector is not a feature that is perculiar to the matrix multiplication example. This is true for any problem that can be solved using a regular iterative algorithm with the requisite properties.

3.3 Efficiency of a Systolic Array

It is well knwon that in the systolic array for matrix multiplication due to Kung and Leiserson, every processor does useful work only once every three cycles. If the fundamental time period of this array is δ, the extrinsic iteration interval (as opposed to the intrinsic iteration-interval to be defined later) is 3δ. (see e.g., Refs. 39 and 40 for more on the iteration intervals and related concepts. In the terminology used by Leiserson et al. [34], the extrinsic iteration interval is the slowdown of the systolic circuit. The extrinsic iteration interval is referred to by Ullman [46] as the period.) Therefore, the efficiency of the Kung-Leiserson array is $1/3$.

Determination of the extrinsic iteration interval

Let I be some index point that is mapped on to the processor at location $P^T I$ in the processor space. The computations corresponding to the iteration unit at this index point are executed by this processor at time step $\underline{t}(I) = \Lambda^T I$. (For the arguments given below, one can assume that $\gamma_j = 0$ for all j. Having nonzero constants does not affect the validity of these arguments, although it increases their notational complexity.) These computations are assumed to be completed by time step $(\Lambda^T I + 1)$ and the processor is now ready to execute the next set of instructions, if necessary. Suppose that J is the index point corresponding to the next set of instructions (i.e., the iteration unit). Then since J is also mapped on to the same processor, one must have

$$J = (I + \alpha U) \qquad\qquad (6.52)$$

Since U is assumed to have coprime elements, α must necessarily be an integer. Now t(J) is given by

$$\underline{t}(J) = t(I) + \alpha(\Lambda^T U) \qquad\qquad (6.53)$$

Therefore, the next set of instrucitons will be executed by the processor at time step $(t(I) \pm \Lambda^T U)$. Hence

$$\text{extrinsic iteration−interval (of a systolic array)} = |\Lambda^T U|\, \delta$$

$$(6.54)$$

and

$$\text{efficiency (of a systolic array)} = \frac{1}{|\Lambda^T U|} \qquad\qquad (6.55)$$

Theorem 9: Efficiency of a systolic array Given Λ^T that defines the linear scheduling function, and U, the iteration vector for a systolic implementation such that the elements of U are mutually coprime, the extrinsic iteration interval for the array is $|\Lambda^T U|\, \delta$, where δ is the fundamental time period of the array. The efficiency of the array is $1/(|\Lambda^T U|)$.

For the systolic arrays shown in Fig. 6.19a, the efficiencies can be calculated to be 1/3, 1/2, 1, and 1 for Kung-Leiserson's, Schreiber's, S. Y. Kung's, and Weiser-Davis's arrays, respectively. This is consistent with the literature. For the new arrays shown in Fig. 6.19b, the efficiencies are 1/2, 1/4, 1, 1, and for (i), (ii), (iii), and (iv), respectively.

One must use extreme caution in applying the efficiency measure for evaluating systolic arrays. Consider, for instance, the following examples:

Example 4: Designing and Optimal Matrix Multiplication Array The problem is to implement the given regular iterative algorithm for matrix multiplication on a systolic array with the following features:

1. All external I/O is handled by the boundary processors.
2. A minimum number of processors is used.
3. The time between the first input and the last output is as short as possible, subject to condition 2.

Solution: The first condition rules out the use of any of the unit vectors as the iteration vector. Hence S. Y. Kung's array cannot be a solution. Furthermore, for any choice of the iteration vector, one must expand the index space so that the inputs and outputs appear on surfaces that are parallel to the iteration vector. This will necessarily increase the range of the linear scheduling functions.

A minimum number of processors must be used. Hence it is advantageous to search for an iteration vector within the set $\{[1 \ 1 \ 0]^T, [1 \ - \ 0]^T\}$. Of course, any cyclic shift of these two vectors is also a possibility. Now, the vector $[1 \ -1 \ 0]^T$ can be eliminated from consideration since it is orthogonal to Λ^T. Hence choose $[1 \ 1 \ 0]^T$ as the iteration vector, resulting in Schreiber's array with $2N^2$ processors. The expanded index space is now described by the constraints (see Fig. 6.20)

$$- N \leqslant (i + j) \leqslant 3N$$

$$- N \leqslant (i - j) \leqslant N$$

$$0 \leqslant k \leqslant N$$

The range of the linear scheduling function within this expanded index space is from $- N$ to $4N$. Hence the time from first input to last output is $5N\delta$, which is slightly longer than the $3N\delta$ achieved by S. Y. Kung's array. Furthermore, since the a and b outputs are not really needed (these are just the inputs propagated to the outputs), one can further restrict the range of the index space to (Fig. 6.20)

$$- N \leqslant (i + j) \leqslant 2N$$

$$- N \leqslant (i - j) \leqslant N$$

$$0 \leqslant k \leqslant N$$

in which case the total time required from first input to last output is only $4N\delta$.

Note that Schreiber's array is a solution to the problem above even though it is only 50% efficient. Now, consider the following variation of the problem.

Example 5 This is the same problem as that of Example 4, with the additional constraint that the efficiency of the systolic array should be 1.

Solution: Once again, a solution must be first sought with the iteration vector being either $[1 \ 1 \ 0]^T$ or $[1 \ -1 \ 0]^T$ or any cyclic shift thereof. The first choice leads to an efficiency of 50% and therefore

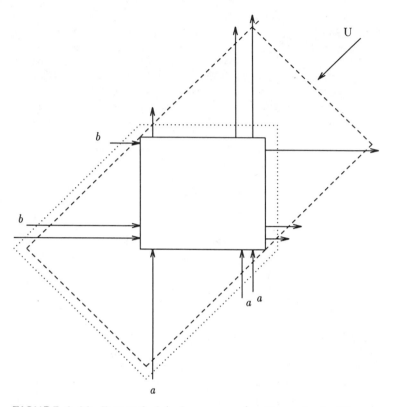

FIGURE 6.20 Expanded index space for Example 4. The necessary expansion is only in the (i,j) planes shown. The solid lines form the boundaries of the real index space, the dashed lines for the expanded index space, and the dotted lines for the expanded index space under the condition that a and b outputs are not needed.

is ruled out. The second vector is orthogonal to $\Lambda^T = [1 \quad 1 \quad 1]$. However, observe that any integer vector Λ^T such that

$$\Lambda^T \geqslant [1 \quad 1 \quad 1]$$

is a valid choice for the linear scheduling function. Hence one can choose

$$\Lambda^T = [2 \quad 1 \quad 1]$$

For this choice of the scheduling function, the iteration vector
[1 −1 0] meets the requirements of nonorthogonality and 100%
efficiency. The expanded index space in this case is given by the set
of constraints

$$0 \leqslant i + j \leqslant 2N$$

$$-2N \leqslant i - j \leqslant 2N$$

$$0 \leqslant k \leqslant N$$

(see Fig. 6.21). The range of the scheduling function [which is now
(2i + j + k)] can be calculated to be from −N to 5N. Hence the total
time required is 6Nδ. Once again, since the a and b outputs are not
of interest, the expanded index space can be restricted to

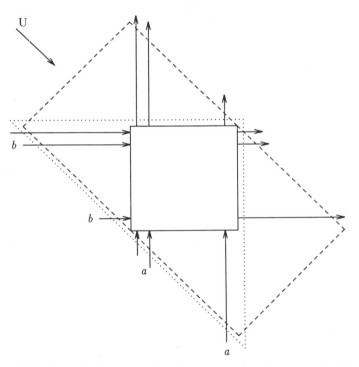

FIGURE 6.21 Expanded index space for Example 5. The solid lines
represent the true index space, the dashed lines for the expanded
index space, and the dotted lines for the expanded index space under
the condition that the a and b outputs are not needed.

$i + j \geqslant 0$

$i, j \leqslant N$

$0 \leqslant k \leqslant N$

in which case the total time required from first input to last output is $5N\delta$. The resulting array is shown in Fig. 6.22.

3.4 Simultaneous Choice of the Iteration Vector and Linear Scheduling Functions for Systolic Arrays

In Example [5] one was fortuitously able to derive a new scheduling function when the choice of the iteration vector conflicted with the original choice of the linear scheduling function. In this section it will be shown that the occurrence of this phenomenon was less due to chance than due to a fundamental property of the RIA: For any choice of the iteration vector U, a linear scheduling function can be found such that $\Lambda^T U \neq 0$, provided that the RIA satisfies the feasibility condition of Eq. (6.31).

First, observe that the feasibility condition, that is, that there exists a constant vector Y^T such that

$$Y^T \begin{bmatrix} \underline{C} \\ \underline{D} \end{bmatrix} = Y^T \underline{G} \geqslant [1 \quad 1 \quad \cdots \quad 1] \tag{6.56}$$

is equivalent to the condition that there exists no solution to the set of constraints

$$\underline{G}Q = 0$$
$$q_i \geqslant 0 \tag{6.57}$$

other than the trivial one $Q = 0$. This is obtained by a direct application of the duality theorem (Theorem 4) in linear programming. Now, suppose that U is a nonzero constant integer vector that one wishes to use as the iteration vector for obtaining a systolic array implementation. Assume that the elements of U are mutually prime so that the GCD of the elements is 1. To apply the synthesis procedure above, it is required that a solution be found for Y^T in Eq. (6.56) such that

$$Y^T = \begin{bmatrix} \Gamma^T \mid \Lambda^T \end{bmatrix}$$

with the usual partitioning, and

$$\Lambda^T U \neq 0 \tag{6.58}$$

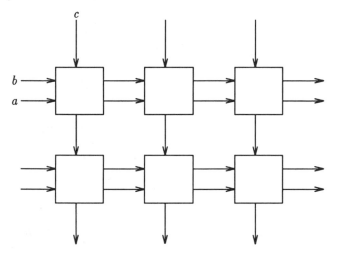

FIGURE 6.22 Processor array obtained for iteration vector $[1 \quad -1 \quad 0]^T$ for the matrix multiplication example. Notice that a shift register is introduced to take into account the fact that $\Lambda^T = [2 \quad 1 \quad 1]$.

Of course, in the interest of ensuring maximum efficiency, one would wish to have $\Lambda^T U$ to be as close to 1 in modulus as possible.

One can now incorporate the constraint in Eq. (6.58) while determining a solution to Eq. (6.56). However, one must first check whether this additional constraint would violate the requirements of the duality theorem, thus rendering the choice of iteration vector undesirable.

The constraint of Eq. (6.58) can be written either as

$$\Lambda^T U \geqslant 1 \tag{6.59}$$

or as

$$\Lambda^T (-U) \geqslant 1 \tag{6.60}$$

Since Λ and U are both constant integer vectors, the constraint of Eq. (6.59) must be equivalent to one of the above. Now one can show that either

$$Y^T \underline{G} \geqslant [1 \quad 1 \quad \cdots \quad 1]$$
$$Y^T \begin{bmatrix} 0 \\ U \end{bmatrix} = \Lambda^T U \geqslant 1 \tag{6.61}$$

or the set of constraints

$$Y^T G \geqslant [1 \quad 1 \quad \cdots \quad 1]$$

$$Y^T \begin{bmatrix} 0 \\ -U \end{bmatrix} = \Lambda^T (-U) \geqslant 1 \tag{6.62}$$

must have a feasible solution.

Suppose that both these sets do not have a feasible solution. Then, by the duality theorem, there must be nontrivial solutions $\{Q_1, \alpha_1\}$ and $\{Q_2, \alpha_2\}$ such that

$$GQ_1 + \begin{bmatrix} 0 \\ U \end{bmatrix} \alpha_1 = 0$$

$$q_{1,i}, \ \alpha_1 \geqslant 0 \tag{6.63}$$

and

$$GQ_2 + \begin{bmatrix} 0 \\ -U \end{bmatrix} \alpha_2 = 0$$

$$q_{2,i}, \ \alpha_2 \geqslant 0 \tag{6.64}$$

In both these solution sets, one must have α_1 and α_2 strictly positive, since it was assumed that there is no nontrivial positive null vector Q such that $\underline{G}Q = 0$ [see Eq. (6.37) and the discussion preceding]. Now consider the vector

$$Q = (Q_1 \alpha_2 + Q_2 \alpha_1) \tag{6.65}$$

This vector must also be nonnegative and cannot be identically zero for the supposition made. However, because of Eqs. (6.63) and (6.64), this vector Q is a solution to the set of constraints in Eq. (6.67), which is impossible. This leads to a contradiciton and thus

Theorem 10 Let the given regular iterative algorithm satisfy the feasibility condition of Eq. (6.56). Let U be an arbitrarily chosen constant integer vector that is required to be used as the iteration vector in Procedure 4 and let the greatest common divisor of the elements of U be 1. Then a constant linear scheduling function can be found such that this implementation procedure results in a systolic array with the desired iteration vector.

This theorem has an important implication: One may now choose the iteration vector in order to achieve the desired data flow in the resulting array, without worrying about the scheduling. Furthermore, since the inner product between the unknown vector Λ^T and the known iteration vector U is a measure of the idleness of the systolic array, it is possible to include an objective criterion for the choice of the linear scheduling function. In this case one needs to find the optimal solution to the linear programming problem

$$\text{minimize } \{Y^T[0 \quad U^T]^T\}$$

$$\text{subject to } Y^T\underline{G}' \geq [1 \quad 1 \quad \cdots \quad 1] \qquad (6.66)$$

3.5 Design of Multirate Systolic Arrays

Multirate systolic arrays are identical to systolic arrays, except for one minor implementational detail. In a systolic array, it is required that the arrival times for the inputs to the iteration unit $\{\tau_{j,in}\}$ and the departure times for the outputs of the same iteration unit $\{\tau_{j,out}\}$ be related as $\tau_{j,out} = (\tau_{j,in} + \delta)$, where δ is the fundamental time period of the array. Furthermore, the arrival times for the inputs to the subsequent iteration unit should also be $\{\tau_{j,out}\}$.

In a multirate systolic array, the departure times for the outputs of an iteration unit are not constrained in any way on the arrival times for the inputs, except that the departure time of an output should be greater than the arrival time for an input, if there is a dependency involved between these two quantities. However, it is required that the arrival times $\{\tau_{j,in}\}$ and the departure times $\{\tau_{j,out}\}$ are such that the set $\{\tau_{j,out} - \tau_{j,in}\}$ is independent of the iteration unit. Furthermore, the arrival times for the inputs to the next iteration unit should be $\{\tau_{j,in} + \delta\}$, where δ is the extrinsic iteration interval of the array.

The removal of this restriction can sometimes result in dramatic improvements in the utilization of the computational resources in the array, as will be shown in what follows. However, the global synchronization of the operation of the array can be a little more difficult to achieve.

For the design of multirate systolic arrays, it is necessary to know more about how the iteration unit is executed by a processor than what was assumed thus far. The intrinsic schedule of a processor (processing element) must be specified before the analysis can be accomplished. This intrinsic schedule consists of an explicit specification of the relative arrival times and departure times for the inputs and outputs, respectively, of an iteration unit computed by a processor. In addition, an intrinsic iteration interval should also be provided. The result of the analysis presented here is the determination of the extrinsic iteration interval for the implementation, and consequently the complete specifications of a time schedule for the operation of the array.

To illustrate the procedures presented below, the matrix multiplication example will be used, as usual. The iteration unit for this example consists of two data-transfer operations and a single multiply-accumulate operation. A multiply-accumulate operation can be executed using a pipelined accumulator that is capable of accepting a new multiply-accumulate operation every clock cycle. On the other hand, it can also be accomplished using a serial accumulator, in which case, the next multiply-accumulate operation can be accepted by the processor only after the current execution is completed. These two extreme situations will be used as examples to demonstrate the various trade-offs that need to be made.

Example [6]: Intrinsic Schedule for a Processor with a Pipelined Multiply–Accumulator for the Matrix Multiplication Example The arrival times for the inputs to the iteration unit, $a(i,j,k)$, $b(i,j,k)$, and $c(i,j,k)$ are all assumed to be the same, say 0. The departure times for the outputs $a(i, j + 1, k)$ and $b(i + 1, j, k)$ are both 1 and that for $c(i, j, k + 1)$ is 16. Due to the pipelined nature of the multiply-accumulator, the next set of data for the next iteration unit can be accepted at time 1. That is, the intrinsic iteration interval for the processor is 1.

Example [7]: Intrinsic Schedule for a Processor with a Serial Multiply–Accumulator for the Matrix Multiplication Example The arrival times for the inputs to the iteration unit, $a(i,j,k)$, $b(i,j,k)$, and $c(i,j,k)$ are all assumed to be the same, say 0. The departure times for the outputs $a(i, j + 1, k)$ and $b(i + 1, j, k)$ are both 1 and that for $c(i, k, k + 1)$ is 16. Due to the serial nature of the multiply-accumulator, the next set of data for the next iteration unit can be accepted at time 16. That is, the intrinsic iteration interval for the processor is 16.

The intrinsic iteration interval, denoted by the symbol δ_{int}, is clearly a measure of the rate at which the processor can operate on the data and depends on how the computational modules within the processor are implemented. The extrinsic iteration interval δ_{ext}, determined below, is a measure of the rate at which the processor is provided with the input data and depends on the precedence constraints within the algorithm and the choice of the iteration vector. Thus it is only natural to define the *percent utilization* of a multirate systolic array as the ratio percent utilization of a multirate systolic array

$$= 100 \times \frac{\{\text{intrinsic iteration interval}\}}{\{\text{extrinsic iteration interval}\}} \tag{6.67}$$

One can, of course, define the percent utilization of a systolic array in a similar fashion. In this case, however, the fundamental time period of the array must first be determined. The fact that the departure times for the outputs of the current iteration unit must coincide with the arrival times for the inputs to the next iteration unit should enable one to do so. For the two examples above, this is seen to be at least 16 time units. Therefore,

percent utilization of a systolic array

$$= 100 \times \frac{\{\text{efficiency} \times \text{intrinsic iteration interval}\}}{\{\text{fundamental time period}\}}$$

$$= 100 \times \frac{\{\text{intrinsic iteration interval}\}}{\{\text{extrinsic iteration interval}\}} \tag{6.68}$$

It can be observed immediately that having a pipelined arithmetic unit does not help at all, in the case of a systolic array implementation. On the other hand, in a multirate systolic array implementation, one can use this feature to advantage under certain circumstances.

As usual, assume that the given RIA is in input-standard form and that it satisfies the feasibility constraints in Eq. (6.31). Therefore, there exist constant row vectors Γ^T and Λ^T such that

$$[\Gamma^T | \Lambda^T] \left[\frac{C}{\underline{D}} \right] \geq H^T \tag{6.69}$$

where H^T is any row vector of appropriate dimensions, with strictly positive integer elements. Now suppose that $(\Lambda^T I + \gamma_j)$ is the time at which variable $x_j(I)$ is computed by some processor in the array. Then if the output $x_j(I + D_j)$ is dependent on $x_k(I)$, from the intrinsic schedule one can deduce that

$$(\Lambda^T I + \Lambda^T D_j + \gamma_j) - (\Lambda^T I + \gamma_k) \geq \tau_{j,\text{out}} - \tau_{k,\text{in}} \tag{6.70}$$

Canceling out the common term $\Lambda^T I$ and stacking these constraints in matrix form

$$[\Gamma^T | \Lambda^T] \underline{G} \geq H^T \tag{6.71}$$

where the row vector H^T is constructed as follows. If the ith arc in
the RDG is from node x_k to node x_j, then $h_i = (\tau_{j,out} - \tau_{k,in})$.
Furthermore, just as in the preceding section, the choice of the
scheduling function and the choice of the iteration vector are constrained by the inequality

$$|\Lambda^T u| \geq \delta_{int} \qquad\qquad (6.72)$$

where δ_{int} is the intrinsic iteration interval for the module.

Once a solution for Λ^T and Γ^T is found, the "excess variables"
determine the amount of external "waits" or shimming delays (see
Fettweis [55]) that must be placed at the inputs of the respective
computational modules. That is, suppose that $x_j(I + D_j)$ is directly
dependent on $x_k(I)$. Then the input $x_k(I)$ must be held for

$$w_{jk} = (\Lambda^T D_j + \gamma_j - \gamma_k - [\tau_{j,out} - \tau_{k,in}]) \qquad\qquad (6.73)$$

clock cycles before it is used by the jth computational module.

Finally, one might wish to maximize the percent utilization while
determining scheduling functions. In this case the objective function
to be minimized is

$$\text{minimize } \{|\Lambda^T u|\} \qquad\qquad (6.74)$$

**Example [8]: Designing a Multirate Systolic Array for the Matrix
Multiplication Problem with Pipelined Processors** The problem is to
implement the given regular iterative algorithm for matrix multiplication
on a multirate systolic array with the following features:

1. All external I/O is handled by the boundary processors.
2. A minimum number of processors are used.
3. The percent utilization of the processors in the array should be
    as large as possible, subject to conditions 1 and 2.
4. The processors are as specified in Example 6.

Solution: Just as in Example 4, the first two conditions limit the
choice of the iteration vector to $U^T = [1 \quad 1 \quad 0]$ or $U^T = [1 \quad -1 \quad 0]$ or
any cyclic shift of these vectors. Thus there are six different possibilities to be examined. Consider, for example, the choice $U^T = [1 \quad 1 \quad 0]$.
The optimization problem can now be written as

minimize $\{\lambda_1 + \lambda_2\}$

subject to $\begin{bmatrix} \lambda_1 \lambda_2 \lambda_3 \end{bmatrix} \begin{bmatrix} 1 & 0 & 0 & 0 & 0 & 1 \\ 0 & 1 & 0 & 0 & 0 & 1 \\ 0 & 0 & 1 & 1 & 1 & 0 \end{bmatrix} \geqslant \begin{bmatrix} 1 & 1 & 16 & 16 & 16 & 1 \end{bmatrix}$

The solution is easily seen to be

$\lambda_1 = \lambda_2 = 1$

$\lambda_3 = 16$

and the percent utilization of the array is 50%. Similarly, the percent utilization can be calculated for the other choices of the iteration vector as

iteration-vector = [0 1 1]   percent utilization = 6%

iteration-vector = [1 0 1]   percent utilization = 6%

iteration-vector = [1 −1 0]   percent utilization = 100%

iteration-vector = [1 0 −1]   percent utilization = 100%

iteration-vector = [0 1 −1]   percent utilization = 100%

Obviously, one of the last three choices of the iteration vector given in the equation above is a solution to the problem. The resulting processor array is shown in Fig. 6.23.

Example 9: Designing a Multirate Systolic Array for the Matrix Multiplication Problem with Serial-Arithmetic Processors Solve Example 8 for processors with serial multiply-accumulators with intrinsic schedule as given in Example 7.

Solution: The solution is similar to the one above, except that the H vector is slightly modified to

$H^T = \begin{bmatrix} 1 & 1 & 16 & 16 & 16 & 16 \end{bmatrix}$

Once again, all six possibilities need to be considered for the iteration vector, with the result

iteration vector = [1 1 0]   percent utilization = 100%

iteration vector = [1 0 1]   percent utilization = 94%

iteration vector = [0 1 1]   percent utilization = 94%

iteration vector = $[1 \;\; -1 \;\; 0]$ percent utilization = 100%

iteration vector = $[1 \;\; 0 \;\; -1]$ percent utilization = 100%

iteration vector = $[0 \;\; 1 \;\; -1]$ percent utilization = 100%

Any of the four choices with 100% utilization is a solution.

One must be careful when interpreting the percent-utilization parameter. It is a measure that truly indicates the utilization of the computational resources in the array when a large number of problems are solved in a contiguous, pipelined fashion on the array. However, if the array is used for solving a single problem, the percent-utilization measure may not be appropriate for evaluating the array. Traditional measures such as the I/O latency are more meaningful in this case.

4   CONCLUDING REMARKS

Thus far, the design of systolic/wavefront array architectures and other processor array architectures has been considered an art, a skill to be learned through intuition, experience, and knowledge of the problem under consideration. Some efforts made by other authors to systematize the design process from the algorithmic level, but these efforts were not completely successful.

In this chapter it was shown that there exists a class of algorithms. itself a subclass of the regular iterative algorithms, that is isomorphic with systolic/wavefront arrays. Given such an algorithm, it is possible to obtain a variety of systolic arrays for implementing the algorithm using the procedure outlined in Sec. 3.

The analysis presented in Sec. 2 shows that given a regular iterative algorithm that is not necessarily in this subclass, one can determine asymptotic lower bounds on the time required for executing the algorithm and on the amount of storage necessary during the execution. The fact that these lower bounds can be met in a processor array implementation of the algorithm was not demonstrated in this chapter; the interested reader can refer to Ref. 2. However, it can be stated that in this case, the resulting processor arrays are not systolic, but could always be described as a regular interconneciton of similar processing elements and variable-length register pipelines operating in a globally synchronous fashion.

All the procedures described in this chapter are executable in constant time, independent of the number of computations in the regular iterative algorithm. Furthermore, the processor arrays derived for implementing the algorithm could be simulated on a fixed torus-connected architecture. Thus it is possible to design a compiler that takes as input a high-level description of the regular iterative algorithm and outputs the program that must be repeatedly executed by each processor in such an architecture.

The usefulness of regular iterative algorithms rests in the fact that many problems in various applications can be solved using them. In most cases, such algorithms could be derived by reformulating existing ones using the techniques described in Refs. 2 and 3. In these references it is also shown that for some problems, such as digital filtering and the problem of determining the transitive closure of a graph, it is advantageous to consider the problem afresh and synthesize regular iterative algorithms for them, from first principles.

In this chapter, the fact that regular iterative algorithms are well matched to mesh-connected and related processor array architectures has been substantiated. As a follow-up on the matter presented here, it is natural to ask whether it is possible to characterize the class of algorithms that is similarly matched to, say, a hypercube, or a shuffle-exchange network, or a butterfly network.

ACKNOWLEDGMENT

The authors would like to thank H. V. Jagadish for many discussions during the early stages of this work.

REFERENCES

1. K. E. Batcher, Design of a massively parallel processor, *IEEE Trans. Comput.*, *C-29*: 836—840 (1980).

2. S. K. Rao, Regular iterative algorithms and their implementations on processor arrays, Ph.D dissertation, Stanford University, (1985).

3. V. P. Roy Chowdury and T. Kailath, Regular iterative algorithms for linear algebraic problems, to be submitted.

4. A. J. Atrubin, A study of several planar iterative switching circuits, S. M. thesis, MIT (1958).

5. E. J. McCluskey, Iterative combinational switching networks— general design considerations, *IRE Trans. Electron. Comput.*, *EC-7*: 285—291 (1958).

6. F. C. Hennie, Analysis of one-dimensional iterative logical circuits, S. M. thesis, MIT (1958).

7. F. C. Hennie, Analysis of bilateral iterative networks, *IRE Trans. Circuit Theory*, *CT-6*: 35—45 (1959).

8. F. C. Hennie, Two-dimensional iterative networks and their decision problems, *General Electric Research Laboratory Report*, General Electric, Schenectady, N.Y. (1960).

9. F. C. Hennie, *Iterative Arrays of Logical Circuits*, MIT Press/ Wiley, New York (1961).

10. F. C. Hennie, *Finite State Models for Logical Machines*, Wiley, New York (1968).

11. S. H. Unger, A computer oriented toward spatial problems, *Proc. IRE, 46*: 1744−1750 (1958).

12. W. M. Waite, Path detection in multidimensional iterative arrays, *J., ACM, 14*: 300−310 (1967).

13. R. McNaughton and H. Yamada, Regular expressions and state graphs for automata, *IRE Trans. Electron. Comput., 9*: 39−47 (1960).

14. R. M. Karp, R. E. Miler, and S. Winograd, The organization of computations for uniform recurrence equations, *J. ACM, 14*: 563−590 (1967).

15. S. K. Rao, H. V. Jagadish, and T. Kailath, Analysis of iterative algorithms for multiprocessor implementations, submitted to *IEEE Trans. Circuits Syst.*, (1985).

16. D. Cohen, Mathematical approach to iterative computation networks, *Proc. 4th Symposium on Computer Architecture*, 226−238 (1978).

17. L. Johnsson and D. Cohen, A mathematical approach to modeling the flow of data and control in computational networks, in *VLSI Systems and Computations*, 509−525 (1979).

18. L. Johnsson and D. Cohen, VLSI approach to real-time computational problems, *Proc. SPIE* 298: 48−58 (1981).

19. L. Johnsson, S. Lennart, U. Weiser, D. Cohen, and A. Davis, *Proc. 2nd Caltech Conference on VLSI*, Toward a formal treatment of VLSI arrays (Jan. 1981).

20. B. Lisper, Description and synthesis of systolic arrays, *TRITA- NA-8318*, The Royal Institute of Technology, Stockholm, Sweden (1983).

21. R. M. King, Research on the synthesis of concurrent computing systems, *Proc. 10th Symposium on Computer Architecture*, 39−46 (1983).

22. R. M. King and E. Mayr, Synthesis of efficient structures for concurrent computation, *VLSI: Algorithms and Architectures*,

(P. Bertolazzi and F. Luccio, eds.), North-Holland, Amsterdam
377–386 (1984).

23. P. Quinton, The systematic design of systolic arrays, *IRISA
International Publication 193* (Apr. 1983).

24. P. R. Capello, VLSI architectures for digital signal processing,
Ph.D dissertation, Princeton University, (1982).

25. P. R. Capello and K. Steiglitz, Unifying VLSI array designs with
geometric transformations, *Proc. International Conference on
Parallel Processing,* Columbus, OH 448–457 (Aug. 1983).

26. P. R. Capello and K. Steiglitz, Unifying VLSI array design with
linear transformations of space-time, *Adv. Comput. Res., 2*:
23–65 (1984).

27. H. T. Kung and W. T. Lin, An algebra for VLSI algorithm
design, *Proc. Conference on Elliptic Problem Solvers,* Monterey,
CA, (Jan. 1983).

28. J. A. B. Fortes, Algorithm transformations for parallel processing
and VLSI architecture design, Ph.D dissertation, University of
Southern California, (1983).

29. R. H. Kuhn, Transforming algorithms for single-stage and VLSI
architectures, *Proc. Workshop on Interconnection Networks for
Parallel and Distributed Processing,* 11–19 (Apr. 1980).

30. G. J. Li and B. W. Wah, The design of optimal systolic arrays,
IEEE Trans. Comput., C-34: 66–77 (1985).

31. W. L. Miranker and A. Winkler, Space-time representation of
computational structures, *Computing 32*: 93–114 (1984).

32. D. I. Moldovan, On the analysis and synthesis of VLSI algorithms,
IEEE Trans. Comput., C-31: 1121–1126 (1982).

33. D. I. Moldovan, On the design of algorithms for VLSI systolic
arrays, *Proc. IEEE 71*: 113–120 (1983).

34. C. E. Leiserson, F. M. Rose, and J. B. Saxe, Optimizing synchronous circuitry by retiming, *3rd Caltech Conference on
VLSI* (R. Bryant, ed.) Computer Science Press, Rockville, MD
87–116 (1983).

35. J. D. Ullman, NP-complete scheduling problems, *J. Comp. Sys.
Sci., 10*: 384–393 (1975).

36. E. Deprettere and P. Dewilde, Orthogonal cascade realization of
real multiport digital filters, *Technical Report,* Network Theory
Section, Dept. of Electrical Engineering, Delft University of
Technology, Delft, The Netherlands (1980).

37. M. Behzad, G. Chartrand, and L. Lesniak-Foster, *Graphs and Digraphs*, Prindle, Wever & Schmidt, Boston (1979).

38. H. E. Mutluay and M. H. Famy, Recursibility of N-dimensional IIR digital filters, *IEEE Trans. Acoust. Speech Signal Process.*, *ASSP-32*: 397—402 (1984).

39. H. V. Jagadish, T. Kailath, J. A. Newkirk, and R. G. Matthews, A study of pipelining in computing arrays, to appear in *IEEE Trans. Comput.*, (1985).

40. H. V. Jagadish, Techniques for the design of parallel and pipe-lined VLSI systems for numerical computations, Ph.D dissertation, Stanford University, (1985).

41. D. E. Heller, Decomposition of recursive filters for linear systolic arrays, *Proc. SPIE* 431: 55—59 (Aug. 1983).

42. S. Y. Kung, VLSI array processors for signal processing, *Conference on Advanced Research in Integrated Circuits*, MIT, Cambridge, MA (Jan. 1980).

43. S. Y. Kung and R. J. Gal-Ezer, Synchronous vs. asynchronous computation in VLSI array processors, *Proc. SPIE* 341: 53—65 (1982).

44. S. Y. Kung, On supercomputing with systolic/wavefront array processors, *Proc. IEEE 72*: 867—884 (July 1984).

45. R. G. Melhem and W. C. Rheinboldt, A mathematical model for the verification of systolic networks, *13*: 541—565 (Aug. 1984).

46. J. D. Ullman, *Computational Aspects of VLSI*, Computer Science Press, Rockville, MD. (1984).

47. R. J. Gal-Ezer, The wavefront array processor and its appli-cations, Ph.D. dissertation, University of Southern California (1982).

48. M. C. Chen and C. A. Mead, Concurrent algorithms as space-time recursion equations, in *VLSI and Modern Signal Processing* (S. Y. Kung, H. J. Whitehouse, and T. Kailath, eds.), Prentice-Hall Englewood Cliffs, NJ 224—240 (1984).

49. L. J. Guibas, H. T. Kung, and C. D. Thompson, Direct VLSI implementation of combinatorial algorithms, *Proc. CalTech Confer-ence on VLSI* (1979).

50. S. Y. Kung and S. C. Lo, A fast systolic algorithm for transitive closure problem, *Proc. International Conference on Acoustics Speech, and Signal Processing* (1985).

51. S. K. Rao, T. K. Citron, and T. Kailath, Mesh connected processor arrays for the transitive closure problem, submitted to *J. ACM* (1985).

52. S. K. Rao, T. K. Citron, and T. Kailath, Mesh connected processor arrays for the transitive closure problem, *IEEE Conference on Decision and Control* (1985).

53. U. Weiser and A. Davis, A wavefront notational tool for VLSI array design, *VLSI Systems and Computations*, (H. T. Kung, R. Sproull and G. Steele, eds.), Computer Science Press, Rockville, MD. 226−234 (1981).

54. H. T. Kung and C. E. Leiserson, Systolic arrays for VLSI, *Sparse Matrix Proceedings*, Real and Applied Mathematics, Philadelphia, 245−282 (1978).

55. A. Fettweis, Realizability of digital filter networks, *Arch. Eleck. Ubertragung, 30*: 90−96 (1976).

7

Algorithms for Integrating Wafer Scale Systolic Arrays

TOM LEIGHTON *Mathematics Department and Laboratory for Computer Science, Massachusetts Institute of Technology, Cambridge, Massachusetts*

CHARLES E. LEISERSON *Laboratory for Computer Science, Massachusetts Institute of Technology, Cambridge, Massachusetts*

1 INTRODUCTION

Very large scale integration (VLSI) technologists are fast developing wafer-scale integration [1]. Rather than partitioning a silicon wafer into chips as is usually done, the idea behind wafer-scale integration is to assemble an entire system (or network of chips) on a single wafer, thus avoiding the costs and performance loss associated with individual packaging of chips. A major problem with assembling a large system of microprocessors on a single wafer, however, is that some of the processors, or *cells*, on the wafer are likely to be defective, or *dead*. In this chapter we present algorithms to construct systolic arrays from the *live* cells of a silicon wafer.

Laser-programming the interconnect of a wafer is one promising means of achieving wafer-scale integration. This technology was pioneered at IBM [2] and pursued in the direction of wafer-scale integration at MIT Lincoln Laboratory [1]. Figure 7.1 shows a scanning electron microscope photograph of a portion of a wafer with programmable interconnect. Laser welds can be made between two layers of metal, and by using the beam at somewhat higher power, wires can be cut. Defective components can thus be avoided by programming connections between only the good components.

Figure 7.2 shows a typical organization of a wafer-scale system with programmable interconnections. The components are organized as a matrix of cells, and between the cells are channels through which the interconnect runs. Figure 7.3 is a close-up of the channel structure. At the intersection of a horizontal and vertical channel, laser-programmable connections can make a horizontal and a vertical wire electrically equivalent. Between two cells, connections can be made from the wires

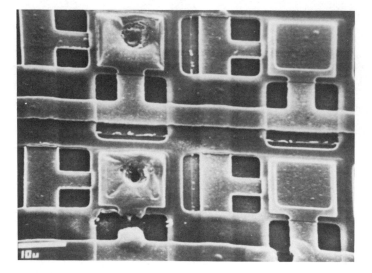

FIGURE 7.1 A close-up of laser-programmable interconnect.

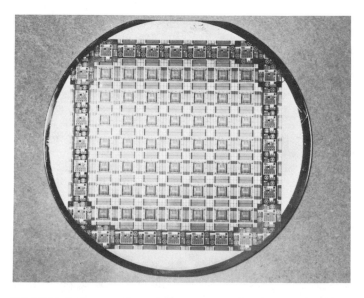

FIGURE 7.2 A wafer-scale system of cells and programmable interconnect.

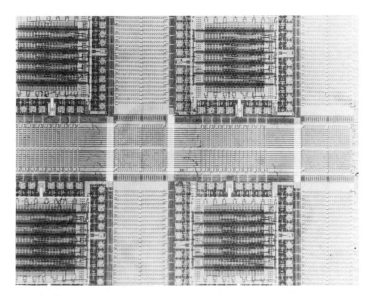

FIGURE 7.3 The channel structure of a wafer-scale system.

in the channel to the inputs and outputs of the two cells. Given that
the interconnect is programmable, we shall adopt a usage of the term
"wire" to mean an electrically equivalent portion of the programmable
interconnect.

Systolic arrays [3–5] are a desirable architecture for VLSI
because all communication is between nearest neighbors. A realization
of a systolic array as a wafer-scale system may lose this advantage if
all nearest neighbors of a processor are dead, however, because a
long wire may be needed to connect electrically adjacent processors.
In general, the longest interconnection between processors is the
communication bottleneck of the system. Of the many possible ways in
which the live cells on a wafer can be connected to form a systolic
array, therefore, the one that minimizes the length of the longest
wire is most desirable.

To illustrate the subtleties inherent in configuring systolic arrays,
consider the problem of constructing a linear (i.e., one-dimensional)
array using all of the live cells in an N-cell wafer. Unfortunately, if
we wish to minimize the length of the longest wire, the problem is
NP-complete [6]. Even more discouraging is that there are some
arrangements of live and dead cells for which even the optimal linear
array has unacceptably long wires. Thus optimal solutions—even if
they could be found quickly—are not always practical.

By assuming a probabilistic model of cell failure, however, many
positive results can be proved. For example, Figure 7.4 illustrates a

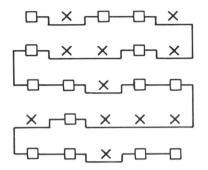

FIGURE 7.4 A simple means of constructing a linear systolic array
from the live cells on a wafer.

possible solution to the problem of connecting the live cells of a wafer
into a linear systolic array. The live cells, which are denoted by small
squares, are connected together, one after another, in a snakelike
pattern. Dead cells, denoted by Xs, are skipped over. With probability
at least $1 - O(1/N)$, the length of the longest wire is $O(\lg N)$, where
N is the number of cells in the wafer and where each cell independently
has a 50% chance of failure.†
     This bound comes from the observation that the length of the
longest wire that connects two cells in the array is just the length of
the longest sequence of dead cells in the snakelike string. For a given
set of k cells, the probability that all are dead is $1/2^k$, and thus the
probability that any set of $2 \lg N$ cells are dead is $1/N^2$. Since there
are less than N sets of $2 \lg N$ consecutive cells, the chances are thus
less than 1 in N of having to skip more than $2 \lg N$ cells in the entire
snakelike path of length N. Hence the maximum wire length is $O(\lg N)$
with probability at least $1 - O(1/N)$.
     To say that "with probability $1 - O(1/N)$ the maximum wire
length is $O(\lg N)$" is a substantially stronger statement than saying
that the expected maximum wire length is $O(\lg N)$. This is because

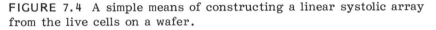

†Here and throughout we use $O(f(N))$ to denote a function that is
bounded above by $cf(N)$ for a fixed constant c and all sufficiently
large N. We also use $\Omega(f(N))$ to denote a function that is bounded
below by $cf(N)$, and Theta($f(N)$) to denote a function that is bounded
above by $c_1 f(N)$ and below by $c_2 f(N)$ for some fixed constants c, c_1,
and c_2, and all sufficiently large N. We also use $\lg N$ to denote $\log_2 N$,
$\lg^2 N$ to denote $(\lg N)^2$, and $\lg \lg^2 N$ to denote $(\lg \lg N)^2$. Finally,
$\lfloor x \rfloor$ denotes the largest integer less than or equal to x, and $\lceil x \rceil$ denotes
the smallest integer greater than or equal to x.

no wire can ever have length greater than $O(\sqrt{N})$, even in the worst case. Hence the expected maximum wire length is at most

$$(1 - O(1/N)) \cdot O(\lg N) + O(1/N) \cdot O(\sqrt{N}) = O(\lg N)$$

Moreover, the chances that the maximum wire length is much greater than $O(\lg N)$ are miniscule. In particular, the probability of having to skip more than $k \lg N$ dead cells at a fixed point in the snakelike path is less than 1 in N^k. Hence every wire has length at most $k \lg N$ with probability at least $1 - 1/N^{k-1}$.

In this chapter we present a survey of algorithms for realizing one- and two-dimensional systolic arrays as wafer-scale systems. Unlike many of the heuristics in the literature, the algorithms here have all been theoretically analyzed, and bounds on their quality have been mathematically proved. The analyses make the assumption that each cell fails independently with probability p, and for simplicity, we assume here that $p = 1/2$. We also assume for ease of explication and analysis that the width of a cell and the width of a wire are each unity. A more complete discussion of the assumptions and their generalizations can be found in [7].

The algorithms are organized to aid an engineer in picking an algorithm for implementation. We try to present enough mathematics to aid the user's intuition, but we do not, for the most part, include the detailed combinatorial arguments appearing in the literature that substantiate the effectiveness of the algorithms. Since programming involves many more "real-world" constraints than can be considered in an algorithmic analysis, we expect that the engineer might choose a less effective algorithm, for example, if it is easier to code. The algorithms here constitute a menu of possibilities to stimulate an intelligent design decision.

The remainder of the chapter is divided into four sections. Section 2 contains basic combinatorial facts underlying the probabilistic analyses used in the literature. Section 3 gives two algorithms for integrating linear arrays. The first algorithm connects all the live cells on a wafer, and the second achieves somewhat shorter maximum wire length by connecting only a large constant fraction of the live cells. Section 4 gives five algorithms for integrating two-dimensional arrays, and includes both worst-case and probabilistic bounds. In Section 5 we provide a summary of the material in this chapter and mention some related work.

2 COMBINATORIAL FACTS

In the preceeding section we showed that with probability at least $1 - O(1/N)$, a sequence of N cells on a wafer contains no more than

O(lg N) dead cells in a row. This type of high probability analysis underlies most of the algorithms in this paper. We shall use the term *high probability* to mean "with probability at least $1 - O(1/N)$," where N is the number of cells on a wafer. We now present some basic facts used in high-probability analyses.

The first fact is the standard definition of independence.

Fact 1 Let A and B be independent random variables. Then

$$Pr\{A \cap B\} = Pr\{A\}Pr\{B\}$$

The second fact bounds the probability of the union of two random events, even if the events are not independent.

Fact 2 Let A and B be random variables. Then

$$Pr\{A \cup B\} \leqslant Pr\{A\} + Pr\{B\}$$

Proof. This fact follows from the principle of inclusion and exclusion. We always have

$$Pr\{A \cup B\} = Pr\{A\} + Pr\{B\} - Pr\{A \cap B\}$$

and since $Pr\{A \cap B\} \geqslant 0$, the result follows.

Fact 2 provides a weak bound if the probabilities involved are large. For example, if the probability of the individual events are each greater than $1/2$, the bound on their union is trivial. When the probabilities are small, however, the bound can be useful.

The next fact bounds a linear function with an exponential. It is most useful when x is near 0.

Fact 3 For all x in the range $-\infty < x < \infty$, we have

$$1 + x \leqslant e^x$$

We now turn to combinatorial theorems that deal more directly with the statistics of faults on wafers. As was mentioned in Sec. 1, we shall typically assume that each cell on the wafer fails independently with probability $1/2$.

Fact 4 With high probability, a given rectangular pattern of live and dead cells of size 2 lg N never appears on an N-cell wafer.

Proof. The proof follows the analysis for the snakelike scheme in Sec. 1, which relies on Fact 2. The generalization from one- to two-dimensional regions is straightforward, as is the generalization from a pattern consisting solely of dead cells to an arbitrary pattern.

Of course, Fact 4 does not imply that no pattern will occur, only that the probability that a given pattern occurs is low. It is like the lottery: somebody will win, but probably not you.

Remarkably, patterns of slightly less than half the size almost always appear on a wafer.

Fact 5 With high probability, a given rectangular pattern of live and dead cells of size $\lg N - 2 \lg \lg N$ appears somewhere on an N-cell wafer.

Proof. Partition the wafer into $N/(\lg N - 2 \lg \lg N)$ rectangular regions of size $\lg N - 2 \lg \lg N$. The probability that a given one of the regions realizes the pattern is

$$2^{-\lg N + 2 \lg \lg N} = 1 - \frac{\lg^2 N}{N}$$

The probability that every region avoids the pattern is therefore

$$\left(1 - \frac{\lg^2 N}{N}\right)^{\frac{N}{\lg N - 2 \lg \lg N}} \leqslant e^{(-\frac{\lg^2 N}{N}) \frac{N}{\lg N - 2 \lg \lg N}}$$

$$= e^{-\frac{\lg^2 N}{\lg N - 2 \lg \lg N}}$$

$$\leqslant e^{-\lg N}$$

$$\leqslant \frac{1}{N}$$

using Facts 1 and 3.

In a region of m cells on a wafer, the expected number of live cells is $(1/2)m$. The actual number will vary, however. The next fact gives tight bounds on the expected deviation.

Fact 6 Let X be the random variable indicating the number of live cells in a region with m cells. Then the expectation of the deviation is

$$E\left(\left|X - \frac{1}{2} m\right|\right) = \Theta(\sqrt{m})$$

Fact 6 tells us that the expected deviation from the mean is $\Theta(\sqrt{m})$. We shall occasionally need to bound the actual probability of some given deviation. The next fact provides such a bound.

Fact 7 Let X be the random variable indicating the number of live cells in a region with m cells. Then for $r \geq 0$, the probability that the deviation exceeds $r\sqrt{m}$ is

$$\Pr\left\{\left|X - \frac{1}{2}m\right| > r\sqrt{m}\right\} = O(e^{-2r^2})$$

We can use Fact 7 to prove a lower bound on the number of live cells in each of a collection of sufficiently large regions. The next fact shows that if each region contains c lg N cells, for some sufficiently large constant c, then with high probability, there are a substantial number of live cells in the each of the regions.

Fact 8 For any c > 4, and for any particular collection of N regions on an N-cell wafer, each with at least c lg N cells, the probability is at least $1 - O(1/N)$ that every region contains $(1/2)c \lg N - \sqrt{c} \lg N$ live cells.

Proof. The probability that a given region does not contain at least $(1/2)c \lg N - \sqrt{c} \lg N$ live cells is $O(e^{-2 \lg N}) = O(1/N^2)$ by Fact 7. By Fact 2, the probability that all the N regions on the wafer, overlapping or not, fail to contain at least $(1/2)c \lg N - \sqrt{c} \lg N$ cells is at most $N \cdot O(1/N^2) = O(1/N)$.

3 INTEGRATING ONE-DIMENSIONAL SYSTOLIC ARRAYS

With high probability, the snakelike scheme described in Sec. 1 connects all the live cells on an N-cell wafer into a linear array with wires of length at most $O(\lg N)$. This section gives two procedures that substantially improve and generalize this bound. The first connects all the live cells on a wafer with wires of length $O(\sqrt{\lg N})$, and the second connects most of the live cells with wires of constant length.

Before presenting the algorithms, we first observe that with high probability, wires of length $\Omega(\sqrt{\lg N})$ are required to connect all the live cells on a wafer. The idea is that somewhere on the wafer, there is a live cell in the center of a square region of $\Omega(\lg N)$ dead cells, an observation that follows directly from Fact 5. (An example of such a region is shown in Fig. 7.5.) Therefore, a wire of length $\Omega(\sqrt{\lg N})$ is required to link the isoalted live cell to any other live cell.

3.1 Patching Method

The first algorithm for integrating a linear systolic array achieves the lower bound of $\Omega(\sqrt{\lg N})$ by partitioning the wafer into squares, forming linear arrays within each square, and then patching together

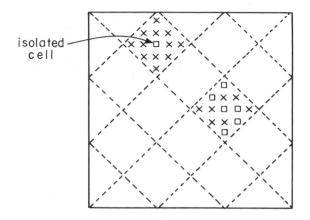

FIGURE 7.5 An example of an isolated cell.

the ends of the small linear arrays to yield a single linear array consisting of all the live cells on the wafer.

More precisely, the method is as follows. Partition the wafer into square regions containing $2 \lg N$ cells each, as is shown by the dashed lines in Fig. 7.6. The probability that each of the $2 \lg N$ cells are dead in one or more of the squares is less then $1/N$ by Fact 4. Thus, with high probability, each of the squares contains at least one live cell.

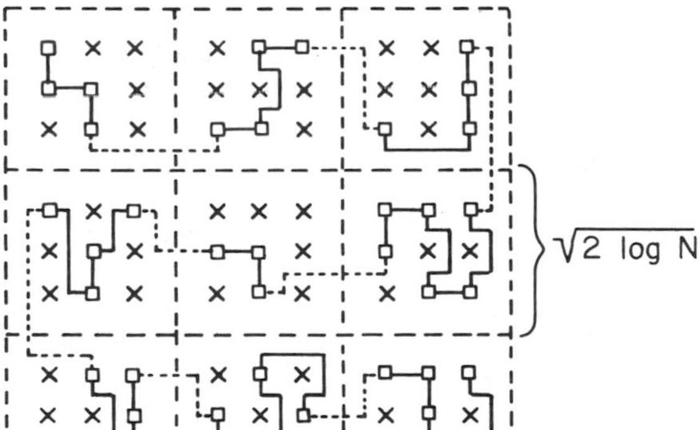

FIGURE 7.6 A scheme for constructing linear arrays from all live cells on a wafer with wires of length $O(\sqrt{\lg N})$ and constant channel widths.

Construct a linear array out of the live cells in each square
using a snakelike scheme on the columns of square, except that when
an empty column is encountered, skip over it. Figure 7.6 shows these
connections with solid lines. Since any pair of cells in the same square
can be linked with a wire of length at most $2\sqrt{2} \lg N$, the wires in each
array have length $O(\sqrt{\lg N})$. Next, add wires, shown by dotted lines
in the figure, to connect the small arrays into one large array. Because each region contains at least one live cell, these connections
can be made with wires of length at most $3\sqrt{2} \lg N$. Thus every wire
in the completed linear array has length $O(\sqrt{\lg N})$ with high probability.

3.2 Tree Method

If all the cells are incorporated in a linear array using the patching
method, the maximum wire length is $\theta(\sqrt{\lg N})$ with high probability.
But the proof of the lower bound suggests that isolated cells induce
the long wires. Instead of insisting that all live cells be incorporated
in the linear array, suppose we only require that most of the live cells
be included. In this section we describe a procedure that can construct
a linear array from almost all of the live cells with constant-length
wires.

The procedure relies on the fact that most live cells on the wafer
are near each other. More specifically, it has been proved [7] that
there exists a positive constant c such that for any d, with probability
$1 - O(1/N)$, at least $1 - O(2^{-cd^2})$ of the live cells on an N-cell wafer
can be connected in a tree using wires of length at most d. Up to
constant factors, this is the best possible bound.

The algorithm consists of two parts. First, a tree T of live cells
is constructed with wires of length at most d, and then the tree is
transformed into a linear array with wires of length at most 6d. (The
constant 6 is due in part to our assumption that the width of a wire
equals the width of a cell. If wire widths are substantially smaller,
the constant shrinks closer to 3.)

The tree T can be constructed by any of the algorithms that
compute the minimum spanning tree of a graph. In particular, Prim's
method [8−10] can be modified to compute the spanning tree in linear
time.

The construction of the linear array from the tree depends on a
result by Sekanina [11], which states that the cube of a nontrivial
connected graph always has a Hamiltonian circuit. Specifically, we now
show that without regard for wire widths, the linear array can be
constructed using wires of length 3d by tracing over wires in the
tree T no more than twice each. Since every wire is traced over at
most twice, the channel widths could (at worst) double in the resulting
wiring, thereby increasing the maximum wire length from 3d to 6d
when wire widths are accounted for.

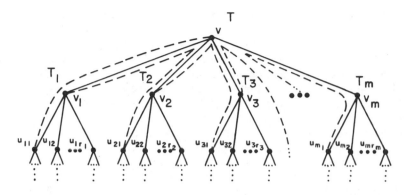

FIGURE 7.7 Constructing a linear array from a spanning tree.

Choose a node v to be the root of T, and let $T_1, T_2, \ldots, T_m$ be the subtrees of v as is shown in Fig. 7.7. (Degenerate cases not like Fig. 7.7 are easily handled, but we do not include the details here.) Recursively construct linear arrays on the nodes of $T_1, T_2, \ldots, T_m$ such that no wire has length greater than 3L, and so that the endpoints of the array in T_i are v_i and u_{i1} for $1 \leqslant i \leqslant m$. Then join the arrays in the subtrees by adding the following wires: (v, u_{11}), $(v_1, u_{21}), (v_2, u_{31}), \ldots, (v_{m-1}, u_{m1})$. (These wires are shown as dashed lines in Fig. 7.7.) Each of these wires has length at most 3L, and the resulting network is a linear array on the nodes of T with endpoints v and v_m. For completeness, we remark that the boundary conditions of the recursion are easily handled.

4 INTEGRATING TWO-DIMENSIONAL ARRAYS

The problem of linking the live cells on a wafer to form a square two-dimensional systolic array is substantially more difficult than the corresponding problem for linear arrays. The main difficulty with constructing two-dimensional arrays is that constant-length wires no longer suffice even if we throw away some of the live cells [12]. In fact, it has been shown [7] that with high probability, every realization of an M-cell two-dimensional array on an N-cell wafer has a wire of length $\Omega(\sqrt{\lg M})$ for all $M = \Omega(\lg^2 N)$. This result means, for example, that wires of length $\Omega(\sqrt{\lg N})$ are required to connect just 1% of the live cells.

For an algorithm to be effective in realizing a two-dimensional array, it must respect the two-dimensional constraints inherent in the problem. For example, consider the following naive algorithm for realizing an M-cell square two-dimensional array from all the live cells of an N-cell wafer. We assume for convenience that $M \approx N/2$ is a perfect square.

Take the top $\sqrt{M}$ live cells on the wafer, breaking ties randomly. These cells, in order left to right, make the first row of the array. Take the top $\sqrt{M}$ cells of the remainder as the second row, in order left to right, and continue similarly to make each row of the array. With high probability no row of the array contains cells from more than three rows of the wafer because Fact 8 guarantees that every row contains nearly $(1/2)\sqrt{N} \approx 0.7\sqrt{M}$ live cells.

At first, this method does not seem so bad because (Fact 5) the horizontal connections among the cells of the array have length $\Theta(\lg N)$. The vertical connections are much worse, however. Consider a vertical line that divides the wafer into left and right halves. Fact 6 says that we can expect that the number of cells in a given row on one side of the dividing line is at least $\Omega(\sqrt{\sqrt{M}}) = \Omega(N^{1/4})$ larger than the number on the other side. Thus, with constant probability, the midpoint of the row is at least $\Omega(N^{1/4})$ cells away from the dividing line. Two consecutive rows have their midpoints on opposite sides of the dividing line half the time, and thus, with constant probability, a wire connecting the two midpoints has length $\Omega(N^{1/4})$. Since there are $\sqrt{M}$ rows, there is a wire of length $\Omega(N^{1/4})$ between two of them with high probability. A bound of $\Theta(N^{1/4}\sqrt{\lg N})$ for the maximum wire length in the resulting array can be shown with more detailed analysis.

4.1 Tree-of-Meshes Method

In this section we present an algorithm that can construct a two-dimensional array from all the live cells of an N-cell wafer if the channels have width $\Omega(\lg N)$. All possible configurations of live and dead cells, however unlikely, can be handled by this technique, but the wire-length bounds are not good. This result will be used as a subroutine in the divide-and-conquer and patching methods to achieve better bounds for wire length on average-case wafers.

We first show how an N-cell wafer with channels of width $\Theta(\lg N)$ can be viewed as an N-leaf tree of meshes [13,16]. The tree of meshes is constructed from a complete binary tree by replacing nodes of the tree with meshes and single edges of the tree with bundles of edges linking the meshes. Figure 7.8 shows a 16-leaf tree of meshes. The root of an N-leaf tree of meshes is a $\sqrt{N}$-by-$\sqrt{N}$ mesh. (We assume for simplicity that $\sqrt{N}$ is a power of 2.) The nodes at the second level are $\sqrt{N}/2$-by-$\sqrt{N}$ meshes, those at the third level are $\sqrt{N}/2$-by-$\sqrt{N}/2$ meshes, and so on until the leaves are replaced by 1-by-1 meshes.

The correspondence between the N-cell wafer and the N-leaf tree of meshes is established as follows. The first step is to construct a lg N-layer three-dimensional layout [17,18] of the tree of meshes. Fold the connections between the root of the tree of meshes and each of

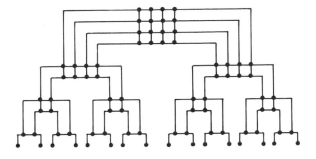

FIGURE 7.8 The 16-leaf tree of meshes.

its two children so that the children fit naturally on a second layer
over the root. Fold the connections to each of the grandchildren so
that they fit naturally over the children on a third layer, and so on.
This procedure generates a lg N-layer three-dimensional layout where
each layer has area N. Next, project the three-dimensional layout
onto a single layer in the manner of [19, pp. 36−38]. Locate cells of
the wafer at the leaves of the tree of meshes. The cross points of
the meshes become programmable switches, and the wires of the meshes
become the wires in lg N-width channels.

   We now wish to make a two-dimensional array from the $M \approx N/2$
live leaves of the tree of meshes. (In general, an exact square array
is not possible, and thus we shall assume the array to be formed is
missing some border cells, as is shown in Fig. 7.9.) We first use divide-
and-conquer to assign each cell a number from 1 to M. We chop the
M-cell array in half vertically into two subarrays with $\lfloor M/2 \rfloor$ and
$\lceil M/2 \rceil$ cells. We recursively assign numbers from 1 to $M/2$ to the first
subarray and numbers from $\lceil M/2 \rceil$ to M to the second subarray, alternating the orientation of the cut between horizontal and vertical at
each recursive step.

   The assignment is now simple. The ith cell of the array is mapped
to the ith live leaf of the tree of meshes counting from left to right.
After swelling the channel capacities by a small constant factor to
accommodate the wires, adjacent cells can be connected by routing
wires through the unique path in the underlying complete binary tree.
Routing through the meshes can be done by treating them as cross-
point switches. The wire lengths are $O(\sqrt{N} \lg N)$ since we need to
route across $O(\sqrt{N})$ channels of width $\Theta(\lg N)$.

   As a practical matter, the tree of meshes need not be used
directly for routing wires. The assignment algorithm can be used to
establish the correspondence between the two-dimensional array and
the live cells of the wafer, and then the wires can be routed using a
standard gate-array routing program. In the case when $\sqrt{M}$ is an

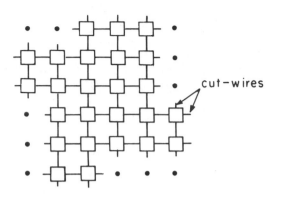

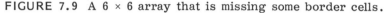

FIGURE 7.9 A 6 × 6 array that is missing some border cells.

exact power of 2, the assignment is particularly simple. The kth live
cell corresponds to the (i,j) position of the array, where i is obtained
by concatenating the even bits of the binary representation of k, and
j is obtained by concatenating the odd bits.

4.2 Divide-and-Conquer Method

The tree-of-meshes algorithm works as well as might be expected
in the worst case, and thus it is natural to wonder how well it works
on average. Unfortunately, the algorithm works poorly in a probabilistic model because the maximum wire length is nearly always large.
In this section we present a similar divide-and-conquer algorithm
which works poorly in the worst case, but which can be proved to
work extremely well on average. With high probability, the algorithm
connects all the live cells of an N-cell wafer with channels of width
O(lg lg N) using wires of length O(lg N lg lg N).

The divide-and-conquer algorithm has two stages. In the first
stage, the wafer is recursively bisected, and the number of live cells
in each half is counted. Based on the count of live cells in each half
of the wafer, the algorithm computes the dimensions of the two subarrays that must be constructed, and then recursively constructs the
subarrays. The two subarrays are then linked together to form the
complete array. The algorithm remains in the first stage as long as
the distribution of cells within the current region of the wafer is good,
which (with high probability) is until subproblems with Θ(lg N) cells
are encountered. Below this point, the distribution of cells can be
arbitrarily bad, and thus the algorithm uses the tree-of-meshes
technique to complete the wiring of a Θ(lg N)-cell subarray. The
exact crossover point between the first and second stages can be set
at subproblems of size c lg N, where c is any constant sufficiently

large to ensure that with high probability, every c lg N-cell region contains $\Omega(\lg N)$ live cells. That such a c exists is a consequence of Fact 8.

Figures 7.10 through 7.13 illustrate the divide-and-conquer procedure. Figure 7.10a shows a 64-cell wafer which contains 36 live cells. In what follows, we step through the algorithm as it constructs a 6-by-6 array, which is identified as the "overall target" in Fig. 7.10b.

The first step is to bisect the wafer vertically, which gives 19 live cells in the left half and 17 in the right. We wish to construct 19-cell subarray in the left-half wafer and a 17-cell subarray in the right-half wafer. Since we want the two subarrays to fit together nicely after they have been constructed, we choose the shapes of the two subarrays that are determined by the partition of the 6-by-6 array shown in Fig. 7.11.

We now invoke the procedure recursively on the two subarrays, but this time we bisect each of the halves horizontally. For example, when the left half wafer is bisected, the 19 live cells are divided into 9 cells above and 10 cells below, as displayed in Fig. 7.12. The algorithm continues in this fashion, alternating between horizontal and vertical divisions, until the wafer and the target have been partitioned into $\Theta(\lg N)$-cell regions, at which point the algorithm proceeds to the second stage, and the tree-of-meshes technique is applied.

```
□  ×  □  □  ×  ×  □  ×
□  □  ×  ×  ×  □  □  ×
×  □  ×  ×  □  ×  □  □
□  □  ×  □  □  ×  ×  ×
□  □  □  ×  ×  □  □  ×
□  ×  ×  □  ×  □  ×  □
□  □  □  ×  ×  □  □  □
×  □  □  ×  □  ×  □  □
```

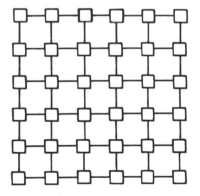

FIGURE 7.10 (a) A 64-cell wafer that contains 36 live cells. (b) The target: a 6-by-6 systolic array.

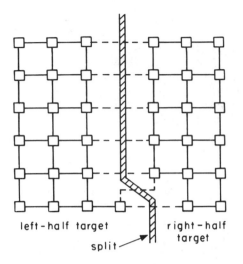

FIGURE 7.11 Partitioning the target.

In this example the number of cells is small enough that the second-stage construction can be performed by inspection. The inspection strategy can be used effectively in practice. Since the second stage operates on regions of size $\Theta(\lg N)$, the routings of this size can conceivably be precomputed. The second stage then consists of a single table look-up.

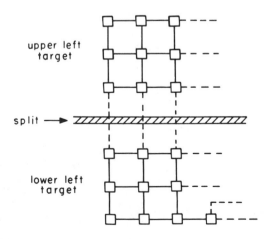

FIGURE 7.12 Partitioning the left target.

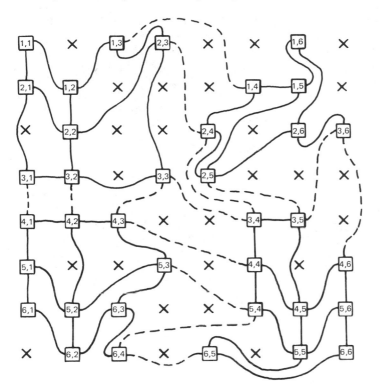

FIGURE 7.13 Completed cell assignment and wiring of the 6-by-6 array.

Figure 7.13 shows the final solution to the problem in Fig. 7.10. For clarity the wires have not been routed within the channels of the wafer. Notice that each quadrant contains the specified targets for the second level of recursion. The dashed lines represent wires that connect cells in different quadrants of the wafer.

With probability $1 - O(1/N)$ the divide-and-conquer method can construct a two-dimensional array from all the live cells on an N-cell wafer using wires of length $O(\lg N \lg \lg N)$ and channels of width $O(\lg \lg N)$. It is not too difficult to see that these bounds hold with probability 1 for the regions of size less than $c \lg N$ that are connected by the tree-of-meshes procedure. Plugging in $c \lg N$ for N in the tree-of-meshes bound yields wires of length $O(\sqrt{\lg N} \lg \lg N)$ and channels of width $O(\lg \lg N)$.

The hard part is showing that the wiring in the upper levels of recursion satisfy the bounds. The analysis, which we briefly sketch, assumes that during the recursion, the channel dividing a subwafer with m > c lg N cells has width $\theta\sqrt{\lg N \lg m}$. Uniform channel widths of lg lg N across the entire wafer can later be obtained by distributing the wider channels across neighboring channels, which does not asymptotically increase the wire lengths in the subsequent analysis.

We begin at the first level of recursion. Consider the wires that link a cell in the left subarray to a cell in the right subarray, as illustrated by the two examples in Fig. 7.14. For the most part, the connecting wires can be routed in the channel that separates the left and right halves of the wafer. The length of the longest wire in the channel is proportional to the longest vertical distance that a single wire must traverse, as is the width of the channel itself.

The length of the longest wire in the center channel depends on the distribution of cells in each quadrant. For example, if we are extremely lucky and the live cells are regularly spaced, the longest wire may have constant length, as in Fig. 7.14a. But if we are very unlucky, half the live cells might occur in the upper right quadrant and the other half in the lower left quadrant (Fig. 7.14b). To connect the two halves in this latter case, some wire must have length $\Omega(\sqrt{N})$.

The length of the longest wire in the center channel can also be influenced by the distribution of cells within a quadrant. For example, if the upper left quadrant contains $\sqrt{N/8}$ live cells (about the right number), but they are distributed as in Fig. 7.15, the center channel still contains a wire of length $\Omega(\sqrt{N})$.

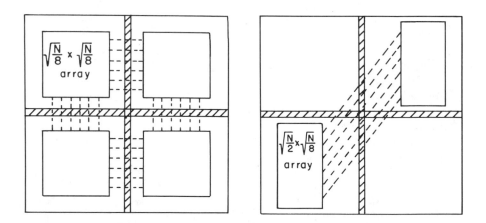

FIGURE 7.14 (a) A distribution of live cells which might allow a narrow center channel. (b) A distribution of live cells which requires a wide center channel.

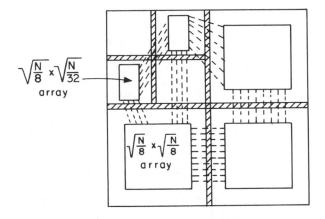

$\sqrt{\dfrac{N}{8}} \times \sqrt{\dfrac{N}{32}}$
array

$\sqrt{\dfrac{N}{8}} \times \sqrt{\dfrac{N}{8}}$
array

FIGURE 7.15 Another distribution of live cells which requires a wide center channel.

Most often, we are not so unlucky that a wire in the center channel has length $\Omega(\sqrt{N})$, but neither are we lucky enough that all wires are constant length. With high probability, we are more lucky than unlucky because the length of the longest wire in the center is $O(\lg N)$. The idea is that the live cells are distributed so evenly that with high probability, the total vertical distortion of the wires in the center channel (over all subproblems of size $\Omega(\lg N)$) is $O(\lg N)$. For channels dividing a subwafer of size $m > c \lg N$, the vertical distortion is $O(\sqrt{\lg m \lg N})$. Thus the channel width bounds assumed earlier suffice.

The wire-length analysis of the divide-and-conquer algorithm is fairly tight. For example, the algorithm requires wires of length $\Omega(\lg N)$ with high probability. Thus if the lower bound of $\Omega(\sqrt{\lg N})$ is to be achieved, a different algorithm must be discovered. It may be possible to improve the channel-width bound, however. For example, any improvement in the worst-case bound given by the tree-of-meshes technique would lead directly to an improvement in the channel-width bounds for the divide-and-conquer algorithm.

4.3 Patching Method

Not surprisingly, we can improve the wire length bounds if we need only construct a two-dimensional array from most of the live cells on a wafer. In particular, we can use a scheme similar to the patching scheme from Sec. 3.1 to construct a two-dimensional array from any constant fraction (less than 1) of the live cells on an N-cell wafer using wires of length $O(\sqrt{\lg N} \lg \lg N)$ and channels of width $O(\lg \lg N)$. These bounds are also achieved with high probability.

The key idea is to partition the wafer into $N/c \lg N$ square regions, each containing $m = c \lg N$ cells. According to Fact 8, we can choose c sufficiently large such that with probability $1 - O(1/N)$, each of the regions contains at least $m' = (1/2)c \lg N - \sqrt{c} \lg N$ live cells. Using the tree-of-meshes technique, we can therefore construct an m'-cell two-dimensional array in each region using wires of length $O(\sqrt{m} \lg m) = O(\sqrt{\lg N} \lg \lg N)$ and channels of width $O(\lg m) = O(\lg \lg N)$. The $N/c \lg N$ two-dimensional arrays are then connected together into one large array with $(1/2)N(1 - 2/\sqrt{c})$ live cells. The added wires also have length at most $O(\sqrt{\lg N} \lg \lg N)$, and can easily fit into the $\Theta(\lg \lg N)$-width channels.

The patching method can be thought of as a refinement of the divide-and-conquer method that throws away a fraction of the cells at each level of recursion. The actual decisions to which cells at a given level are thrown away can be postponed until lower in the recursion, but it is important that at each level, every region of the wafer have exactly the same number of live cells.

4.4 Greene's Method

The next method, due to Greene [20], also connects any constant fraction of the live cells on an N-cell wafer into a two-dimensional array. With high probability, it uses wires of length $\Theta(\sqrt{\lg N})$ and channels of constant width, thus achieving the lower bound for integration of two-dimensional arrays. It is similar to the algorithm presented at the beginning of this section in that it creates rows of the array, but it is considerably more clever. The algorithm that determines the rows and columns of the array is based on network flow techniques, but we present it in a manner that does not require a knowledge of combinatorial optimization.

Greene's algorithm can construct a $(1 - e)\sqrt{N}$-by-$(1/2)(1 - e)\sqrt{N}$ array, for any constant $e > 0$. For any such e, we require the N-cell wafer to have channels of width w, where w is a sufficiently large constant that depends on e. The higher the percentage of cells we wish to integrate into an array, the wider we must make the channels.

Partition the wafer as shown in Fig. 7.16 into blocks of size 1-by-$c_1\sqrt{\lg N}$ such that there are $\sqrt{N}$ rows of blocks and $\sqrt{N}/c_1\sqrt{\lg N}$ columns of blocks, where c_1 is a constant depending on e. Mark a block as *bad* if it contains fewer than t live cells, and *good* otherwise, where t is also a constant depending on e. For the exact values of constants, we refer the reader to [20].

The first part of the algorithm determines tentative rows for the array. We divide the w vertical tracks between blocks on the wafer into two bundles, each consisting of $w/2$ tracks. For this part of the

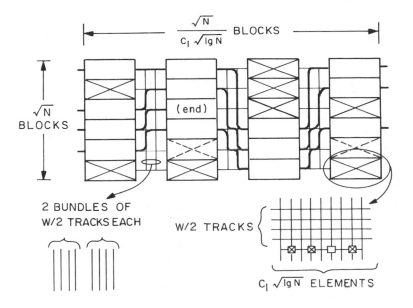

FIGURE 7.16 Forming the tentative rows in Greene's method. Blocks containing fewer than t live cells are marked with solid X's. Blocks marked as bad during the scan are marked with dashed X's.

algorithm, we will treat the two bundles as two routing tracks. Later, we will need to reexpand the capacity of the two tracks by $w/2$ each.

The algorithm first determines $(1 - e)\sqrt{N}$ horizontally running chains from the left edge of the wafer to the right edge through the good blocks. The chains must satisfy the constraint that no wire is longer than $c_2\sqrt{\lg N}$, for some constant c_2 depending on e. The algorithm determines the chains in the following manner. Scan the columns of blocks left to right. For each column, proceed through the blocks from top to bottom. At each point, if the current block is good, we attempt to connect it to a good block on the left. This connection is made to the uppermost good block within distance $c_2\sqrt{\lg N}$, up or down, from the current block that has not yet been connected to a block in the current column. It must also satisfy the constraint that the routing does not exceed the channel capacity of 2. If such a connection cannot be made, we mark the current block as bad. Block $(5,2)$ in Fig. 7.16 is marked bad for this reason. Some chains are terminated by this procedure—for example, the chain ending in block $(3,2)$ of the figure. With high probability, however, this procedure establishes $(1 - e)\sqrt{N}$ horizontal chains, each with $\sqrt{N}/c_1\sqrt{\lg N}$ blocks.

The horizontally running chains can be viewed conceptually as shown in Fig. 7.17. We now expand the blocks in the chains to see

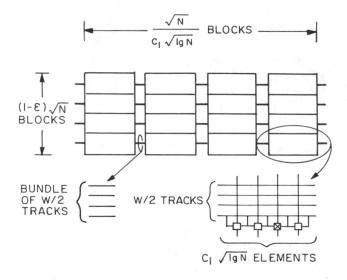

FIGURE 7.17 Normalized view of the rows of blocks.

their internal structure, as shown in Fig. 7.18. The horizontal tracks in Fig. 7.18 actually correspond to sections of both horizontal and vertical tracks in Fig. 7.16 because the chains run both horizontally and vertically. The horizontal channels in Fig. 7.18 have $w/2$ tracks, and thus the two vertical tracks between blocks in Fig. 7.16 must each be expanded by $w/2$ to accommodate the wires we shall now route to make the vertical connections.

We establish the vertically running chains by essentially the same procedure as before, except we scan top to bottom and route through horizontal channels of width $w/2$. With high probability, the algorithm constructs $(1/2)(1 - e)\sqrt{N}$ vertical chains. The horizontal chains are now modified to include only those cells used in the vertical chains, which completes construction of the $(1 - e)\sqrt{N}$-by-$(1/2)(1 - e)\sqrt{N}$ array. All channels are constant width w, and it turns out to be the case that all wire lengths are $O(\sqrt{\lg N})$.

Greene's method generates a rectangular array with aspect (length to width) ratio 2, but we may wish to realize a square array without throwing away half the cells. by embedding a $(1 - e)\sqrt{N/2}$-by-$(1 - e)\sqrt{N/2}$ square array into a $(1 - e)\sqrt{N}$-by-$(1/2)(1 - e)\sqrt{N}$ rectangular array so that adjacent cells of the square array are constant distance away in the rectangular array, we can use Greene's method directly. The first row of the square is embedded in the first two rows of the rectangle such that all the first row of the rectangle is used and an evenly spaced portion of the second row is used. We

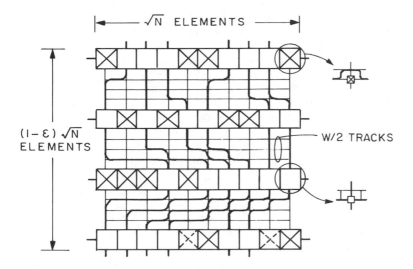

FIGURE 7.18 Forming the columns in Greene's method. Dead cells are marked with solid Xs. Cells marked as bad during the scan are marked with dashed Xs.

connect the cells of the first row of the square linearly left to right in the rectangle. The second row of the square is embedded linearly in the second and third rows of the rectangle using all the remaining cells in the second row and a uniformly spaced portion of cells in the third row. The third row of the square uses all the remaining cells in the third row of the rectangle, all the cells in the fourth row, and a uniformly spaced portion of cells from the fifth row. We continue in this fashion until the embedding is completed. Every adjacent pair of cells in the square array is within horizontal and vertical distances of four cells in the rectangular array. This procedure can be generalized to construct any rectangular array of any aspect ratio.

4.5 Matching Method

We conclude with a method whose proven bounds are not as good as those presented thus far, but which is nevertheless interesting. In the case of widthless wires, this method, which is based on bipartite matching in a graph, can integrate all the cells on an N-cell wafer with wires of length $O(\lg^{3/4} N)$. When we consider the normal case of unit width wires, however, we could conceivably need channels of width $\Theta(\lg^{3/4} N)$, and because the wires would need to cross these channels, wires of length $O(\lg^{3/2} N)$. This algorithm is certainly worth considering when wire widths are small because the $O(\lg^{3/4} N)$

wire-length bound is better than the bound of $\Theta(\lg N)$ which the divide-and-conquer method yields for widthless wires. Moreover, the true performance of the matching method might be better than that suggested by the upper bound for unit-width wires. In comparison, the divide-and-conquer method has a hard lower bound of $\Theta(\lg N)$ even for widthless wires. In addition, the algorithm is easily tailored to handle the situation when we wish to integrate any constant fraction of the live cells, in which case the widthless wire bound shrinks to $\Theta(\sqrt{\lg N})$, which is optimal.

The first step of the matching method is to determine the number M of live cells on an N-cell wafer. Then we pick a target wire length d that we hope to achieve. The algorihtm now determines the locations of points in a uniform $\sqrt{M}$-by-$\sqrt{M}$ grid superimposed on the wafer. It then constructs a bipartite graph between the grid points and the live cells of the wafer with an edge between a grid point and a live cell if the distance between them is at most d. Then, using a bipartite matching algorithm [9], the procedure determines whether every grid point can be matched one-to-one with a live cell. If a perfect matching exists, we know a routing of the corresponding assignment with widthless wires has maximum edge length d.

It is possible to show [21,22] that if $d = \Theta(\lg^{3/4} N)$, the matching succeeds with high probability. As a practical matter, it is better to search for the smallest d that works for a given wafer using exponential search. Try $d = 1, 2, 4, 8, \ldots$ until a value of d is found that results in a perfect mathcing, and then binary search to find the exact value.

The same technique can be applied to construct a two-dimensional array from any number $m < M$ of the M live cells by using a $\sqrt{m}$-by-$\sqrt{m}$ grid. For the case when $m = (1 - e)M$, it can be shown that wires have length $O(\sqrt{\lg N})$ with high probability.

Table 7.1 Bounds for One-Dimensional Arrays

Method	Portion of cells used	Maximum wire length	Maximum channel width
Patching	All	$\Theta(\sqrt{\lg N})$	$\Theta(1)$
Optimal	All	$\Theta(\sqrt{\lg N})$	$\Theta(1)$
Tree	99%	$\Theta(1)$	$\Theta(1)$

Table 7.2 Bounds for Two-Dimensional Arrays

Method	Maximum wire length for widthless wires	Maximum channel width	Maximum wire length for unit-width wires
Worst-Case Wafer, Using All Live Cells			
Tree of meshes	$\Theta(\sqrt{N})$	$O(\lg N)$	$O(\sqrt{N} \lg N)$
Optimal	$\Theta(\sqrt{N})$	$\Omega(1)$	$\Omega(\sqrt{N})$
Average-Case Wafer, Using All Live Cells			
Divide and conquer	$\Theta(\lg N)$	$O(\lg \lg N)$	$O(\lg N \lg \lg N)$
Matching	$O(\lg^{3/4} N)$	$O(\lg^{3/4} N)$	$O(\lg^{3/2} N)$
Optimal	$\Omega(\sqrt{\lg N})$	$\Omega(1)$	$\Omega(\sqrt{\lg N})$
Average-Case Wafer, Using 99% of the Live Cells			
Patching	$\Theta(\sqrt{\lg N})$	$O(\lg \lg N)$	$O(\sqrt{\lg N} \lg \lg N)$
Greene	$\Theta(\sqrt{\lg N})$	$\Theta(1)$	$\Theta(\sqrt{\lg N})$
Matching	$\Theta(\sqrt{\lg N})$	$O(\sqrt{\lg N})$	$O(\lg N)$
Optimal	$\Theta(\sqrt{\lg N})$	$\Theta(1)$	$\Theta(\sqrt{\lg N})$

5 SUMMARY AND CONCLUSIONS

The content of this chapter is taken primarily from Ref. 7 and somewhat from Refs. 20 and 12. The algorithms presented are summarized
in Tables 7.1 and 7.2. The literature contains many more techniques
for integrating systolic arrays. Manning [23,23], Hedlund and Snyder
[25], Koren [26], and Fussell and Varman [27] look at the basic
problem of constructing arrays from wafers containing faulty cells.
Rosenberg [28,29], Chung et al. [30,31], and Bhatt and Leighton
[13] have also investigated fault tolerance.

ACKNOWLEDGMENTS

We would like to thank the members of the MIT Lincoln Laboratory
Restructurable VLSI project for acquainting us with the details of their
work and for providing the photographs in Figs. 7.1, 7.2, and 7.3.

This research was supported in part by the Defense Advanced Research Projects Agency under Contract N00014-80-C-0622 and in part by the Air Force under Contract OSR 82-0326. Tom Leighton and Charles Leiserson are both supported in part by NSF Presidential Young Investigator Awards. Portions of this work are based on the paper "Wafer-scale integration of systolic arrays" by Leighton and Leiserson, which appeared in *IEEE Transactions on Computers,* Vol. C-34, No. 5, pp. 448—461, May 1985, © 1985 IEEE.

REFERENCES

1. J. I. Raffel, On the use of nonvolatile program links for restructurable VLSI, *Proc. CalTech Conference on VLSI,* 95—104, (Jan. 1979).

2. J. Logue, W. Kleinfelder, P. Lowy, J. Moulic, and W. Wu, Techniques for improving engineering productivity of VLSI designs, *Proc. IEEE International Conference on Circuits and Computers* (1980).

3. H. T. Kung, Why systolic architectures, *Computer* (IEEE), *15*: pp. 37—46 (Jan. 1982).

4. H. T. Kung and C. E. Leiserson, Systolic arrays (for VLSI), in *Sparse Matrix Proc.* (I. S. Duff and G. W. Stewart, eds.), Society for Industrial and Applied Mathematics, Philadelphia, PA 256—282 (1978).

5. C. E. Leiserson, *Area-Efficient VLSI Computation,* MIT Press, Cambridge, MA (1983).

6. A. Itai, C. H. Papadimitriou, and J. L. Szwarcfiter, Hamiltonian paths in grid graphs, *SIAM J. Comput., 11*: 676—686 (1982).

7. F. T. Leighton and C. E. Leiserson, Wafer-scale integration of systolic arrays, *IEEE Trans. Comput., C-34*: 448—461 (1985).

8. A. V. Aho, J. E. Hopcroft, and J. D. Ullman, *Data Structures and Algorithms,* Addison-Wesley, Reading, Mass. (1983).

9. S. Even, *Graph Algorithms,* Computer Science Press, Rockville, Md. (1979).

10. R. C. Prim, Shortest connection networks and some generalizations, *Bell Syst. Tech. J., 36*: 1389—1401 (1957).

11. M. Sekanina, On an ordering of the set of vertices of a connected graph, *Publications of the Faculty of Science, University Brno,* Brno, Czechoslovakia, No. 412, 137—142 (1960).

12. J. W. Greene and A. El Gamal, Configuration of VLSI arrays in the presence of defects, *J. ACM 31*: 694–717 (1984).

13. S. N. Bhatt and F. T. Leighton, A framework for solving VLSI graph layout problems, *J. Comput. Syst. Sci.*, *28*: 300–343 (1984).

14. F. T. Leighton, *Complexity Issues in VLSI: Optimal Layouts for the Shuffle-Exchange Graph and Other Networks*, MIT Press, Cambridge, MA (1983).

15. F. T. Leighton, New lower bound techniques for VLSI, *Math. Sys. Theory*, *17*: 47–70 (1984).

16. F. T. Leighton, A layout strategy for VLSI which is provably good, *Proceedings of the 14th ACM Symposium on Theory of Computing*, pp. 85–98 (May 1982).

17. F. T. Leighton and A. L. Rosenberg, Three-dimensional circuit layouts, *SIAM J. Comput.*, *15*(3): 793–813 (1986).

18. A. L. Rosenber, Three-dimensional integrated circuitry, *VSLI Systems and Computations* (H. T. Kung, R. Sproull, and G. Steele, eds.) Computer Science Press, Rockville, MD. 69–80, (Oct. 1981).

19. C. D. Thompson, A complexity theory for VLSI, Ph.D. dissertation, Carnegie-Mellon University, (1980).

20. J. W. Greene, Configuration of VLSI arrays in the presence of defects, Ph.D. dissertation, Stanford University, (1983).

21. F. T. Leighton and P. W. Shor, Tight bounds for minimax grid matching, with applications to the average-case analysis of algorithms, *Proc. 18th ACM Symp. on Theory of Computing*, pp. 91–103 (1986).

22. P. W. Shor, Average-case analyses for bin packing and planar problems, Ph.D. dissertation, MIT (1985).

23. F. Manning, Automatic test, configuration, and repair of cellular arrays, Ph.D. dissertation, MIT (1975).

24. F. Manning, An approach to highly integrated, computer-maintained cellular arrays. *IEEE Trans. Comput.*, *C-26*, 536–552 (1977).

25. K. Hedlund and L. Synder, Wafer-scale integration of configurable, highly parallel (CHiP) processors, *Proc. IEEE International Conference on Parallel Processing*, 262–264 (1982).

26. I. Koren, A reconfigurable and fault-tolerant VLSI multiprocessor array, *Proc. 8th Annual IEEE/ACM Symposium on Computer Architecture*, 425–431, (May 1981).

27. D. Fussell and P. Varman, Fault-tolerant wafer-scale archi-
     tectures for VLSI, *Proc. 9th Annual IEEE/ACM Symposium on
     Computer Architecture,* 190—198 (Apr. 1982).

28. A. L. Rosenberg, The Diogenes approach to testable fault-
     tolerant networks of processors, *Technical Report CS-1982-6.1,*
     Department of Computer Science, Duke University, Durham,
     N.C. (May 1982).

29. A. L. Rosenberg, On designing fault-tolerant arrays of proceeors,
     Technical Report CS-1982-14, Duke University, Durham, N.C.
     (1982).

30. F. R. K. Chung, F. T. Leighton, and A. L. Rosenberg,
     Diogenes: a methodology for designing fault-tolerant VLSI
     processor arrays, *Proc. IEEE Symposium on Fault-Tolerant
     Computing,* 26—31, (June 1983).

31. F. R. K. Chung, F. T. Leighton, and A. L. Rosenberg, Em-
     bedding graphs in books: a layout problem with applications to
     VLSI design, *SIAM J. Alg. & Disc. Meth.,* *8*(1): 33—58 (1985).
     (1985).

8

Fault Tolerant Computing Approaches

MARIAGIOVANNA SAMI and RENATO STEFANELLI
Department of Electronics, Politecnico di Milano, Milano, Italy

1 INTRODUCTION

Historically, fault-tolerant techniques have been adopted in discrete-component computing systems as a consequence of a particular application requirement. It can actually be claimed that fault tolerance itself, as a subject of research, has been originated when applications in the aerospace, telecommunications, or other mission-critical areas have been envisioned. When complex very large scale integration (VLSI) architectures have become feasible, new problems—mostly application independent—have made it mandatory to adopt fault-tolerance techniques for system design.

In fact, the very intrinsic complexity of a VLSI system—whatever its specific nature and applications—creates a number of problems that require use of fault-tolerance techniques. We list briefly the most important ones:

1. Testing is a first, basic issue; both structural complexity and relative scarcity of control and observation points make it very difficult to adopt in a straightforward manner the "classical" approaches to digital testing. It becomes necessary to introduce design-for-testability techniques (see, e.g., Ref. 1) or even to provide the system with self-testing capacity.

2. High reliability and long lifetime are increasingly requested even in areas that are not mission critical, due to high maintenance costs. On the other hand, as is well known, device complexity actually decreases intrinsic reliability. So-called "fault-avoidance" approaches would be unrealistic; rather, it is necessary to use design techniques that through the introduction of suitable redundancies allow us to overcome faults and to survive possibly without performance degradation or, at least, with predetermined "graceful" degradation.

3. High production yield is mandatory for economic reasons, but again, clashes with device complexity. Already for some particular devices (typically, memories) it has been proved that adoption of internal redundancies and *production-time* reconfiguration allows to reach satisfactory yield figures.

In the present chapter we do not deal explicitly with the complex testing problem. We assume that self-testing is available in the VLSI architectures that we consider, and we examine only one possible technique as strongly related to the fault-tolerance ones we discuss in detail. Problems considered will then be the ones regarding production yield and device reliability.

Classical, general approaches such as triple modular redundancy (TMR) are scarcely applicable in this context. In fact, silicon-area requirements would increase more than threefold with respect to the nominal device area, with a consequent relevant decrease in device mean time between failures (MTBF), while in the worst case, only a single error could be corrected and a second one detected. Such techniques may become interesting only for particular devices in which, due to functions implemented or to specific structural characteristics, better error-recovery performances can be achieved (an example is offered by arithmetic-logic units (ALUs) [2]).

Fault-tolerance approaches for VLSI devices may follow two main lines: they can be function dependent or architecture dependent. Let us explore briefly the two alternatives.

1. *Function-dependent approaches*. In this case, the particular technique adopted (use of specific coding, use of some form of information compression for the self-test phase, etc.) are strongly related to the specific function implemented by the device or, more generally, by the single section of a more complex device. The main scope is to achieve detection and correction of (possibly large) fault distributions while limiting redundancy. Thus function-dependent techniques may be considered in three different contexts:

a. *Single-function devices*. The best known instance is offered by memories [in particular, random access memories (RAMs)]. An extensive literature is available; most solutions are based on the use of suitable coding techniques, adapted to the error model best suited to a particular chip implementation [3—5]. A number of the largest memory chips present on the market use such techniques principally to achieve acceptable yields (in this case, testing is performed at production time and faulty cells are discarded by a static technique such as laser cutting). A more recent solution, aiming at wafer-scale memory devices, is independent of coding techniques; rather, the entire memory is considered as a long "string" of cells that can be connected in suitable order to obtain a working device.

Another interesting class of devices is that of arithmetic units. Again, coding has been suggested by various authors as a possible solution [6,7]; use of particular algebras, foremost among them residue arithmetics, has been explored extensively as an alternative solution [8−11]. These proposals are based on error detection through duplication and comparison of results. To limit the cost increase (here evaluated as silicon-area requirements), whereas the "main" arithmetic unit is implemented in conventional arithmetics, the alternative one is implemented in residue arithmetics. Residues of the inputs are applied to it, and its results are compared with the residue of the main ALU's result, suitably evaluated. If residues are computed with respect to more than one base, the difference between the results of the alternative unit and the residue of the main ALU result may carry sufficient information to allow not only detection, but also correction of one (as in Ref. 10) or two (as in Ref. 11) faults.

A final class of devices that has recently received attention is that of microprogrammed control units; again, the immediate solution (use of massive replication) would not only involve an excessive amount of added silicon area, but also actually check—or even correct—a limited number of possible faults. Proposals in the recent literature make use of "signature analysis" techniques to implement error-checking mechanisms effective against the largest class of error sources (including decoding and sequencing) [12]. Considering the relevant incidence of transient and intermittent faults on the general statistics of run-time faults in VLSI devices, a proposal has recently been published [13] based on the previous checking technique and upon software fault-tolerance criteria, leading to fault tolerance through backward-error-recovery schemes.

b. *Multiple-function devices.* In this case, different techniques may have to be adopted for the different functional units. A case in point is that of microprocessors; some examples have been presented in the literature [14,15], generally aiming at self-test rather than complete fault tolerance. A peculiar solution (oriented to wafer-scale integration) was that proposed by Triology, in which large faulty areas were presumed so that conventional techniques could not be adopted; actually, however, the proposal falls under the massive redundancy approach.

c. *Array devices, or more generally, architectures built up of a number of identical processing elements.* In this case function-level fault tolerance may be kept at the local self-testing level, so that each processing element will provide information on its own working/faulty status.

2. *Architecture-dependent approaches.* In this case fault-tolerance techniques exploit the particular characteristics of the architecture in order to reach high yield and/or reliability. Typically, such an approach has been adopted in the past for multiprocessor or multicomputer systems; most solutions proposed in the environment of discrete-processor systems are, in anycase, scarcely applicable to

VLSI architectures. In a discrete-processor environment, 100% spares utilization is regarded as essential. The complexity of the interconnection network is considered as a figure of merit insofar as it involves the complexity of the related control circuits and switching structures, not as regards length or topology [except when this affects the processors themselves, due to input/output (I/O) ports or similar]; redundancy is limited only by processor-cost considerations.

When a VLSI multiple-processing element structure is considered (typically, a VLSI array architecture) new figures of merit acquire dominant importance. *Regularity* of the architecture and of interconnection topology is a basic factor; *locality* of internconnections is also a primary requirement. This leads us to consider new strategies for reconfiguration after fault. To begin with, the "direct substitution" strategy (by which a faulty element is immediately substituted by any available spare) cannot be afforded, as it would require a complex and irregular interconnection network. Although high spares utilization is, of course, an important aim, it may become acceptable to achieve less than 100% spares utilization provided that the previous requirements are satisfied: in fact, a compromise is usually sought between the complexity of the reconfiguration algorithm (thence of the related control circuitry), the complexity of the interconnection network, and the effectiveness of the reconfiguration algorithm. Complexity in this context involves another figure of merit, silicon-area requirements; it has been proved by various authors [16,17] that excessive increase in the silicon area ultimately leads to overall reliability decrease outstripping fault-tolerance effectiveness.

This chapter is dedicated to *architecture-related fault-tolerance techniques for VLSI arrays of processing elements* (hereafter called *cells*). The importance of processing arrays as VLSI-oriented architectures has been stressed in the earlier chapters, so we do not deal with it further. Here we are interested primarily in the particular aspects that make such architectures well suited to self-testing and reconfiguration techniques. Basically, criteria introduced in the literature aim at keeping the regularity of the structure as high as possible; spares are added as regular patterns of additional processing elements, and the augmented interconnection structure is also very regular.

Some preliminary considerations are necessary concerning the error model hereafter adopted. "Classical"—stuck-at—fault models are scarcely applicable, owing both to the complexity even of the single cells and to the fact that such models are strictly technology dependent [and even, as for complementary metal-oxide semiductors (CMOS's), not well suited to given technologies]. We prefer to consider functional *cell-level errors*, which allow us to distinguish simply a "faulty" or "working" state for any given cell.

A further consideration regards allowable fault distributions. Basically, we may distinguish *random* distributions of mutually independent faults and *cluster* distributions of faults presumably deriving from one common cause. We assume the first distribution to be acceptable for VLSI devices; there, large fault clusters affecting a number of neighboring cells, as well as the interconnection lines in the same area, would lead us to discard the entire device rather than attempting some form of recovery. Cluster distributions are reasonable assumptions for wafer-scale integration (see Ref. 18). Here we restrict our study to the first type of fault distribution.

Even prior to any analysis of the *reconfiguration algorithms*, it is possible to distinguish three approaches to reconfiguration— approaches that influence, first, the implementation technology of the reconfigurable array.

1. Reconfiguration can be performed at *production time,* with the aim of increasing production yield; we speak then of "static" reconfiguration. Testing is in this case performed externally; reconfiguration is uniquely determined at production time, and it is actuated mainly by irreversible actions (e.g., by laser cutting). The reconfiguration algorithm is driven by an external device; no on-chip controlling circuits are needed. Quite obviously, complexity of the reconfiguration algorithm is not a critical issue; rather its efficiency is of the utmost importance, since it directly influences final yield (on the other hand, complexity of the augmented structure has to be considered). Some examples of static reconfiguration are given in Refs. 19—21. A technology allowing "semistatic" reconfiguration, by means of EPROM cells acting as switches, has recently been proposed.

2. Alternatively, reconfiguration can be performed at *run time,* with the aim of increasing device lifetime; these solutions are termed "dynamic reconfigurations." In turn, we have two alternatives:

a. *Self-reconfiguration.* Circuits controlling reconfiguration (on the basis of error information available) are on the chip itself. Such a solution allows us to reach very low error latency before reconfiguration is performed; basically, reconfiguration may take place as soon as the error is detected. Here complexity of reconfiguration-controlling algorithms becomes a main issue, since it influences both silicon requirements and reconfiguration speed (which must be completed in the time slot between two subsequent array processing steps). The literature available on this subject is by now quite extensive; a number of techniques can be found, for example, Refs. 22—32.

b. *Host-driven reconfiguration.* In this case, routing devices (however simple) in the interconnection network are controlled by a suitable "reconfiguration protocol" generated by the external host on the basis of available error information. In this case, reconfiguration actions will not, as a general rule, be undertaken by the host at each processing step, so that the problem of error latency may become

important. (Some examples are discussed in Refs. 33–36.) Although
problems such as locality and complexity of interconnections and
complexity of routing devices are again very important, complexity
of the algorithm proper is not essential since the host machine is
usually a powerful one; thus it is possible to aim for very high probability of survival through complex reconfiguration algorithms.

   Although techniques described here are best suited to dynamic
self-reconfiguration, the most efficient ones (in terms of spare
utilization) also lend themselves to static reconfiguration.

2 ANALYSIS OF SOME PREVIOUS RECONFIGURATION PROPOSALS

Before discussing in detail our algorithmic approach to reconfiguration
of VLSI arrays, we analyze a very few of the many proposals presented in the literature in this context. The ones we consider are
closest to our own technique in that they strive to optimize (at least
in part) the same figures of merit that we consider: locality of interconnections, simplicity of algorithm, and overall regularity.

   A widely discussed example is constituted by the "Diogenes
approach" 23,24. The Diogenes philosophy aims at implementing a
high number of alternative interconnection topologies on a given
(basically, linear) layout consisting of processing elements (PEs),
bundles or "stacks" of wires and switches. Consider the elementary
structure of a single PE with an associated stack of wires and switch
in Fig. 8.1; such a structure corresponds to a depth-3 complete

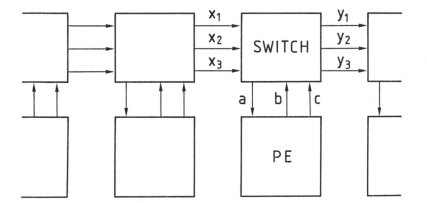

FIGURE 8.1 Diogenes approach: basic interconnection structure for
implementation of fault-tolerant binary trees.

Table 8.1 Switch Operation for the System of Fig. 8.1

PE faulty	PE working as leaf	PE working as nonleaf node
x_1 to y_1	x_1 to y_2	x_2 to y_1
x_2 to y_2	x_2 to y_3	x_3 to a
x_3 to y_3	x_3 to a	b to y_2
	0 to y_1	c to y_3

binary tree. Operation of the switch is summarized briefly in Table 8.1. The second and third columns of the table represent the rules for implementation of the binary tree, while the first column provides for bypass of a faulty cell; the tree will be fault tolerant if more than seven PEs are present. Different architectures involve greater numbers of stacks of wires of greater depth (and also a higher number of connection points for the PEs). The Diogenes approach assumes that switches and wires are *always* correctly working, and it guarantees 100% spares utilization; on the other hand, the length of interconnecting wires may easily become very high for contiguous-fault distributions, and the area occupied by the bundles of wires will also increase with the envisioned architecture complexity. Thus two of the figures of merit (spares utilization and structure regularity) are optimized; the solution is not as satisfactory for the other two (locality and silicon requirements).

Length of interconnections and locality of the reconfiguration are, on the contrary, main figures of merit in Ref. 32. The approach deals explicitly with systolic arrays, and it considers possibly very large numbers of faulty cells (in fact, it deals with wafer-scale integration). The aim is mainly that of high production yield, so that the complexity of algorithms is relatively less important than in other papers (the algorithm being implemented by an external machine, without affecting the array's processing time). The approach by Leighton and Leiserson is based on a probabilistic fault model, giving each cell in the array, independently a 50% chance of failure, and it shows that with high probability a two-dimensional array can be constructed on an N-cell wafer with wires of limited length (very close to $O(\sqrt{\lg N})$). This result is reached by a "divide-and-conquer" philosophy; here the aim is not so much that of providing an X × X working array through introduction of Y spare elements, as that of obtaining an array exploiting (almost) all working cells on a given wafer.

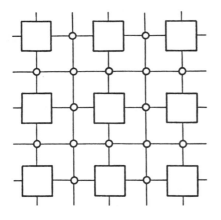

FIGURE 8.2 CHiP: basic array.

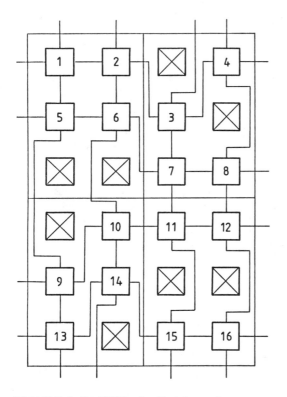

FIGURE 8.3 CHiP: fault-tolerant array.

The last case that we deal with briefly is that of the CHiP architecture [34]; it is also the one that comes closer to our global approach. Here, locality, regularity, and low silicon requirements are the main figures of merit, while a less-than 100% spares utilization is accepted. The basic (non-fault-tolerant) architecture is given in Fig. 8.2; to reach fault tolerance, the number of buses in each "channel" between a pair of adjacent rows or columns of PEs is augmented to two (all interconnections are properly switched) and spare PEs are added [30,37,38]. Locality is achieved by restricting reconfiguration possibilities inside subarrays properly defined; the cxample in Fig. 8.3 involves subarrays of 2×3 PEs that are capable of reconfiguring so as to give a 2×2 subarray for any set of two faults. Reconfiguration is then very simple, as is the interconnection network; on the other hand, the number of spares added is of the order of N^2 if the array is a square $N \times N$ one, and the probability of survival is certainly less than 100% (In the example of Fig. 8.3 it is obvious that any distribution of three faults in a subarray will lead to "fatal failure," that is, the impossibility of reconfiguration.)

3 RECONFIGURATION THROUGH INDEX MAPPING: BASIC DEFINITIONS

From now on, our considerations will be restricted to rectangular, $N \times N$ arrays of identical processing elements or "cells." Cells are of a combinatorial nature or at least "externally" they behave as combinatorial, not keeping memory beyond an array processing step (this is valid, e.g., for serial arithmetic devices, that while being intrinsically of sequential nature behave externally—with regard to cooperation with other parts of the array—as if they were combinatorial). Information flows following a "wavefront computation" technique (i.e., along one direction only of each interconnection axis). The assumptions implicitly adopted for the definition of reconfiguration techniques can be summarized as follows:

1. *Cells are provided with self-testing capacity.* This can derive from cell-level self-checking design or from an array-level self-testing approach. Whatever the choice, any cell is associated with its own error signal (true if the cell is faulty, false otherwise), kept stable in a local 1-bit memory device. To allow for transient faults, it is sufficient that memory devices storing such fault information be cleared periodically. Since our reconfiguration techniques do not require any assumptions, such as single fault, error sequencing in time, and so on, the foregoing solution is perfectly acceptable.

2. *Faults are associated with processing cells.* Failures in the inter-
    connection network are (as far as possible) assimilated to cell
    failures. Reconfiguration-controlling circuits are self-checking,
    so as not to constitute a bottleneck for the array's reliability.
3. *Fault tolerance is provided by array-level reconfiguration, making*
    possible the use of a number of spare cells. "Spares" has the
    meaning "processing element not otherwise used during the par-
    ticular operation phase considered." This allows to consider two
    alternatives:
a. *Space redundancy.* Physical spare cells are substituted for faulty
    ones; array-level operation time is kept essentially unchanged
    even after reconfiguration, until a "fatal failure" condition is
    reached.
b. *Time redundancy.* Correctly working cells substitute for faulty
    ones during multiple operation phases; no added processing cells
    are required (a limited space redundancy exists, due to added
    interconnection links and control circuits), but array-level
    processing time increases.

In the present chapter we deal explicitly with the space-redundancy
approach, better suited to devices with very stringent speed requirements; time redundancy has been treated, for example, in Refs. 29
and 39. Spares are arranged in regular patterns: as an added row
and/or an added column of spare cells. A formal frame useful for subsequent algorithm definition is the following:

1. Each cell is identified by a pair of physical indices, denoting row
    and columns; thus it can be uniquely denoted as cell (i,j). The
    whole array is then represented by a "physical index" matrix PH,
    whose entries are the physical coordinates.
2. At operation time, each cell (i,j) is identified by the functions it
    performs in the working array; these are represented by a pair
    of "logical indices" $LG(i,j) = i',j'$. For any cell that is not working
    (either because it is faulty or because it is an unused spare) at
    least one of the associated logical indices is set to 0. In the absence
    of faults, it is

$$LG(i,j) = PH(i,j) \quad \text{for } 1 \leqslant i, j \leqslant N$$

$$0,0 \quad \text{for all spares}$$

(see the example in Fig. 8.4).
3. As a consequence of reconfiguration after fault, entries in the
    logical index matrix may in general differ from the corresponding
    entries of the physical index matrix; it will be $LG(i,j) = i',j'$ with
    $i' \neq i$ and/or $j' \neq j$. The *locality* of the algorithms proposed guaran-

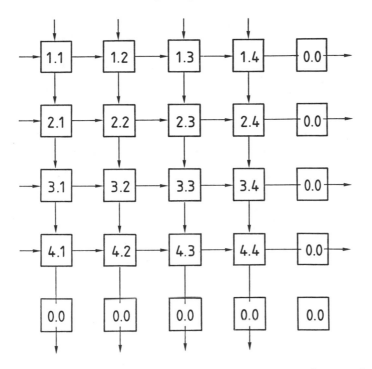

FIGURE 8.4 Sample 4*4 array with one row and one column of spares. Logical indices are indicated inside the cells.

tees that the distance between logical and physical indices will in any case, be strictly limited (e.g., restricted to the values +1, 0, −1).

4. Correct reconfiguration takes place iff the logical index matrix LG produced is, in turn, correct; namely:
   a. Any pair of entries (h,k) $1 \leqslant h, k \leqslant N$ appears once and only once in LG;
   b. If the structure is provided with x spares, there are x zero entries in LG.
5. Additional conditions for correct reconfiguration are:
   a. Locality (as defined above) is satisfied.
   b. All faulty cells are associated with 0 entries in the final logical matrix.

Before entering into detail concerning our formal reconfiguration approach, it may be useful to consider intuitively a very simple reconfiguration technique. Assume that spares are organized simply along one (rightmost) column (we then have N spares) and that reconfiguration is performed only along the rows. Reconfiguration rules (algorithm 1) can be stated as follows: For any given value of i, scan the row for increasing values of j:

1.   Until no faulty cell is found, it is i' = i, j' = j.
2.   As soon as a faulty cell (i,k) is found, its logical indices are set to 0 and for any cell (i,j), j > k, it is i' = i, j' =j − 1.
3.   Whenever two or more cells in one row are faulty, reconfiguration is impossible and fatal failure is declared.

An example is given in Fig. 8.5; quite obviously, both augmented-interconnection-network and reconfiguration-controlling circuits are very simple (they are detailed in Ref 25); on the other hand, the

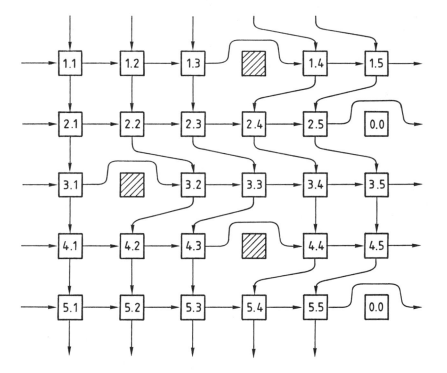

FIGURE 8.5 Example of reconfiguration following Alg. 1; one column of spares is added to the 5*5 basic array.

probability of survival is not very high, since the fatal failure condition is quite common. (For a 20 × 20 array, the probability of survival to six faults is only 50%.)

Getting back now to the formalization of our algorithmic approach, two distinct theoretical steps may be identified as concurring to perform reconfiguration.

1. A reconfiguration operator RO is defined, capable of producing a new logical matrix LG starting from an initial logical matrix LG1 and from a deformation matrix DG (of the same dimensions as PH and LG) whose entries give, for the corresponding cells, all information necessary to perform the given deformation. It is then

$$LG2(i,j) = RO(DG(i,j),LG1(i,j))$$

2. The deformation matrix is computed on the basis of a fault matrix F (again, of the same dimensions as all others) whose entries have nonzero value iff the corresponding cells are faulty. (Thus F gives all information on fault distribution.)

Step 1 may be seen as defining the reconfiguration methodology (basically, the number and sequence of operators identify the complexity of the interconnection network, organization of switches or other devices routing the information through the working cells, "depth" of locality, etc.), while step 2 actually identifies the specific reconfiguration algorithm (and therefore, if self-reconfiguration is envisioned, also the circuits controlling it). The algorithm must be capable of creating a correct matrix DG, that is, a matrix such that starting from a correct logical matrix LG1 (as is the initial logical matrix in the absence of faults), a correct logical matrix LG2 is produced.

In the simplest instance, the initial logical matrix coincides with the physical index matrix; we speak then of a "one-layer" reconfiguration, since as will be seen, the interconnection network around each cell involves one "layer" or "shell" of simple devices. It is obviously possible to iterate the procedure. For example, a two-layer reconfiguration algorithm is defined by a sequence

$$LG2(i,j) = RO1(DG1(i,j),LG1(i,j))$$

$$LG3(i,j) = RO2(DG2(i,j),LG2(i,j))$$

With this algorithm there will be associated a two-layer interconnection network, with each "shell" corresponding to one reconfiguration operator. In the next section, the general reconfiguration operator is defined and some specific instances are analyzed in detail.

```
      9  A              2  1  8

6  5  4  B           C  3  0  7

7  0  3  C           B  4  5  6

8  1  2              A  9

      (a)                (b)
```

FIGURE 8.6. (a) Example of adjacency domain; (b) corresponding
inverse adjacency domain.

4 GENERAL RECONFIGURATION OPERATOR

To define a general reconfiguration operator correctly, we introduce
the concept of "adjacency domain" for any given cell (i,j), as consisting
of a limited number of cells at restricted distance from (i,j) [and of
(i,j) itself]. The specific dimensions and the set of cells constituting
the domain are related to a particular algorithm (or set of algorithms).
In the very simple example seen in the preceding section, this domain
consisted of the two cells $(i, j - 1)$ and (i,j). Another possible
instance is given in Fig. 8.6a [where, for simplicity, cell (i,j) is
denoted as cell 0, and its adjacent cells are suitably numbered].

Now we define the *general reconfiguration operator RO* as an
operator that, given two adjacent cells (i,j) and (i'',j''), transfers
into (i,j) the pair of logical indices (i',j') associated with (i'',j''). It
is then possible to define the *inverse adjacency domain* as consisting,
for any given cell (i,j), of all cells into which the logical indices
associated with it can be transferred (see Fig. 8.6b).

The concept of locality can now be stated formally as follows:
Given the inverse domain of adjacency for the cells in an array, two
physically adjacent cells [i.e., cells connected by a direct interconnection link, e.g., (i,j) and $(i + 1, j)$] may as a consequence of
reconfiguration see their logical indices transferred only to two cells
lying in their respective inverse domains of adjacency; the destination
cells will be "logically adjacent." Given a domain of adjacency of
limited dimensions, the distance between the two logically adjacent
cells will be limited. As an example, in Fig. 8.7 the inverse adjacency
domains for two physically adjacent cells a, b are outlined, and the
two positions at maximum distance within them are marked.

Obviously, larger domains may allow more complex reconfigurations
and greater spares utilization; on the other hand, silicon requirements may correspondingly rise beyond acceptable limits. We will
restrict the adjacency domain to cells 1 to 8; thus, as stated previously,

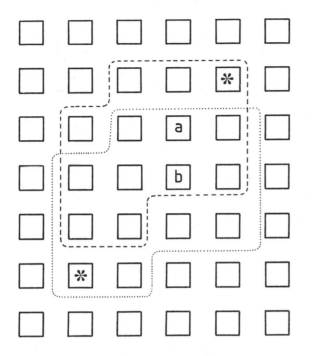

FIGURE 8.7 Inverse adjacency domains of cell a (dashed line) and b (dotted line) as defined in Fig. 6. Cells marked with a star denote the maximum-distance position to which logical indices of a and b can be moved as a consequence of reconfiguration.

locality is such that the difference between i, j and, respectively, i', j', may have one out of three values: +1, 0, −1. (For some particular algorithms, an even more limited domain will be considered.) We will use the "coding" suggested in Fig. 8.6 for identification of the various adjacent cells; thus it will be LG(0) = LG(i,j) and LG(8) = LG(i + 1, j − 1).

RO can now be defined through the actions it performs:

1. If DG(0) = k, then LG2(0) = LG1(k) $0 \leqslant k \leqslant 8$. Here each entry DG(i,j) denotes the cell k in the adjacency domain of (i,j) such that entry LG1(k) consists of the pair of logical indices to be transferred into LG2(i,j) [e.g., if DG(i,j) = 3, then LG2(i,j) = LG1(i, j + 1)]. (Note that, by this definition, it is impossible that two different cells of LG1 be transferred into the same cell of LG2.)
2. If DG(0) = 0 and a value k exists such that DG(k) = k, then LG2(0) = 0,0. This happens if LG1(i,j) is transferred into

LG2(i1,j1) but no entry is transferred into LG2(i,j) (a consequence of a fault in cell (i,j) or, in particular algorithms, of some
fault distribution leading to consider (i1,j1) as a "virtual fault").

The definition above is not sufficient to guarantee that LG2 will
be correct; in particular, the following two conditions might appear:

1. One entry LG1(i,j) is duplicated into two different entries of LG2.
    This condition would be verified iff there were two values k1, k2,
    $1 \leqslant k1$, $k2 \leqslant 8$, such that $DG(k1) = k1$, $DG(k2) = k2$.
2. One (or more) pairs of logical indices appearing in LG1 "disappear"
    from LG2; this instance is present if, for a given (i,j), it is
    $DG(k) \neq k$ for all $0 \leqslant k \leqslant 8$.

Both instances may be avoided only by suitable computation of
the deformation matrix DG (i.e., by a correct definition of the reconfiguration algorithm proper). Thus deformation matrices cannot be
built arbitrarily, but they must be built so as to guarantee such
correctness.

We can now make visible the actual operation of our general reconfiguration operator RO. For this purpose we define a *deformation
line* as a sequence of index pairs (i_1,j_1), . . . , (i_n,j_n) such that:

1. For any k, $1 \leqslant k < n$, LG1(i_k,j_k) is transferred into LG2(i_{k+1},
    j_{k+1}).
2. No pair of logical indices is transferred from LG1 into LG2(i_1,j_1);
    rather, a zero entry is written in this position: LG2(i_1,j_1) = (0,0).
3. LG1(i_n,j_n) "disappears"; to allow correct reconfiguration, entry
    LG1(i_n,j_n) must of necessity be a zero entry.

Consider the example in Fig. 8.8, where LG1 is assumed to be the
initial logical matrix, while entries of LG2 are written inside the cells
and values of DG are written at the right of each corresponding cell.
Nonzero values in DG identify a deformation line starting from cell
$(i_1,j_1) = (7,2)$ and ending on cell $(i_n,j_n) = (2,7)$.

If the conditions given previously for correct definition of RO are
satisfied (in particular, the ones forbidding replication of a given
entry in LG2), it follows that DG may originate more than one deformation line (as it is actually necessary when several faults are present),
but that all are disjoint (i.e., no physical cell may belong to two or
more deformation lines).

Reconfiguration operators can be composed into sequences, each
operator making use of its associated deformation matrix; the "inner"
operator acts on the initial logical matrix and provides a new logical

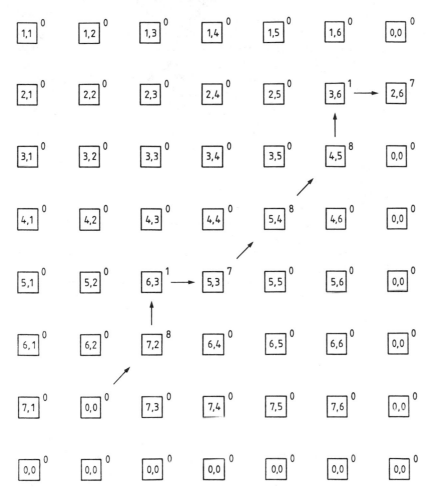

FIGURE 8.8 Deformation line starting from cell in row 7, column 2 and ending on spare cell in row 2. Logical indices following reconfiguration are written inside the cells. Entries of matrix DG are written beside the cells.

matrix on which the second operator acts, and so on. Thus in the most general case, we have

$$LG_p = RO_{p-1}(DG_{p-1}, RO_{p-2}(DG_{p-2}, (\cdots RO_1(DG_1, LG_1)))) $$

In particular, we consider sequences of two operators (longer sequences, while theoretically possible, would lead to impractical requirements in terms of silicon area and circuit complexity). Consider two deformation matrices DG_1 and DG_2. Assume that there is a deformation line $(i_1,j_1), \ldots, (i_r,j_r)$ in DG_2 and another one $(i_r,j_r), \ldots,$ (i_s,j_s) in DG_1; the union of these two lines is called a "complete line" $(i_1,j_1), \ldots, (i_s,j_s)$. Then:

1. As a consequence of application of the first operator, logical pair (i_s,j_s) disappears, and zero indices are associated with (i_r,j_r).
2. As a consequence of application of the second operator, the zero logical indices associated with (i_r,j_r) disappear and zero logical indices are associated with (i_1,j_1).

Actually, the effect of sequential application of two operators is equivalent to that of one single operator whose associated deformation matrix DG_3 contains a deformation line identical to the complete line. Increased complexity of reconfiguration algorithms, interconnection network and related control circuits, and reconfiguration-controlling circuits is balanced by greater freedom in definition of complete lines; as will be shown later, they may be nondisjoint. Statistical results, as regards probability of survival, are far more satisfactory.

Having thus completed the introduction of formal tools, in the next sections we will examine a number of actual algorithms and evaluate their comparative requirements and performances.

5 RECONFIGURATION ALGORITHMS

We define here algorithms making use of one operator or of a sequence of two operators. As already stated, while longer sequences of operators are in theory possible, they are actually impractical.

Whatever the algorithm type, we could define a theoretical optimum as an algorithm achieving the highest limit to reconfiguration capacity within a given number of operators and a given domain of adjacency for each of them. Such optimum algorithms would, anway, be fairly complex and thus suited to host-driven reconfiguration rather than to self-reconfiguration.

On the other hand, we can consider suboptimum algorithms allowing simple computation of reconfiguration-controlling signals; most algorithms described here belong to this second class.

In *single-operator algorithms*, the algorithm, given the fault-distribution matrix F (already defined), creates a deformation matrix DG:

$$DG(i,j) = f(F(i,j))$$

In *two-operator algorithms*, instead, given the fault matrix F, the algorithm creates *two* deformation matrices DG_1 and DG_2.

A deformation algorithm must create a deformation matrix having as many deformation lines as there are faulty cells. If a multiple-operator algorithm is adopted, complete lines—deriving from composition of the various deformation matrices—must be as many as faulty cells. The following conditions must be met:

1. Target points of the deformation lines must coincide with spare cells.
2. Origin points of the deformation lines must coincide with faulty cells.

In fact, a deformation line leads to the disappearance of a pair of logical indices in the target cell (that, as a consequence, must initially be void) while it creates zero indices in the origin cell, which therefore must coincide with a fault. Thus it can be said that each deformation line creates a logical correspondence between a fault and a spare; note that such correspondence is *not* a substitution [as well evidenced by Fig. 8.8, where (7,2) is the faulty cell and (2,7) the spare one in correspondence with it]. In this section we consider first a number of algorithms with two-operator sequences (comprising an optimum one) and then some single-operator algorithms.

5.1 Algorithms Based on Two-Operator Sequences

In all algorithms of this class, the complete line creating a correspondence between one fault and one spare is made up as the composition of a first line segment starting from the faulty cell c_1 and of a second line segment ending on the spare cell c_s. The effect of the first operator application is transfer of logical indices as follows:

$$c_{s-1} \to c_s$$

$$c_{s-2} \to c_{s-1}$$

$$\cdots \cdots \cdots$$

$$c_v \to c_{v+1}$$

$$0,0 \to c_v$$

so that c_v acts as if it were a "virtual fault." Subsequently, the second operator effects the following transfer of logical indices:

$$c_{v-1} \rightarrow c_v$$

$$c_{v-2} \rightarrow c_{v-1}$$

. . . .

$$c_1 \rightarrow c_2$$

$$0,0 \rightarrow c_1$$

so that here c_v acts as if it were a "virtual spare."

A basic rule is that no other cell in the second line, besides c_1, can be faulty. In fact, consider the two lines above and assume that there is a cell c_h, $1 < h \leqslant v$, which is faulty. As a consequence of the second deformation, c_h will receive the logical indices associated with c_{h-1}; but, being faulty, it will not be capable of performing the functions related to this pair of indices. The only acceptable condition would be that during the first deformation, c_{h-1} received in turn zero logical indices; but then c_{h-1} would be the origin of a new line c_{h-1}, c'_h, . . . , c'_s. This actually coincides with the existence of two distinct pairs of lines, namely,

$$c_1, \cdots, c_{h-1} \quad \text{and} \quad c_{h-1}, c'_h, \cdots, c'_s$$

$$c_h, \cdots, c_v \quad \text{and} \quad c_v, \cdots, c_s$$

This second solution involves the same cells as the first one, but it follows the rules that each fault corresponds to a spare by means of a complete line starting from the fault itself, and that the second segment does not contain any other faulty cell.

On the contrary, it is possible for the first segment to contain a faulty cell. Let c_k be this fault; in the total deformation, c_1 is put in correspondence with c_s via the two lines

$$c_1, \cdots, c_v \quad \text{and} \quad c_v, \cdots, c_{k-1}, c_k, c_{k+1}, \cdots, c_s$$

while c_k is put in correspondence with a cell c'_s via two lines,

$$c_k, c'_{k+1}, \cdots, c'_v \quad \text{and} \quad c'_v, \cdots, c'_s$$

The first deformation transfers logical indices according to the following table:

$$c_{s-1} \to c_s \qquad c'_{s-1} \to c'_s$$

$$\cdots\cdots\cdots \qquad c'_{s-2} \to c'_{s-1}$$

$$c_k \to c_{k+1} \qquad \cdots\cdots\cdots$$

$$c_{k-1} \to c_k \qquad c'_v \to c'_{v+1}$$

$$\cdots\cdots\cdots \qquad 0,0 \to c'_v$$

$$c_v \to c_{v+1}$$

$$0,0 \to c_v$$

The second deformation, in turn, transfers logical indices as by the following table:

$$c_{v-1} \to c_v \qquad c'_{v-1} \to c'_v$$

$$\cdots\cdots\cdots \qquad \cdots\cdots\cdots$$

$$c_1 \to c_2 \qquad c_{k-1} \to c'_{k+1}$$

$$0,0 \to c_1 \qquad 0,0 \to c_k$$

As a consequence, the composition of the two deformations is correct and both faulty cells are duly substituted by working ones.

In the same way, the third basic property can be proved, namely, that two lines may intersect or even share common segments, provided that they belong to different deformations.

We describe now five algorithms (Algorithm 2, 3, 4, 5, 6) suitable for run-time self-reconfiguration and a sixth (optimum) one better suited to host-driven reconfiguration (Algorithm 7).

Algorithm 2: Simple fault-stealing reconfiguration, fixed choice This algorithm can be introduced by considering first the main cause of low performances for Algorithm 1; there each fault invoked reconfiguration to the right along the row, involving the spare in column $N + 1$, and the presence of two faults in one row originated fatal failure. Algorithm 2 modifies this technique by allowing "stealing" of spares from adjacent rows; consider the simple example in Fig. 8.9. The rightmost fault in row i invokes reconfiguration to the right, the other one transfers its logical indices to the cell in the same column belonging to row i + 1, "stealing" it. In turn, this second cell acts "as if" it were faulty, and it invokes reconfiguration to the right (see the arrow markers on the figure).

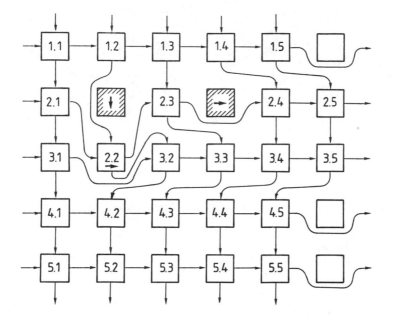

FIGURE 8.9 Simple fault-stealing: basic reconfiguration approach.
Cell in the third row, second column is "stolen" by the faulty cell
above it.

Two deformations are used, a horizontal one with alternatives
0,7 and a vertical one with alternatives 0,5; for the example in Fig.
8.9, the deformation matrices are:

DG1	DG2
0 0 0 0 0 0	0 0 0 0 0 0
0 0 0 0 7 7	0 0 0 0 0 0
0 0 7 7 7 7	0 5 0 0 0 0
0 0 0 0 0 0	0 0 0 0 0 0
0 0 0 0 0 0	0 0 0 0 0 0

The complete definition of the algorithm is as follows:

1. Rows are scanned for increasing values of index i; assume that
    in row i (i < N) there are s faulty or stolen cells (i, k_1), . . . ,
    (i, k_s), with $k_1 < \cdots < k_s$, while cells $(i + 1, k_1)$, . . . ,$(i + 1,$
    $k_{s-1})$ are not faulty; then:

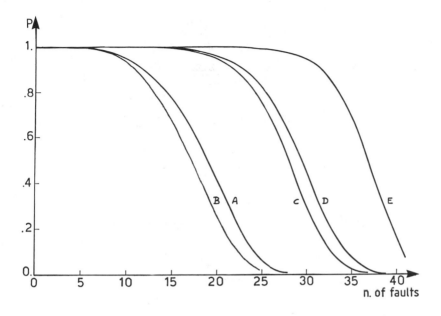

FIGURE 8.10 Probability of survival for algorithms 2 to 7, given a 20*20 basic array.

  a. Cell (i,k_s) invokes reconfiguration along row i; DG1(i,j) = 7, $j > k_s$.
  b. Cells (i,k_1), . . . ,(i,k_{s-1}) steal the positions of cells $(i + 1, k_1)$, . . . ,$(i + 1, k_{s-1})$; DG2$(i + 1, j)$ = 5 for all $j \in \{k_1, \ldots, k_{s-1}\}$.
2. A fatal failure condition is reached if:
  a. At least one cell $(i + 1, h)$ with $h \in \{k_1, \ldots, k_{s-1}\}$ is faulty.
  b. If only N spares are present (one column of spares), the second condition is due to the presence of two or more faulty or stolen cells in row N.

To overcome the last-named restriction, it is sufficient to insert also one row of spares (row N + 1); statistical results for this case are given by curve A in Fig. 8.10. In Fig. 8.11 a complex example solved by this case is presented.

Algorithm 3: Restricted fault stealing, fixed choice This algorithm gives performance inferior to those of Algorithm 2; on the other hand (as will be seen when implementation of interconnection networks will be presented), it can be implemented by the augmented bus structure

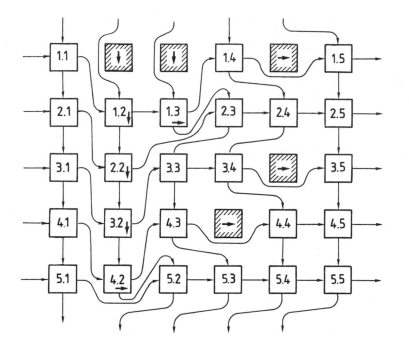

FIGURE 8.11 Simple fault-stealing, fixed choice: example of reconfiguration.

presented in Ref. 30 requiring only two buses, while Algorithm 2 requires higher redundancy. The reduction in silicon-area requirements makes it reasonable to consider this structure. The rules are the following (we consider directly $(N + 1) \times (N + 1)$ structures, with one row and one column of spares):

1. Rows are scanned for increasing values of index i; assume that in row i there are s faulty or stolen cells, (i,k_1), . . . ,(i,k_s), with $k_1 < \cdots < k_s$, while cells $(i + 1, k_1)$, . . . ,$(i +1, k_{s-1})$ are not faulty; then:
    a. If (i,k_s) is stolen and $(i - 1, k_s)$ is faulty and $(i + 1, k_s)$ is not faulty then:
        (1) No cell invokes reconfiguration along row i.
        (2) All cells (i,j), $j \in \{k_1, \ldots ,k_s\}$ steal the position of the corresponding cells $(i + 1, j)$; $DG2(i + 1, j) = 5$ for all $j \in \{k_1, \ldots ,k_s\}$.

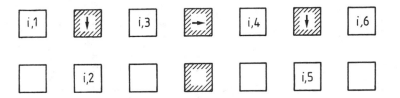

FIGURE 8.12 Simple fault-stealing, variable choice: reconfiguration policy for a case not solvable with fixed-choice.

   b. Otherwise:
      (1) Cell (i,k_s) invokes reconfiguration along row i; DG1(i,j) = 7, $j > k_s$.
      (2) Cells (i,k_1), . . . ,(i,k_{s-1}) steal the position of cells $(i + 1, k_1)$, . . . ,$(i + 1, k_{s-1})$; DG2(i + 1, j) =5 for all $j \in \{k_1, \ldots ,k_{s-1}\}$.
2. Fatal failure condition is the same as for Algorithm 2. Statistical results are given by curve B in Fig. 8.10.

Algorithm 4: Simple fault stealing, variable choice Predetermining reconfiguration along the row for the rightmost fault, in fact, restricts fault distributions for which fault stealing allows survival; by introducing an additional degree of freedom—variable choice for the cell that invokes reconfiguration along the row—better probability of survival can be achieved. Consider the example in Fig. 8.12; fixed-choice fault stealing would lead to declaring fatal failure. Instead, we propose that reconfiguration along the row is invoked by the faulty or stolen cell (if any) that could not steal the cell in the lower row; in the example, this happens for the cell in the first row, fourth column. All other cells (including the rightmost one) steal positions from the lower row. Such modification is possible because reconfiguration along the row is effected as the first deformation (DG1) and, as stated before, cells along the corresponding deformation line may be faulty.

    Formal rules are as follows (again, we consider directly the (N + 1) × (N + 1) structure):

1. Rows are scanned for increasing values of index i; assume that in row i (i < n) there are s faulty or stolen cells (i,k_1) $\cdots$ (i,k_s), with $k_1 < \cdots < k_s$, while cells $(i + 1, k_1)$ $\cdots$ $(i + 1, k_s)$ are not faulty except cell $(i + 1, k_h)$, $k_h \in \{k_1 \cdots k_s\}$; then:
    a. Cell (i,k_h) invokes reconfiguration along row i; it is DG1(i, j) = 7 for $j > k_h$.

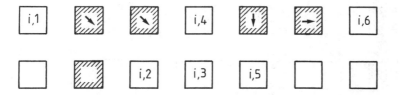

FIGURE 8.13 Complex fault-stealing, fixed choice: example of reconfiguration policy.

    b. Cells $(i,k_1) \cdots (i,k_{h-1})$ and $(i,k_{h+1}), \ldots ,(i,k_s)$ "steal"
        the positions of cells $(i + 1, k_1) \cdots (i + 1, k_{h-1})$ and (i,k_{h+1}),
        $\ldots ,(i,k_s)$; it is $DG2(i + 1, j) = 5$ for all $j \in \{k_1, \ldots ,k_{h-1}$
        $k_{h+1}, \ldots ,k_s\}$.
2. A "fatal failure" condition is reached if at least two cells $(i + 1,$
    h1) and $(i + 1, h2)$, with h1 and h2 $\in \{k_1 \cdots k_s\}$, are faulty.

Performances of this algorithm are given by curve C in Fig. 8.10; the
greater flexibility offered by this more complex algorithm is reflected
in a higher probability of survival.

Algorithm 5: Complex fault stealing, fixed choice This algorithm is
in practice an alternative to Algorithm 4 as regards restrictions on the
"stolen" cell. Consider the example in Fig. 8.13: Algorithm 2 would
exclude reconfiguration, since the cell in row 1, column 2 would require
stealing from a *faulty* cell. Rather than leaving freedom for the choice
of the cell invoking reconfiguration along the row, we introduce a
"skew" stealing, as shown by the arrow markers in the figure [cells
that can be used for stealing by (i,j) can be found in the set $(i + 1, j)$,
$(i + 1, j + 1)$]; thus, actually, a three-way choice is allowed for the
first deformation. Horizontal deformation matrix DG1 has entries 0,7,
while the vertical deformation one has entries 0,5,6.
    Formal rules are as follows:

1. Rows are scanned for increasing value of index i; assume that in
    row i there are s faulty or stolen cells $(i,k_1), \ldots ,(i,k_s)$,
    $k_1 < \cdots < k_s$. Cell (i,k_s) invokes reconfiguration along the row;
    it is $DG1(i,j) = 7$ for $j > k_s$.
2. For each h $\in \{k_1, \ldots ,k_{s-1}\}$:
    a. If $(i + 1, h)$ is not faulty or already stolen, (i,h) steals it:
        $DG2(i,h) = 5$.
    b. If $(i + 1, h)$ is faulty or already stolen, but $(i + 1, h + 1)$ is
        neither faulty nor stolen, (i,h) steals it and it is $DG2(i,h) = 6$.
    c. If $(i + 1, h)$ and $(i + 1, h + 1)$ are both faulty or stolen, fatal
        failure is declared.

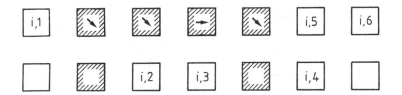

FIGURE 8.14 Complex fault-stealing, variable choice: example of reconfiguration policy.

Statistical performance of this algorithm is given by curve D in Fig. 8.10.

Algorithm 6: Complex fault stealing: the variable choice This algorithm merges the variable-choice feature of Algorithm 4 with the complex stealing function of Algorithm 5; it is well exemplified by Fig. 8.14. Briefly, given a row i in which two or more faulty and/or stolen cell exist, it is necessary to decide which one invokes reconfiguration along the row. We choose to this end a cell that could not otherwise invoke stealing from row i + 1; if no such cell exists, the rightmost one is elected. Cell (i,2) in Fig. 8.14 makes use of complex fault stealing since (i + 1, 2) is faulty; so does (i,3), since (i +1, 3) has already been stolen by (i,2); (i,4) cannot invoke stealing since (i + 1, 4) has been stolen and (i + 1, 5) is faulty. Thus (i,4) invokes reconfiguration to the right. The complete algorithm is then:

1. Each row (for increasing values of row index) is scanned from left to right: let (i,k) be the rightmost faulty or stolen cell.
2. Given a cell (i,h), h < k, faulty or stolen, it invokes stealing from (i + 1, h) if this cell is not faulty or stolen; otherwise, it invokes stealing from (i + 1, h + 1). If this attempt also fails, (i,h) invokes reconfiguration along the row and all other faulty or stolen cells (i,j), h < j ⩽ N + 1, invoke stealing from row i + 1; this, of course, is true also for (i,k).

Condition for fatal failure, given one column and one row of spares, is now presence in row i of two or more cells that must invoke reconfiguration along row i. A complete example is given in Fig. 8.15, for a 5 × 5 array with one spare row and one spare column. The probability of survival is given by curve E in Fig. 8.10.

Although definitely quite complex, Algorithm 6 gives 50% probability of survival to a number of faults equal to 90% of available spares.

FIGURE 8.15 Complex fault-stealing, variable choice: example of reconfiguration.

Algorithm 7: Host-driven optimum reconfiguration If host-driven
reconfiguration is considered, far more complex and time-consuming
algorithms are acceptable provided that they optimize spare utilization.
This does not imply that complex deformations are considered. In
fact, we still limit possible alternatives, respectively, to (0,7) and
(0,5) for DG1 and DG2.

Refer again to the $(N + 1) \times (N + 1)$ array with one spare row and
one spare column. A logical correspondence may be created between
any given faulty cell (i,j) and one of the following spares:

1. $(i, N + 1)$ as a consequence of one, horizontal deformation,
    described by

    $$DG1(i,k) = 7 \quad j \leqslant k \leqslant N$$

2. (N + 1, j) as a consequence of one, vertical deformation, described by

   $$DG2(k,j) = 5 \quad i \leqslant k \leqslant N$$

3. (r, N + 1) with $i < r \leqslant N$ as a consequence of one horizontal and one vertical deformation, described by

   $$DG1(r,k) = 7 \quad j \leqslant k \leqslant N$$
   $$DG2(k,j) = 5 \quad i \leqslant k < r$$

   provided that $F(k,j) = 0, \ i < k \leqslant r$.

It is then possible to define a "correspondence table" having one row for each fault (obviously, the number of rows cannot exceed the number of spares; otherwise, reconfiguration can immediately be stated as impossible) and one column for each spare (in our example, $2 \times N + 1$ columns); entry (h,k) in the table is marked if logical correspondence can be created between spare k and fault h, void otherwise. The correspondence table is actually a coverage table in the classical sense: A reconfiguration algorithm effects a coverage operation on this table, meaning that a set of marks has to be found such that there is one mark in each row (each fault finds a corresponding spare) and no more than one mark in each column (no spare can correspond to more than one fault). If the coverage algorithm is optimum, that is capable of finding a solution if any exists [40], reconfiguration will also be optimum.

An example of this approach is given in Table 8.2. Part (a) presents matrix F; cells 1 · · · E are spares; cells a · · · h are faulty; in part (b) the association table is shown; a possible coverage is represented by Xs. Note that the fault pattern in Table 8.2 could be reconfigured only by Algorithm 6, which required, in turn, more complex operators than the ones assumed for Algorithm 7. This proves, among other things, that other reconfiguration algorithms, although very efficient in terms of complexity, are not optimum.

5.2 Single-Operator Algorithms

In this second class of algorithms, the line creating a correspondence between faulty cells and spares is generated by application of one operator only. Even so, to obtain high probability of survival, it is necessary to define algorithms more complex than the previous ones, with deformation matrices allowing a wider set of possible deformations. Deformation matrices examined here will have entries chosen among a set of three or even four alternatives.

Table 8.2   (a) Matrix F; (b) The Association Table

```
.  .  .  .  .  .  . 1          1 2 3 4 5 6 7 8 9 0  A B C D E

. c  .  b  .  a . 2       a . x x x X x x . . x  . . . . .

. ~  .  .  .  .  . 3       b . X x . . . . . . .  . . . . .

. f  e  d  .  .  . 4       c . x X . . . . . . .  . . . . .

.  .  . h  g  . . 5        d . . . X . . . . . .  . . . . .

.  .  .  .  .  .  . 6       e . . . x x X x . . .  . . x . .

.  .  .  .  .  .  . 7       f . . . x x x X . . .  . . . x .

E D  C  B  A  0 9 8        g . . . . x x x . . .  X . . . .

                           h . . . . x x x . . .  . X . . .
        (a)
                                            (b)
```

Given such complexity, identification of optimum algorithms based on a correspondence table, as in Algorithm 7, is not easily accomplished. In fact, in that case the only correspondence between a faulty cell (i_1,j_1) and a spare cell (i_S,j_S) completely defines the deformation line as composed of a vertical segment (i_1,j_1), . . . ,(i_S,j_1) and a horizontal one (i_S,j_1), . . . ,(i_S,j_S). When a more complex single-operator algorithm is used, more than one path exists between the faulty cell and the spare cell; a correspondence matrix could then be defined with as many rows as the number n_f of faults and as many columns as the number of all the allowable paths.

Algorithms can no longer proceed by simple row-by-row analysis; rather the method proposed is based on propagation of request signals and of answering grant or block signals. As will be seen, this allows us to create correct logical matrices.

Algorithm 8: Deformation 0,5,7   The aim of the algorithm is identification of a set of n_f completely disjoint lines (assuming n_f faults), each creating a correspondence between an origin faulty cell and a target spare cell. By definition, a single operator creates only simple lines that can at most fan out into a set of segments, each ending on a different target cell. As already said, this would correspond to a noncorrect deformation matrix: to avoid it, algorithms must guarantee that each simple line originating from a faulty cell does not split and that it ends on one target spare cell.

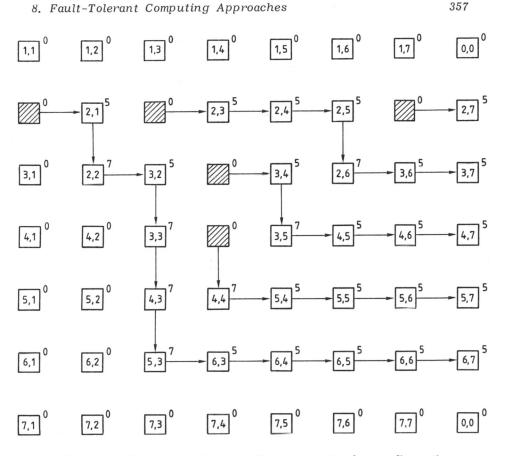

FIGURE 8.16 Single-operator algorithm: example of reconfiguration.

In Algorithm 8, each deformation line is made up only of horizontal segments oriented to the right, and of vertical segments oriented downward (we assume, again, a spare row and a spare column, respectively, at the bottom and at the right). A faulty cell originates a request for creation of a deformation line, and it is attempted to propagate the line until a spare cell is found.

As a general rule, each deformation line must propagate, as far as possible, in the highest rows of the array; to this end, it propagates horizontally until a block signal is received. Entries in the deformation matrix may have values 0 (no deformation), 5 (vertical deformation downward), or 7 (horizontal deformation rightward).

An example of reconfiguration by this algorithm is given in Fig. 8.16. Deformation lines are marked by arrows; entries of DG are

indicated besides the corresponding cells, while final logical indices
are written in the cells. Data interconnections between cells may easily
be derived from the logical pair of indices.

In the array structure corresponding to this class of algorithms,
each cell is provided with a "propagation circuit" performing the
functions requested by the request-acknowledge technique adopted.
In the case of Algorithm 8, the propagation circuit performs the following steps:

1. Faulty cell (i,j) transmits to one of its neighbors (i, j + 1) and
   (i + 1, j) a request to propagate a deformation line ending on a
   spare cell. Given the basic rule of priority, horizontal segments
   are favored over vertical ones.
2. Cells receiving the request attempt, in turn, to propagate it, first
   to the horizontal neighbor (DG = 7). Only if this cell answers by
   a block signal (either because it is faulty or because it has already
   been used in a vertical segment of a different line) is the request
   propagated to the vertical neighbor (DG = 5).
3. Propagation is continued until either an unused spare cell is
   reached or two block signals, one horizontal from cell (i, j + 1)
   and one vertical from cell (i + 1, j), propagate to the faulty cell
   (i,j), thus forbidding the creation of a deformation line. In this
   last case, fatal failure is declared.

To perform the steps listed above, the propagation circuit of any
cell (i,j) must have four inputs and four outputs (see Fig. 8.17):
8.17):

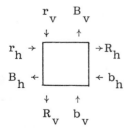

FIGURE 8.17 I/O organization of propagation circuit.

They are, respectively:

    Inputs: r_h horizontal request [from (i, j − 1)]

               r_v vertical request [from (i − 1, j)]

               b_h horizontal block [from (i, j + 1)]

               b_v vertical block [from (i + 1, j)]

    Outputs: R_h horizontal request [to (i, j + 1)]

               R_v vertical request [to (i + 1, j)]

               B_h horizontal block [to (i, j − 1)]

               B_v vertical block [to (i − 1, j)]

Moreover, the propagation circuit of (i,j) receives the error signal $r_f = e(i,j)$, stating the faulty/working state of the cell itself. The function of the propagation circuit can be stated as follows:

1. If $r_f = 1$, then B_h and B_v are set to 1.
2. If $z_f + z_h + z_v = 1$ and $b_h = 0$, a horizontal request $R_h = 1$ is propagated.
3. If $r_f + r_h + r_v = 1$ and $b_h = 1$, $b_v = 0$, a vertical request $R_v = 1$ is propagated and B_h is set to 1.
4. If $b_h = 1$ and $b_v = 1$, B_h and B_v are both set to 1: cell (i,j) cannot propagate the line because neither one of its neighbors can be part of it.

The logic scheme of the propagation circuit is shown in Fig. 8.18. The probability of survival with this algorithm is given by curve A in Fig. 8.19.

Algorithm 9: Deformation 0,6,7 The only modification with respect to Algorithm 8 consists of the set of neighbors, which are now (i, j + 1), (i + 1, j + 1); entries of DG may then have one of three values 0 (no deformation), 7 (horizontal deformation rightwards), or 6 ("skew" deformation rightward and downward); skew deformation has priority over the horizontal one.

    The complexity of the algorithm, of the hardware propagation circuit, and of reconfiguration circuits is similar to the one of Algorithm 8; so is the probability of survival.

Algorithm 10: Deformation 9,5,6,7 This is the most complex one; the set of neighbors that can be considered for line propagation consists of three cells: (i, j + 1) (horizontal rightward deformation, DG = 7), (i + 1, j + 1) (skew deformation, DG = 6), and (i + 1, j) (vertical downward deformation, DG = 5). Priority is of horizontal segments over the skew ones and of the skew segments over the vertical ones.

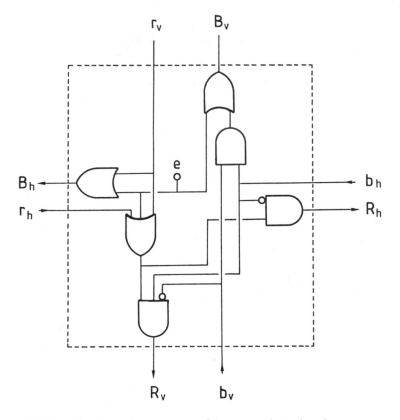

FIGURE 8.18 Logic diagram of propagation circuit.

The propagation circuit now has six inputs and six outputs, respectively:

r_h, r_s, r_v (requests for propagation, inputs)

b_h, b_s, b_v (blocks for propagation, inputs)

R_h, R_s, R_v (requests for propagation, outputs)

B_h, B_s, B_v (blocks for propagation, outputs)

Functions of the propagation circuit are an immediate extension of the ones derived for Algorithm 8; propagation possibilities are higher,

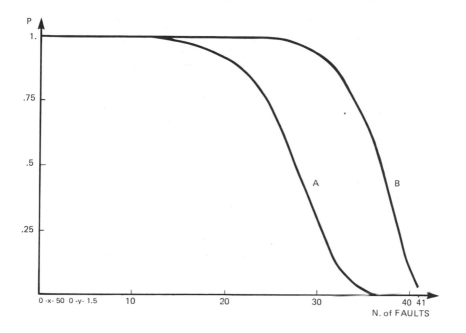

FIGURE 8.19 Probability of survival for algorithms 8 and 10, given a 20*20 array.

and so is the probability of survival (curve B in Fig. 8.19); on the other hand, both propagation circuit and interconnection network are more complex.

6 CIRCUIT IMPLEMENTATION OF RECONFIGURATION ALGORITHMS

The "nominal" array without any capacity for fault tolerance basically consists of N × N processing cells and of data paths connecting them.† This theoretical structure is augmented in a relevant way when we introduce fault tolerance through reconfiguration. We now have

---

† Power supply, synchronization signals, and control signals from an external host device driving the array are not explicitly considered here since they are not in any way modified by fault-tolerance characteristics.

1. A measure of structure redundancy guaranteeing detection and
    identification of faulty cells.
2. A number of alternative data paths, allowing us to implement the
    interconnections required by reconfiguration and constituting the
    global interconnection network.
3. Circuits associated with each cell (i,j) and allowing us to select
    the particular data paths to be activated in the presence of a
    given fault distribution. These may be (depending on the imple-
    mentation philosophy) multiplexers, demultiplexers, or switches.
4. Transmission paths for control signals activating the circuits
    above.
5. Circuits computing reconfiguration signals and generating the
    control signals above. If reconfiguration is host driven, the last-
    named circuits will obviously be absent.

In the present chapter, we do not make any assumption regarding
the functions performed by the processing cells. These are clearly
application dependent (in most cases, the cell will perform a suitable
set of arithmetic operations); for our purposes, the cell is simply a
square "black box."

Alternative data paths and related selection circuits depend on
the *reconfiguration operators* adopted; from these operators, or more
precisely, from the associated adjacency domain, it is possible to
derive the set of cells that can be connected as horizontal or vertical
input and output to any cell (i,j), given any allowable reconfiguration.
Such a definition leads to the statement that the interconnection network proper depends on the reconfiguration procedure adopted
(namely, the sequence of operators and adjacency matrix of each
operator) but not on the specific algorithm as represented by the
deformation matrix. This is "in principle" true; actually, since a
specific algorithm may not fully use all deformation possibilities
offered by the operators, the algorithm itself may influence the interconnection network by allowing us to simplify it with respect to the
"theoretical" one. The specific algorithm, on the other hand, is
essential for definition of the reconfiguration-controlling circuit, which
actually constitutes the hardware implementation of the algorithm itself.

Comparative evaluation of the various reconfiguration proposals
made in the preceding section must be performed on a cost-performance
basis. Although the main performance envisioned is, as already said,
probability of survival as related to a given number of spares, cost is
basically seen as related to silicon-area requirements. To this purpose,
we make some simplifying assumptions:

1. The cost is due primarily to the area increase deriving from
    alternative data paths and selection circuits.

2. The cost of a circuit's computing reconfiguration signals and of the related transmission paths is assumed to be much lower than for the previous one, and therefore is not taken into account.

Such assumptions are consistent with experimental layouts completed for some designs related to parallel data transfers (designs were carried out for 8-bit-wide data paths). They could become invalid in the case of serial data transfers, where actually the second cost figure could become preeminent with respect to the first one (arrays of serial processing elements are widely used for image processing applications [41]).

In the sequel we consider three alternative implementation philosophies for interconnection network and selection circuits. They correspond to:

1. A *theoretical structure*, derived directly from the definition of a reconfiguration operator or of a sequence of operators; each operator implies a layer of "shell" made up of interconnection paths and selection circuits around each processing cell. Although formally elegant and easily defined on the basis of theory introduced in the preceding section, this structure may be suited to actual implementation for simple instances involving one layer of shell only; otherwise, it becomes silicon consuming in addition to creating excessive delays along the interconnection paths.

It can be modified, by a sequence of simplifications, into a

2. *Multiplexer-based compact structure.* The basic technique consists of inserting a multiplexer on each cell input, which receives data sent from all possible "input adjacents," while outputs of the cell are "broadcast" to all "output adjacents." Selection circuits are reduced to their simpler form; alternative data paths are increased with respect to the previous structure, since all possible interconnections are directly present.

The structure lends itself easily to array-level self-testing policies, because it adapts to a simple global fault model; it is, in fact, immediate to associate multiplexers and interconnections as "belonging" to a cell, and therefore to treat their own faults as if they were due to cell failure.

3. Finally, the *switched bus structure* will be considered; it belongs to the same class as that of the fault-tolerant CHiP architecture. The interconnection network no longer consists of direct interconnection links but of bidirectional buses with a number of bidirectional 2 × 2 switches. The silicon area occupied by the interconnection network is reduced with respect to the previous solutions (a reduction often compensated by the augmented complexity of the selection circuits).

As discussed in Ref. 30, a further cause of interest of this structure is constituted by its dynamic adaptability not only due to

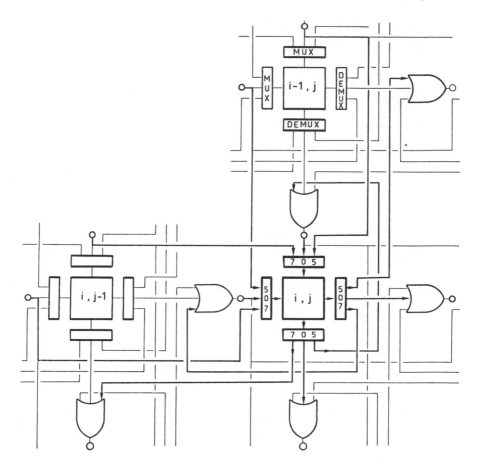

FIGURE 8.20   Theoretical interconnection structure: basic shell (one layer).

fault tolerance, but also to array-level functional reconfiguration (implementation of different classes of architectures onto the same array).

6.1 Theoretical Structure

We examine first a single-operator reconfiguration with deformation alternatives $(0,5,7)$. The example has already been discussed with reference to reconfiguration possibilities; extension to different cases is immediate. Then a two-layer structure corresponding to a two-operator sequence will be examined.

One layer of shell; deformation (0, 5, 7)

Definition of the algorithm can be summarized as follows:

1. If DG(i,j) = 0, logical and physical indices coincide.
2. If DG(i,j) = 5, cell (i,j) performs the functions of cell (i − 1, j).
3. If DG(i,j) = 7, cell (i,j) performs the functions of cell (i, j − 1).

The basic "shell" is represented in Fig. 8.20. Each processing cell is enclosed by a shell consisting of two multiplexers selecting the proper inputs, and two demultiplexers distributing the outputs of the cell.

Around each shell, four large dots mark the points where correct data will be available, even after reconfiguration. In fact, the two input multiplexers to cell (i,j) [directly controlled by entry DG(i,j)] provide to the cell itself the correct input data, derived from correct input points to cell (i,j) (selection signal 0), (i − 1, j) (selection signal 5), and (i, j − 1) (selection signal 7). The results produced by cell (i,j) are then deviated to the correct output points through the two demultiplexers on the outputs of (i,j), controlled once again by entries DG(i,j).

Data provided by the three adjacent cells foreseen by the algorithm are ORed. If DG is correct, one input only to any OR gate will be active at any time. Thus signals present at the terminals of the shell are actually the correct data corresponding to that array location.

Some details on layouts implemented for this theoretical structure may be of interest. The design was carried out for CMOS $\lambda = 1\ \mu m$ transmission-gate technology and for 8-bit-wide data paths. With these assumptions, unselected outputs from demultiplexers are "open" and OR gates can be substituted by wired-ORs, so that the simplified structure in Fig. 8.21 is obtained.

Each multiplexer and demultiplexer is 900 λ long and 94 λ wide. Thus for the technique to be cost-effective, the minimum dimensions for the processing cell are 900 × 900 λ. If this lower limit is considered, given 69 λ as the width of each data path, the interconnections and selection circuits require 50% of the total area of the chip. The relative cost of fault tolerance gets lower for larger processing cells and if more advanced technologies (e.g., multiple metal layers) are adopted. Transmission delay from cell to cell has been estimated, by simulation, to be about 18 ns for both a unreconfigured and a reconfigured array.

Two layers of shell

We consider the composition of two simple, single-direction deformations (as is the case with many of the algorithms examined, e.g., all algorithms from 2 to 7). As a test case, we assume that DG1 = {0, 7} and DG2 = {0,5}.

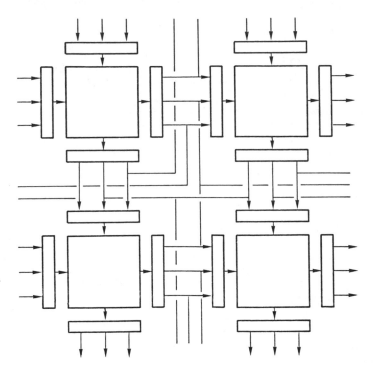

FIGURE 8.21 Simplified structure with one-layer shell.

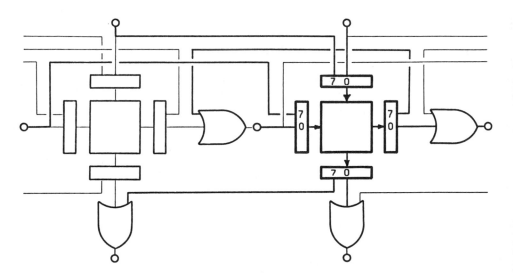

FIGURE 8.22 Elementary shell for single-direction deformation.

For one single-direction deformation, the elementary shell is provided by the structure shown in Fig. 8.22. (Again, large dots mark the points where correct data, associated as outputs to the "nominal" logical indices identical to the physical ones, are available.) Then composition of two deformations can be implemented by super-imposing two layers of shell as shown in Fig. 8.23. The dashed line enclosing the inner shell may be seen as creating a "virtual processing element" to which a second identical deformation shell is added.

The outer shell corresponds to the first deformation, the inner one to the second operator application. "Identical" refers only to the structure; control signals selecting multiplexers inputs and demulti-

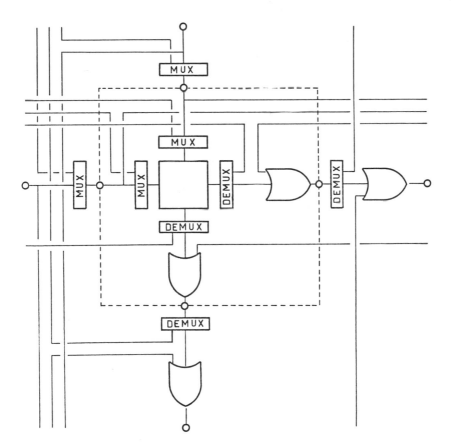

FIGURE 8.23 Two-layer shell made up from composition of two elementary shells.

plexer outputs are obviously different, corresponding to entries of different deformation matrices (note that this technique may be iterated for longer sequences of operators).

Although formally elegant, from a practical point of view this structure is less attractive. As can easily be deduced, it is fairly silicon consuming; moreover, the multiple stages of multiplexing or demultiplexing circuits introduce delays that for the high-throughput application envisioned can easily become unacceptable. In any case, this solution is still interesting in that it provides the basis for the second approach.

6.2 Multiplexer–Based Compact Structures

From the theoretical routing scheme defined previously, a simpler inter-connection structure can actually be devised as follows:

1. For the vertical input to cell (i,j), all alternative paths coming from outputs of different cells are followed; the "input shell" structure to (i,j) is substituted by one multiplexer having as many inputs as there are different paths, controlled by a function derived from the control signals to the multiplexers and demulti-plexers found on the various paths.
2. The same approach is followed for the horizontal input to cell (i,j)

The backtracking procedure leads to the creation of a "tree" of inter-connections such as the ones given in Ref. 28.

If the theoretical structure in Fig. 8.23 is considered, which is related to two layers of shell with deformations of 0,5 and 0,7, re-spectively, the set of cells whose outputs are connected to inputs of cell (i,j) is presented in Table 8.3. A schematic layout corresponding to the interconnections presented in Table 8.3 is shown in Fig. 8.24.

Some algorithms set limits on the possible values of reconfiguration signals of adjacent cells. This allows further simplification of the cor-

Table 8.3 Connectivity table for the structure of Fig. 8.23

Horizontal input	Vertical input
$(i-1, j-2)$, $(i-1, j-1)$	$(i-2, j-1)$, $(i-2, j)$
$(i, j-2)$, $(i, j-1)$	$(i-2, j+1)$, $(i-1, j-1)$
$(i+1, j-2)$, $(i+1, j-1)$	$(i-1, j)$, $(i-1, j+1)$
	$(i, j-1)$, $(i, j+1)$

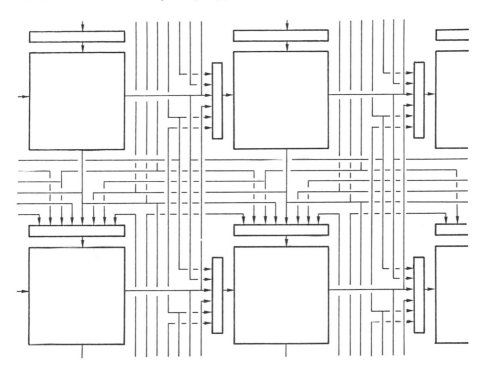

FIGURE 8.24 Compact structure deriving from the one in Fig. 23, for algorithms 4 and 7.

responding interconnection structure. For example, in Algorithms 3 and 4 the layout of Fig. 8.24 can be simplified by substituting, respectively, the eight and six-input multiplexers by a six- and five-input multiplexers and by reducing the number of alternative interconnection paths (from seven to five along the horizontal axis, from six to five along the vertical one). The set of all alternative inputs for all algorithms analyzed in the preceding section is given in Table 8.4.

Schematic layouts are presented in Fig. 8.24 for Algorithms 4 and 7; in Fig. 8.25 for Algorithms 2 and 3; in Fig. 8.26 for Algorithms 5 and 9; in Fig. 8.27 for Algorithm 6; in Fig. 8.28 for Algorithm 8.

6.3 Switched Bus Structure

As a third possible implementation alternative, we consider an interconnection network consisting of a reduced number of bidirectional

Table 8.4 Connectivity Table for Alternative Algorithms of Sec. 5

Algorithm	Horizontal input	Vertical input
	Deformation $0,7-0,5$	
Simple F-S fixed choice	$(i - 1, j - 1)$, $(i, j - 1)$ $(i, j - 2)$, $(i + 1, j - 1)$ $(i + 1, j - 2)$	$(i, j - 1)$, $(i - 1, j - 1)$ $(i - 1, j)$, $(i - 1, j + 1)$ $(i - 2, j)$, $(i - 2, j + 1)$
Simple F-S variable choice	$(i - 1, j - 2)$, $(i - 1, j - 1)$ $(i, j - 2)$, $(i, j - 1)$ $(i + 1, j - 2)$, $(i + 1, j - 1)$	$(i - 2, j - 1)$, $(i - 2, j)$ $(i - 2, j + 1)$, $(i - 1, j - 1)$ $(i - 1, j)$, $(i - 1, j + 1)$ $(i, j - 1)$, $(i, j + 1)$
	Deformation $0,7-0,5,6$	
Complex F-S fixed choice	$(i - 1, j - 2)$, $(i - 1, j - 1)$ $(i, j - 2)$, $(i, j - 1)$ $(i + 1, j - 2)$, $(i + 1, j - 1)$ $(i + 1, j)$	$(i - 2, j - 1)$, $(i - 2, j)$ $(i - 2, j + 1)$, $(i - 1, j - 1)$ $(i - 1, j)$, $(i - 1, j + 1)$ $(i, j - 1)$
Complex F-S variable choice	$(i - 1, j - 3)$, $(i - 1, j - 2)$ $(i - 1, j - 1)$, $(i, j - 3)$ $(i, j - 2)$, $(i, j - 1)$ $(i + 1, j - 2)$, $(i + 1, j - 1)$ $(i + 1, j)$	$(i - 2, j - 2)$, $(i - 2, j - 1)$ $(i - 2, j)$, $(i - 2, j + 1)$ $(i - 1, j - 2)$, $(i - 1, j - 1)$ $(i - 1, j)$, $(i - 1, j + 1)$ $(i - 1, j + 2)$, $(i, j - 1)$ $(i, j + 1)$, $(i, j + 2)$
	One Layer of Shell	
$0,5,7$	$(i - 1, j - 1)$, $(i - 1, j)$ $(i, j - 2)$, $(i, j - 1)$ $(i + 1, j - 2)$, $(i + 1, j - 1)$	$(i - 2, j)$, $(i - 2, j + 1)$ $(i - 1, j - 1)$, $(i - 1, j)$ $(i - 1, j + 1)$, $(i, j - 1)$
$0,5,6,7$	$(i - 1, j - 1)$, $(i - 1, j)$ $(i, j - 2)$, $(i, j - 1)$ $(i + 1, j - 2)$, $(i + 1, j - 1)$ $(i - 1, j - 2)$	$(i - 2, j)$, $(i - 2, j + 1)$ $(i - 1, j - 1)$, $(i - 1, j)$ $(i - 1, j + 1)$, $(i, j - 1)$ $(i - 2, j - 1)$

buses and of the switches controlling them. As already said, this solution derives from the CHiP architecture [34,30].

To obtain fault-tolerance capabilities, a number of alternative buses, running into each channel between a pair of rows or columns, are added to the basic interconnection structure presented in Fig. 8.2. A suitable number of 2 × 2 switches is inserted at some intersection between two buses or between an intercell link and a bus.

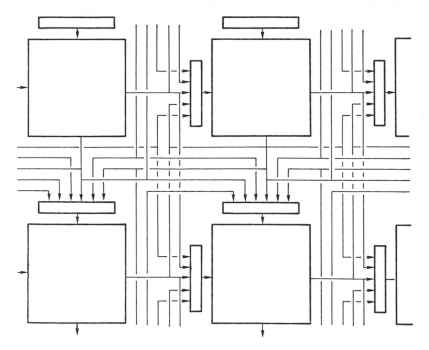

FIGURE 8.25 Compact structure for algorithms 2 and 3.

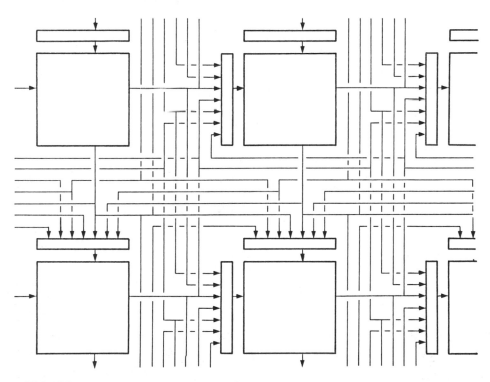

FIGURE 8.26 Compact structure for algorithms 5 and 9.

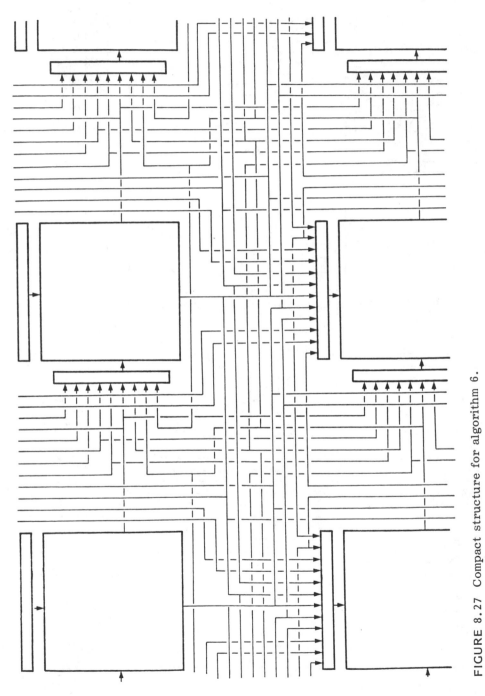

FIGURE 8.27 Compact structure for algorithm 6.

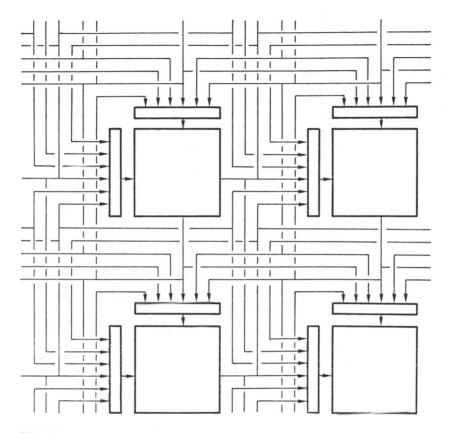

FIGURE 8.28 Compact structure for algorithm 8.

The number of buses depends on the reconfiguration algorithm;
three horizontal buses and two vertical ones are required if Algorithm
2, 4, or 7 is adopted; two horizontal and two vertical buses are used
for Algorithm 3. A more complex structure (which allows us to implement fairly complex reconfiguration strategies, such as the ones
adopted in Algorithm 5 and 6) requires three buses in each channel
(and a correspondingly increased set of switches).

It should be pointed out that these interconnection networks are
more fragile than the previous ones; failure of a bus segment or of a
switch can forbid completion of a high number of alternative connections,
and thus it can lead more rapidly to fatal failure.

Mapping of reconfiguration algorithms onto this structure is best
performed by starting from Table 8.4. We will briefly discuss the
approach for some sample algorithms previously analyzed. (Further
details are given in Ref. 42.)

6.4 Triple-Bus Solution

We consider first a structure that is valid for a large class of reconfiguration algorithms (Algorithms 2, 4, and 7). The interconnection
network is based on three horizontal buses between any two rows
(i, i + 1) and two vertical buses between any two columns (j, j + 1).
Buses are in a way "dedicated" to a particular class of data path; one
vertical and one horizontal bus are reserved to routing of reconfigured
horizontal interconnections, while the other vertical bus and the
remaining two horizontal buses are reserved to routing of reconfigured
vertical interconnections. (In the absence of faults, no bus is active.)
If we restrict the reconfiguration possibilities of the network to purely
fault-tolerance ones, switches are inserted only at the crossing of
links and/or buses reserved to the same class of data paths. The
switch distribution concerning the outputs of any cell (i,j) is given
in Fig. 8.29; routing functions required to a switch are specified in
Fig. 8.30.
       To identify the path activated in the switched structure by any
possible reconfiguration, and to prove as a consequence that the
structure itself is capable of supporting the reconfiguration algorithms
indicated above, we refer to the structure presented in the previous
subsection and in particular to Table 8.4, giving, for each algorithm,
the set of possible vertical and horizontal predecessors. Thus,
referring in particular to Algorithms 4 and 7 (more complex than
Algorithms 2 and 3), these sets lead to identification of the sets of

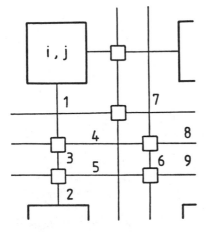

FIGURE 8.29 Triple-bus structure: switch distribution.

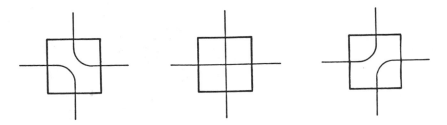

FIGURE 8.30 Routing functions of switches.

alternative paths represented in Fig. 8.31. On this basis, application of Algorithm 4 leads—for a sample fault distribution—to the reconfiguration shown in Fig. 8.32.

It is obviously necessary to prove that the bus structure envisioned is "sufficient," meaning that for no fault distribution allowing reconfiguration, a conflict for access to any path segment would arise. A

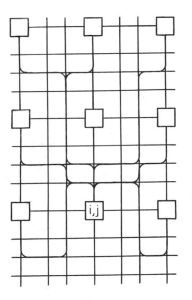
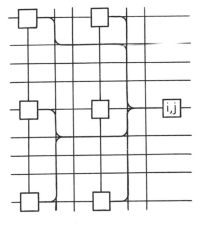

FIGURE 8.31 Triple-bus structure: alternative interconnection paths.

FIGURE 8.32 Triple-bus structure: example of reconfiguration with
simple fault-stealing, variable choice algorithm.

complete proof can easily be deduced following the outline given below
for vertical interconnection paths (refer to Fig. 8.29):

1. Segment 1 does never originate a conflict since it is engaged only
    by vertical output of cell (i,j); thus it is never required by two
    distinct paths in the same instance of reconfiguration.
2. Segment 2 is engaged only by vertical input to $(i + 1, j)$, so that
    the above holds in this case also.
3. Segment 3 may be required by two paths only:

    from (i,j) to $(i + 1, j)$

    from $(i - 1, j)$ to $(i + 1, j)$

Since these end on the same cell, they are mutually exclusive.

4. Segment 5 is engaged only by two alternative paths, both entering $(i + 1, j)$, which again are mutually exclusive.
5. Segment 9 is engaged only by two alternative paths entering $(i + 1, j + 1)$, which again are mutually exclusive.

As regards the remaining segments engaged by vertical inter-connection paths, it is necessary to refer not only to the topology of the structure (as was done above) but also to the reconfiguration algorithms:

6. Segment 7 may be engaged by five different reconfiguration paths, i.e.,:

    a. From $(i - 1, j)$ to $(i + 1, j + 1)$
    b. From $(i - 1, j + 1)$ to $(i + 1, j)$
    c. From $(i - 1, j + 1)$ to $(i + 1, j + 1)$
    d. From (i,j) to $(i, j + 1)$
    e. From $(i, j + 1)$ to (i,j)

    Some of the possibly conflicting pairs above are automatically excluded because they originate from the same cell (pair b,c) or end on the same cell (pair, a,c) or create a loop (pair d,e). As for the other possible conflicts, alternative a requires that $(i - 1, j)$ has not undergone any displacement, while alternative b imples a displacement of $(i - 1, j)$ into $(i - 1, j + 1)$; the alternatives are clearly mutually exclusive. Similar considerations allow us to exclude all other conflicts
7. The same line of reasoning excludes conflicts for segments 4, 6, and 8.

6.5 Double-Bus Structure

This solution is valid only for Algorithm 3, which, as already stated, is simpler than Algorithm 2 or 4. As given in Table 8.4, the set of alternative inputs for a given cell is reduced with respect to the other algorithms, and therefore possible alternative paths and consequent conflicts are reduced.

It can be proved that for this algorithm, the switched-bus solution can be implemented with only two buses in each channel; one vertical and one horizontal buses are reserved for reconfigured vertical inter-connections, the other pair is reserved for reconfigured horizontal interconnections. As before, considering only fault-tolerance-related reconfigurations, the switches required are only those at the inter-sections of links and buses dedicated to the same class of intercon-nection paths; thus they reduce to four switches only (see Fig. 8.33).

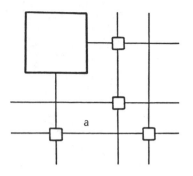

FIGURE 8.33 Double-bus structure: switch distribution.

Alternative paths adopted for possible reconfigurations are shown in Fig. 8.34. Once again it is possible to prove that the structure proposed is actually sufficient to implement without any conflicts all reconfigurations allowed by Algorithms 3. The proof can be derived along the same lines adopted for the previous structure; without entering it in detail, we can note that the only segment from which

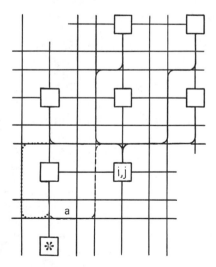
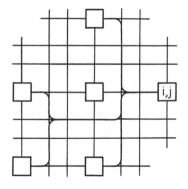

FIGURE 8.34 Double-bus structure: alternative interconnection paths.

conflicts may arise is segment a. The conflict can anyway be solved
by the following criteria:

1. Introduce an alternative path from (i, j − 1) to (i,j) (dashed line
   in Fig. 8.34); the path to be activated will depend on the direction
   from which input to (i + 1, j − 1) comes.
2. Since it can be proved that conflicts arise when vertical deformations lines have length 1, a way of avoiding them is to require
   vertical deformations to have length no less than 2, as stated by
   Algorithm 3.

A complete example of reconfiguration is given in Fig. 8.35.

7 SOLUTION FOR ARRAY-LEVEL SELF-TESTING

Up to now, we have simply assumed that processing cells were provided
with self-testing capacities, without making any further assumption
as regards the means by which such a result could be achieved.
"Local techniques" could belong to a functional self-test class if the
processing cell can be identified as a single-function device (examples
were given in Section 1); otherwise, general techniques such as
double-rail logic and self-checking checkers, or straightforward duplication and comparison, should be used. In such instances, area
redundancy becomes very high, and, of course, it adds to that required by reconfiguration policies. Moreover, it should be kept in
mind that area increase due to self-testing actually increases the
possibility of failure for the single cell, all considerations that make
the above-named techniques a very critical issue in the present context.† This most favorable aspect is the very low error latency they
afford. The error is detected at the very moment that faulty information would be transmitted to other units, so that error confinement
may be effectively achieved. Still, in a number of applications, such
complete confinement may not be required, in the sense that a limited
error propagation followed by periodical detection and confinement
would be acceptable. A periodical self-check philosophy can be en-
couraged further by statistics on fault distribution in time.

---

† Although this problem is specifically related to dynamic self-or host-
driven reconfiguration, it is actually important also for static production-time reconfiguration. The complexity of a processing array, as
stated earlier, also make self-testing supports very attractive for
initial testing.

We present here, as an alternative to individual cell-level self-testing, an array-level technique based on "time redundancy" that still allows us to detect and correctly locate individual faulty cells. A limited structure redundancy is still required. Increases in processing time become very high (twofold) if the same error latency guaranteed by cell-level self-testing is required; otherwise, if only periodical test actions are foreseen, it can be kept to a reasonable level. The time-redundancy approach to array self-testing is fully discussed in Ref. 43; here, we outline it briefly to provide some insight into this alternative self-testing technique.

The functional fault model already adopted is further detailed with the following assumptions:

1. A faulty cell always provides faulty horizontal and vertical outputs.
2. A correctly working cell, whenever it receives at least one faulty input, provides faulty horizontal and vertical outputs.
3. Whether faulty or correctly working, a cell produces different outputs given different input data.

The technique described here is dedicated strictly to run-time testing; therefore, it is assumed that initially the whole array is working correctly. We consider first the simplest request, identification of the first fault.

Basically, the technique requires addition of N cells organized as the (N + 1)th column of the array. [We will see that actually an (N + 1)th cell also has to be added in column 1]. The interconnection network also has to be augmented and completed by two-way multiplexers on both inputs of row and column 1 and on the vertical inputs of column 2, as well as of two-way demultiplexers on the horizontal outputs of column N and on the vertical outputs of row N. Finally, a set of 2 × N comparators has to be inserted on the array's horizontal and vertical outputs; a schematic design is given in Fig. 8.36 (where all multiplexers and demultiplexers are implicitly incorporated into the corresponding cells, for greater simplicity).

During a self-testing cycle, two operation phases $\mathscr{P}_0$ and $\mathscr{P}_1$ must be activated as follows:

Phase $\mathscr{P}_0$. A set I of input data is applied to the "virtual" array consisting of columns 1 to N [the first column consisting of cells (1,1) to (N, 1).]
Phase $\mathscr{P}_1$. The same set I is applied to the virtual array consisting of columns 2 to N + 1; moreover, horizontal inputs are also applied to cells (2,1) to (N + 1, 1).

Thus each elementary computation is repeated by at least two different processing cells, and each cell—during the two phases—

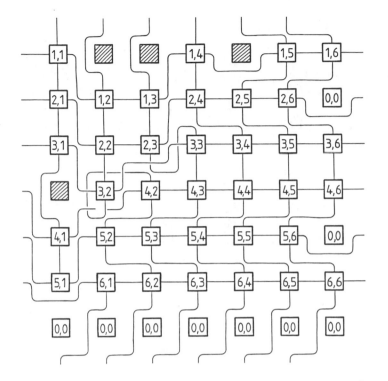

FIGURE 8.35 Double-bus structure: example of reconfiguration with restricted fault stealing algorithm.

operates on different input data. A fault is detected if at least one comparator detects a difference between the results of phases $\mathscr{S}_0$ and $\mathscr{S}_1$.

In Fig. 8.36, "logical" indices indicating functions performed during phase $\mathscr{S}_0$ (upper indices) and $\mathscr{S}_1$ (lower indices) are associated with each cell. Assume that the cell in row 2, column 3, is faulty; fault is propagated to the other cells as indicated by thick lines. Then all comparators marked with a "D" show disagreement, while the other ones (marked with "A") show agreement. Identification of the faulty cell is reached by looking for:

1. The first comparator from the top, on "row" outputs, showing disagreement (this identifies the row in which there is a faulty cell).
2. The first comparator from the left, on "column" outputs, showing disagreement (this identifies the column index of the faulty cell).

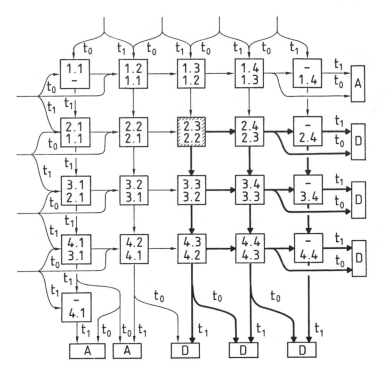

FIGURE 8.36 Array-level self-testing.

Cell (N + 1, 1) also allows us to locate correctly faults in column 1.
    Some further considerations are possible as regards this technique:

1. If a more complex fault model is adopted, allowing a faulty cell
    to produce an error on one input only, a three-phase operation
    becomes necessary [43];
2. If the interconnection network is suitably augmented, a reconfigu-
    ration phase may follow detection and identification of the first
    fault; then, externally, the array operates as if it were correct.
    Logical indices—with the meaning adopted in the previous
    sections—are now associated with its cells. A new self-testing
    action can detect a second (subsequent) fault; this time, identi-
    fication of the faulty cell gives the *logical* indices of the cell itself.
    Thus the technique is valid also for multiple faults, provided that
    they appear sequentially in time.

3. While cell-level testing allows us to provide for transient faults by periodically clearing all fault information, here this is not possible since it would invalidate the sequentiality assumption for fault appearance. It would be possible to take care of possible transient faults by clearing error information for one single cell at a time, and going in this way through all faulty cells.

4. Considering the self-reconfiguring array, failure of a multiplexer on a cell input or of a line is detected "as if" it were a cell failure. The cell will be confined from subsequent operation in the usual way, so that a failure in the interconnection or selection circuits will not affect proper operation of the whole array. When a self-level approach to self-testing is chosen, this result can be achieved only by making multiplexers also self-checking and by inserting some form of redundancy as regards data transmission. (All this, obviously, requires further area increase.)

8 CONCLUDING REMARKS

The algorithmic approach presented in this chapter allows to give a common, formal definition to a number of reconfiguration procedures for processing arrays. Regular reconfiguration structures correspond to these algorithms; networks are characterized by locality and simplicity, and the strict limits given to maximum distances between logically adjacent guarantee that added silicon area and time delays will also be limited.

Since interconnection networks and selection circuits can be directly derived from the algorithms via well-defined steps, the technique is well suited to computer-aided design. Even the simplification and compaction procedures leading from the theoretical structure to less space-consuming ones are suitable for automatic implementation.

A further cause of interest is due to the fact that the same interconnection network can be kept valid for a number of reconfiguration algorithms; only circuits computing the control signals have to be modified. If self-reconfiguration is envisioned, this allows us to keep a relevant part of the layout unchanged for different algorithm implementations. Even better, in the case of host-driven (or possibly microprogram-driven) reconfiguration, the identical array can be driven through different reconfiguration algorithms, even dynamically modificable as a consequence of varying specifications.

REFERENCES

1. T. W. Williams and K. P. Parker, Design for testability, a survey, *IEEE Trans. Comput.*, C-31: 2—15 (1982).

2. R. Stefanelli, Multiple error correction in arithmetic units with triple redundancy, *Proc. EUROMICRO 84,* (B. Myrhaugh and D. R. Wilson, eds.), North-Holland, Amsterdam pp. 205–216 (1984).

3. B. F. Fitzgerald and E. P. Thoma, Circuit implementation of fusible redundant addresses on RAMs for productivity enhancement, *IBM J. Res. Dev.,* 24: 291–298 (1980).

4. G. A. Papachristou and N. B. Sahgal, An improved method for detecting functional faults in semiconductor random access memories, *IEEE Trans. Comput., C-24:* 110–116 (1985).

5. R. A. Rutledge, Models for the reliability of memory with ECC, *Proc. Annual Reliability and Maintainability Symposium,* Philadelphia, PA 57–62 (Jan. 1985).

6. H. L. Garner, Error codes for arithmetic operations, *IEEE Trans. Comput., C-15:* 763–770 (1966).

7. T. R. N. Rao and K. Vathanvit, A class of $A(N + C)$ codes and its properties, *Proc. 7th Symposium on Computer Arithmetic,* Urbana, IL 293–295 (June 1985).

8. A. Avizienis, Low-cost residue and inverse residue error detecting codes for signed-digit arithmetic, *Proc. 5th IEEE Symposium on Computer Arithmetic,* 165–168, (May 1981).

9. A. Avizienis, Arithmetic algorithms for operands encoded in two-dimensional low-cost arithmetic error codes, *Proc. 7th Symposium on Computer Arithmetic,* Urbana, IL 285–292 (June 1985).

10. T. R. N. Rao, Biresidue error-correcting codes for computer arithmetic, *IEEE Trans. Comput., C-19:* 398–402 (1970).

11. M. Annaratone and R. Stefanelli, A multipler with multiple error correction capabilities, *Proc. 6th IEEE Symposium on Computer Arithmetic,* Aarhus (1983).

12. S. M. Thatte, D. S. Ho, H. T. Yuan, T. Sridhar and T. J. Powell, An architecture for testable VLSI processors, *Proc. IEEE Test Conference,* 484–492 (1982).

13. A. Antola, R. Negrini, M. G. Sami and N. Scarabottolo, Transient fault management in microprogrammed units: a software recovery approach, *Proc. EUROMICRO '85,* Bruxelles (Sept. 1985).

14. C. P. Disparte, A design approach for an electric engine controller self-checking microprocessor, *Proc. EUROMICRO 81,* Paris (1981).

15. M. P. Halbert and S. M. Bose, Design approach for a VLSI self-checking MIL-STD-1750A microprocessor, *Proc. FTCS, 14*: 254–259 (June 1984).

16. T. M. Mangir and A. Avizienis, Fault-tolerant design for VLSI: effect of interconnection requirements on yield improvement of VLSI design, *IEEE Trans. Comput., C-31*: 609–615 (1982).

17. I. Koren and M. A. Breuer, On area and yield considerations for fault-tolerant VLSI processor arrays, *IEEE Trans. Comput., C-33*: 21–27 (1984).

18. R. Negrini and R. Stefanelli, Algorithms for self-reconfiguration of wafer-scale regular arrays, *Proc. ICCAS*, Beijing (1985).

19. J. Grinberg, G. R. Nudd and R. D. Etchells, A cellular VLSI architecture, *Computer* (IEEE) *17*(1): 69–81 (Jan. 1984).

20. F. T. Leighton and C. E. Leiserson, Wafer-scale integration of systolic arrays, *2nd IEEE Symposium on Foundations of Computer Science* (Oct. 1982).

21. H. T. Kung and M. S. Lam, Fault-tolerance and two-level pipelining in VLSI systolic arrays, *Proc. MIT Conferences on Advanced Research in VLSI* MIT, Cambridge, MA, 74–83, (Jan. 1984).

22. I. Koren, A reconfigurable and fault-tolerant VLSI multiprocessor array, *Proc. 8th IEEE Symposium on Computer Architecture*, (1981).

23. F. R. K. Chung, F. T. Leighton and A. L. Rosenberg, Diogenes: a methodology for designing fault-tolerant VLSI processor arrays, *Proc. IEEE Symposium on Fault-Tolerant Computing*, Milano (June 1983).

24. A. L. Rosenberg, The Diogenes approach to testable fault-tolerant VLSI processor arrays, *IEEE Trans. Comput., C-32*: 902–910 (1983).

25. M. G. Sami and R. Stefanelli, Reconfigurable architectures for VLSI implementation, *Proc. NCC 83, AFIPS*, Los Angeles (May 1983).

26. J. Fried, Wafer-scale integration of pipelined processors, *Proc. ICCD 84*, New York, 611–615 (Oct. 1984).

27. D. S. Fussell and P. J. Varman, "Designing systolic algorithms for fault-tolerance", *Proc. ICCD 84*, New York 623–628 (Oct. 1984).

28. R. Negrini, M. G. Sami and R. Stefanelli, Fault-tolerance approaches for VLSI/WSI arrays, *Proc. IEEE Conference on Computers and Communication*, Phoenix, AZ (1985).

29. M. G. Sami and R. Stefanelli, Fault-tolerance of VLSI processing arrays: the time-redundancy approach, *Proc. 1984 IEEE Real-Time Systems Symposium*, Austin TX 200−207 (Dec. 1984).

30. K. S. Hedlund and L. Snyder, Wafer-scale integration of configurable highly parallel (CHiP) processor, *Proc. IEEE International Conference on Parallel Processing*, 262−264 (1982).

31. K. S. Hedlund and L. Synder, Systolic arrays: a wafer-scale approach, *Proc. ICCD 84*, New York, 604−610 (Oct. 1984).

32. T. Leighton, C. E. Leiserson, Wafer-scale integration of systolic arrays, *IEEE Trans. Comput.*, *C-34*: 448−461 (1985).

33. K. E. Batcher, Architecture of a massively parallel processor, *Proc. 7th IEEE Symposium on Computer Architecture*, 168−173 (May 1980).

34. L. Snyder, Introduction to the configurable, highly parallel computer, *Computer* (IEEE), *15*(1): 47−56 (1982).

35. F. B. Manning, An approach to highly integrated, computer-maintained cellular arrays, *IEEE Trans. Comput.*, *C-26*: 536−552 (1977).

36. R. C. Aubusson and I. Catt, Wafer-scale integration: a fault-tolerant procedure, *IEEE, J. Solid-State Circuits*, *SC-13*: 339−344 (1978).

37. W. R. Moore, A review of fault-tolerant techniques for the enhancement of integrated circuit yield, *GEC J. Res.* *2*(1): 1−15 (1984).

38. W. R. Moore, Switching circuits for yield-enhancement of array chips, *Electron. Lett.*, *20*: 667−669 (1984).

39. R. Negrini and R. Stefanelli, Time redundancy in WSI arrays of processing elements, *First International Conference on Supercomputing Systems*, St. Petersburg, Fla. (Dec. 1985).

40. E. L. Lawler, *Combinatorial Optimization, Networks and Matroids*, Holt, Rinehart and Winston, New York (1976).

41. T. J. Fountain, A survey of bit-serial array processor circuits, *Computing Structures for Image Processing*, Academic Press, London, pp. 1−13 (1983).

42.   V. N. Doniants, V. G. Lazarev, M. G. Sami and R. Stefanelli, Reconfiguration of VLSI arrays: a technique for increased flexibility and reliability, *11th Symposium on Microprocessing and Microprogramming,* Brussels (Sept. 1985).

43.   M. G. Sami and R. Stefanelli, Self-testing VLSI array structures, *Proc. ICCD '84, New York,* (Oct. 1984).

Index